Airborne Biocontaminants and Their Impact on Human Health

Airborne Biocontaminants and Their Impact on Human Health

Edited by

Rajeev Singh
Department of Environmental Science, Jamia Millia Islamia, Delhi, New Delhi, India

Anamika Singh
Maitreyi College, University of Delhi, Delhi, New Delhi, India

WILEY

Published by John Wiley & Sons, Inc., Hoboken, New Jersey.
Published simultaneously in Canada.

Library of Congress Cataloging-in-Publication Data Applied for:

Hardback ISBN: 9781394178933

Cover Design: Wiley
Cover Image: © Sandipkumar Patel/Getty Images

Set in 9.5/12.5pt STIXTwoText by Straive, Pondicherry, India

This book is dedicated to our beloved parents:

Late Shri Keshav Singh	*Shri Paras Nath Singh*
Smt Keshara Singh	*Smt Shailja Singh*

Contents

About the Editors

Dr. Rajeev Singh is Associate Professor at Department of Environmental Science, Jamia Millia Islamia, New Delhi. He is working in the area of environmental health. He is Founder Secretary of the Bio-electromagnetic Society of India. Dr. Singh is member of several scientific and academic bodies. He has organized and participated in several international and national conferences and delivered invited talks. He has published around 100 research papers, chapters in books, conference proceedings, etc., of international repute. He is also a principal investigator of several major research projects funded by various government agencies. He has published a good number of research articles in reputed journals of high impact, namely *Seminars in Cancer Biology* (impact factor – 17), *Scientific Reports* (impact factor – 4.9), *Bioresource Technology* (impact factor – 11.8), *Science of the Total Environment* (Impact factor – 10.7), *Environmental Pollution* (impact factor –9.98), *Cells* (impact Factor – 7.6), *Chemosphere* (impact factor – 8.9), *Environmental Research* (impact factor – 8.4), *Fuel* (impact factor – 8.0), *Biomedicine and Pharmacotherapy* impact factor – 7.4), *International Journal of Hydrogen Energy* (impact factor – 7.1), *Journal of Cancer Research and Clinical Oncology* (impact factor –4.5), *Air Quality Atmosphere and Health* (impact factor –5.8), *Aerobiologia* (impact factor –2.7), *Reproductive Toxicology* (impact factor –3.4), and *International Journal of Environmental Health Research* (impact factor –4.4), among several others. He has edited a number of books published by Springer Nature. He is also a reviewer of several high-impact journals like *Environmental International, PLOS One*, *Free Radical Research*, *Mutation Research*, etc. Dr. Singh is recipient of Young Scientist Award of Federation of European Microbiological Societies (FEMS), Gold Medal for best oral presentation, etc.

Dr. (Mrs.) Anamika Singh is currently working as Assistant Professor (Selection grade) in Maitreyi College, University of Delhi. She holds a PhD degree in bioinformatics from the Indian Institute of Information Technology, Allahabad (IIIT-A). She has 12 years of teaching and research experience. Her areas of research interest are genetic engineering, bio-informatics, Phytomolecular analysis, biological databases, phylogeny, molecular modeling, drug designing, computer-based drug designing, computational biology, etc. Dr. Singh is member of the Indian Bioinformatics Society, APBioNet, the European Federation of Biotechnology, and IAENG. She is also actively involved in the review of research articles for various journals. She has organized national and international conferences, faculty development programs, and workshops in different platforms. She has published several research articles in reputed journals and book chapters with leading publishers. She has also authored a book in biochemistry and medicinal botany.

List of Contributors

Sujata Adhana
Bhaskaracharya College of Applied Sciences, University of Delhi, Delhi
New Delhi, India

Anamika
Department of Zoology, Ramjas College, University of Delhi, Delhi
New Delhi, India

Sadaf Anwar
Department of Biochemistry, College of Medicine, University of Ha'il, Ha'il
Saudi Arabia

Khencha Aran
Department of Botany, Cotton University, Guwahati, Assam, India

Rajnish Awasthi
Faculty in University College of Medical Sciences, University of Delhi
Delhi, New Delhi, India

Amrita Bakshi
Department of Zoology, Ramjas College, University of Delhi, Delhi
New Delhi, India

Palak Balyan
Environmental Pollution Laboratory
Department of Environmental Studies
University of Delhi, Delhi
New Delhi, India

B. D. Banerjee
Faculty in University College of Medical Sciences, University of Delhi
Delhi, New Delhi, India

Anamika Bhardwaj
Department of Microbiology
Guru Nanak Dev University, Amritsar
Punjab, India

Pankaj Bhatt
Lloyd Institute of Management and Technology, Greater Noida
Uttar Pradesh, India; Department of Pharmaceutical Sciences, Gurukul Kangri (Deemed to be University)
Haridwar, Uttarakhand, India

Dhitri Borah
Department of Zoology
Biswanath College, Biswanath Charali
Assam, India

Maneet Kumar Chakrawarti
Antimicrobial Research Laboratory
School of Environmental Sciences
Jawaharlal Nehru University
New Delhi, India

Atavika Chandan
Department of Zoology, Hindu
College, University of Delhi, Delhi
New Delhi, India

Shilpa Samir Chapadgaonkar
Department of Biosciences and
Technology, Dr. Vishwanath Karad
MIT World Peace University, Pune
Maharashtra, India

Anchal Chaudhary
Department of Microbiology
Guru Nanak Dev University, Amritsar
Punjab, India

Twinkle Chetia
Department of Botany, Cotton
University, Guwahati, Assam, India

Shukla Das
Faculty in University College of
Medical Sciences, University of Delhi
Delhi, New Delhi, India

Bitupan Deka
Department of Botany, Cotton
University, Guwahati, Assam, India

Chirashree Ghosh
Environmental Pollution Laboratory
Department of Environmental Studies
University of Delhi, Delhi
New Delhi, India

Anita Girdhar
Department of Environmental
Sciences, J.C. Bose University of
Science and Technology, YMCA
Faridabad, Haryana, India

Priti Giri
Department of Botany, Maitreyi
College, University of Delhi
New Delhi, India

Ravi Kumar Goswami
Department of Zoology, Hindu
College, University of Delhi, Delhi
New Delhi, India

Shiwangi Gupta
Department of Environmental Studies
Satyawati College, University of Delhi
Delhi, New Delhi, India

Jaldhi
Biochemistry and Molecular Biology
Laboratory, Department of Zoology
Faculty of Science, University of
Delhi, Delhi, New Delhi, India

Kunal Kanojia
Department of Pharmaceutics, KIET
Group of Institutions (KIET School of
Pharmacy), Muradnagar, Ghaziabad
Uttar Pradesh, India

Vinita Katiyar
Regional Services Division, Indira
Gandhi National Open University
(IGNOU), Maidan Garhi, Delhi
New Delhi, India

Manandeep Kaur
Department of Microbiology
Guru Nanak Dev University, Amritsar
Punjab, India

Simran Kaur
School of Biosciences, Apeejay Stya University, Gurugram, Haryana, India

Mohd Adnan Kausar
Department of Biochemistry, College of Medicine, University of Ha'il, Ha'il Saudi Arabia

P. Kavita
Department of Botany, Maitreyi College, University of Delhi New Delhi, India

Devendra Kumar
Department of Environmental Science, Ramanujan College University of Delhi, Delhi New Delhi, India

Mohit Kumar
Department of Zoology, Hindu College, University of Delhi, Delhi New Delhi, India

Pradeep Kumar
Department of Environmental Studies Satyawati College, University of Delhi Delhi, New Delhi, India; Department of Agricultural and Biosystems Engineering, South Dakota State University, Brookings, SD, USA

Umesh Kumar
Department of Biosciences, Institute of Management Studies Ghaziabad (University Courses Campus) Ghaziabad, Uttar Pradesh, India

Vipin Kumar
Department of Pharmaceutical Sciences, Gurukul Kangri (Deemed to be University), Haridwar Uttarakhand, India

Pratibha Kumari
Department of Botany, Daulat Ram College University of Delhi, Delhi New Delhi, India

Anirban Kundu
Department of Zoology, Laboratory of Helminth Parasitology, Visva-Bharati University, Shantiniketan West Bengal, India

Shalini Kaushik Love
Department of Botany, Hansraj College, University of Delhi, Delhi New Delhi, India

Sudeshna Mandal
Department of Zoology, Laboratory of Helminth Parasitology, Visva-Bharati University, Shantiniketan West Bengal, India

Shashank Kumar Maurya
Biochemistry and Molecular Biology Laboratory, Department of Zoology Faculty of Science, University of Delhi, Delhi, New Delhi, India

Anil Kumar Mavi
Department of Botany, Sri Aurobindo College, University of Delhi, Delhi New Delhi, India

Samina Mazahar
Department of Botany, Dyal Singh College, University of Delhi, Delhi New Delhi, India

Arti Mishra
Department of Botany, Hansraj College, University of Delhi, Delhi New Delhi, India; Umeå Plant Science Center, Department of Plant Physiology, Umeå University Umeå, Sweden

Chandrani Mondal
Department of Zoology, Laboratory of Helminth Parasitology, Visva-Bharati University, Shantiniketan West Bengal, India

Kasturi Mukhopadhyay
Antimicrobial Research Laboratory School of Environmental Sciences Jawaharlal Nehru University New Delhi, India

Mohammad Zeeshan Najm
School of Biosciences, Apeejay Stya University, Gurugram, Haryana, India

Vinay Mohan Pathak
Department of Microbiology University of Delhi (South Campus) Delhi, New Delhi, India; Department of Botany and Microbiology, Gurukul Kangri Deemed to be University Haridwar, Uttarakhand, India

K. Priti
Department of Microbiology Gurukula Kangri (Deemed to be University), Haridwar Uttarakhand, India

Samir Qiblawi
Medical and Laboratory Sciences Department, Palestine Technical University - Kadoorie, Tulkarm Palestine.

Garima Rathi
Delhi Public School Ghaziabad Ghaziabad, Uttar Pradesh, India

Chanchal Rawat
Department of Biotechnology School of Engineering & Technology Manav Rachna International Institute of Research and Studies, Faridabad Haryana, India

Varunendra Singh Rawat
Department of Zoology, Hindu College, University of Delhi, Delhi New Delhi, India

Mou Singha Ray
Department of Zoology, Laboratory of Helminth Parasitology, Visva-Bharati University, Shantiniketan West Bengal, India

Sadaf
Department of Biotechnology Jamia Millia Islamia, Delhi New Delhi, India

Kalpana Sagar
Department of Microbiology Gurukula Kangri (Deemed to be University), Haridwar Uttarakhand, India

Arun Kumar Sharma
Faculty in University College of Medical Sciences, University of Delhi
Delhi, New Delhi, India

Nikhil Sharma
Department of Microbiology
Guru Nanak Dev University,
Amritsar Punjab, India

Abhilasha Shourie
Department of Biotechnology
School of Engineering & Technology
Manav Rachna International Institute of Research and Studies, Faridabad
Haryana, India

Shweta
Biochemistry and Molecular Biology Laboratory, Department of Zoology
Faculty of Science, University of Delhi, Delhi, New Delhi, India

Madhuri Singh
Antimicrobial Research Laboratory
School of Environmental Sciences
Jawaharlal Nehru University
New Delhi, India

Rajeev Singh
Department of Environmental Science, Jamia Millia Islamia, Delhi
New Delhi, India

Rajiv Kumar Singh
Primary Health Center (PHC)
Government of Uttar Pradesh, Pali
Hardoi, Uttar Pradesh, India

Sevaram Singh
Translational Health Science and Technology Institute, NCR Biotech Science Cluster, Faridabad, Haryana
India; Jawaharlal Nehru University
Delhi, New Delhi, India

Suruchi Singh
Department of Pharmacology
Accurate College of Pharmacy
Greater Noida, Uttar Pradesh, India

Kriti Singhal
Department of Biotechnology
School of Engineering & Technology, Manav Rachna International Institute of Research and Studies, Faridabad
Haryana, India

Anurag Thapliyal
Biochemistry and Molecular Biology Laboratory, Department of Zoology
Faculty of Science
University of Delhi, Delhi
New Delhi, India

Ekta Tyagi
Department of Bioscience and Biotechnology, Banasthali University
Banasthali, Rajasthan, India

Arun Kumar Yadav
Environmental Pollution Laboratory
Department of Environmental Studies
University of Delhi, Delhi
New Delhi, India

Himanshi Yadav
Biochemistry and Molecular Biology Laboratory, Department of Zoology, Faculty of Science
University of Delhi, Delhi
New Delhi, India

Preface

The primary objective of this book is to understand the indoor contaminants and why they are important with reference to their impact on human health. As this topic is very broad and every small scientific studies related to contaminants are very important, we are trying to focus on different parameters (from basic to advanced concepts). There are many expert contributors to this book who are all well known in the field. With the advancement of this topic as it is completely research based, we have touched upon different areas that helps readers understand the subject at its basic level.

Environmental studies related to human health itself is a broad term and includes all disease-related studies as well as environmental distortion that may cause diseases in animals and humans. In addition to this, consider the reciprocal impact between humans and the environment. Since the beginning of environmental research, outdoor contaminants and greenhouse gases have always been hot topics of study, but under adverse climatic conditions (COVID, pollution, etc.), we spent most of the time in our homes, so again this question arises: We are safe or not in our homes, or are we still under high risk due to some other pollutants present in our homes. The answer is yes. Pollutants of the second category, that is indoor pollutants, may be bacteria, virus, fungal spores, dust, etc. This book gives a clear insight of household or indoor pollutants and their impact. The book also attempted to address historical elements concerning contaminants, alongside their various types and methods of treatment. The connection between mental health and environmental contaminants, as well as the correlation between our professions and exposure to pollutants, is noteworthy. Exploring the impact of these associations on our well-being is essential.

In the last chapter, we have compiled the government guidelines and precautions for indoor contaminants. With this, all we wish is that the readers definitely get a clear-cut idea about the contaminants. This book also suggests some novel and untouched areas of environmental research.

1

Biocontaminants in Indoor Environments

Occurrence, Spread, and Prevention

Abhilasha Shourie[1], *Anita Girdhar*[2], *Kriti Singhal*[1], *Chanchal Rawat*[1], *and Shilpa Samir Chapadgaonkar*[3#]

[1] *Department of Biotechnology, School of Engineering & Technology, Manav Rachna International Institute of Research and Studies, Faridabad, Haryana, India*
[2] *Department of Environmental Sciences, J.C. Bose University of Science and Technology, YMCA, Faridabad, Haryana, India*
[3] *Department of Biosciences and Technology, Dr. Vishwanath Karad MIT World Peace University, Pune, Maharashtra, India*

1.1 Introduction

In the contemporary realm of civilization, a significant portion of human existence is now spent within the confines of homes. The advent of the pandemic further accelerated this trend, prompting several companies to offer remote work arrangements. As a result, individuals find themselves predominantly indoors, leading to a noteworthy escalation in indoor environmental contaminants. Microbes, omnipresent from the kitchen to the bedroom, assume diverse forms, some innocuous while others perilous. Among these, mites and bacteria tend to adhere to clothing and the food procured from markets, subsequently infiltrating indoor spaces and exerting an influence on personal health.

Particularly noteworthy among these indoor contaminants are certain molds, such as *Aspergillus* and *Penicillium*, which frequently find refuge within furniture upholstery. These molds cause skin afflictions and infections, thereby underscoring the intricate relationship between health and environmental conditions. The term "biocontaminant" encapsulates the panorama of toxic pollutants and microorganisms – viruses, fungi, bacteria, algae, mites, and insects – pervading the indoor sphere. The genesis of these biological pollutants can be traced to sources as

Corresponding Author

Airborne Biocontaminants and Their Impact on Human Health, First Edition.
Edited by Rajeev Singh and Anamika Singh.

diverse as household pets, construction sites, building materials, dust particles, plants, and domestic furnishings, compounded by human occupancy.

Research conducted by Li et al. (2007) has shed light on a conspicuous linkage between indoor contamination and the efficacy of ventilation systems. Given that the transmission of disease-causing agents predominantly occurs through airborne pathways, the quality of indoor air is inextricably tied to the performance of AC units, fans, and the influx of external air. Globalization, with its dynamic movement of people and products, has engendered a surge in airborne diseases, notably respiratory disorders, and asthma. These health concerns are augmented by either shifting climate patterns or the extensive traveling across regions.

Even within meticulously sanitized environments such as hospitals – where emphasis on cleanliness is paramount – biocontaminants continue to thrive. This is attributed to the airborne nature of infectious agents, which proliferate through patients carrying communicable or noncommunicable diseases. A study by Schilling et al. (2014) underscores the persistence of biocontaminants within healthcare settings, further attesting to the intricate challenges inherent in infection control. The amalgamation of altered lifestyles, work paradigms, and pervasive indoor living necessitates comprehensive strategies for upholding indoor air quality (IAQ). This demands an integration of advanced ventilation and filtration mechanisms, meticulous sanitation practices, and a holistic understanding of the microbial ecology within closed environments. This chapter thus endeavors to unravel the nuanced tapestry of indoor contaminants, their far-reaching implications, and the strategies imperative for maintaining a wholesome indoor milieu.

Biocontaminant exposures commonly arise from allergens present in house dust, furry pets, rodents, and cockroaches as well as endotoxins from Gram-negative bacteria and the inflammatory compound beta (1,3) D-glucan found in microbial cells abundant within buildings. These contaminants, as studied by Kader et al. (2018), Murrison et al. (2019), and Sun et al. (2022), can lead to serious health issues including allergic diseases, skin conditions, asthma, and acute respiratory disorders, as observed by Miller et al. (2007), and Sun et al. (2022). Odors in houses are often attributed to mold and fungal growth in concealed areas, affecting both adults and children and causing health problems such as rhinitis, as reported by Caillaud et al. (2018), Chew et al. (2016), and Moses et al. (2019). The impact of biocontaminants on human body, causing different diseases and illness, is depicted in Figure 1.1.

Heating, ventilation, and AC systems impact indoor environments by harboring microbial contaminants such as *Actinobacteria*, *Firmicutes*, and *Cyanobacteria*, which significantly deteriorate air quality (Watanabe et al. 2022). Indoor pollutants, acting as allergens, carcinogens, neurotoxins, immunotoxins, or irritants, contribute to sick building syndrome (SBS). This syndrome, associated with nasal, ocular, and general health issues, notably presents symptoms like dry eyes, dry throat, and headaches in 70% of affected individuals, as reported by

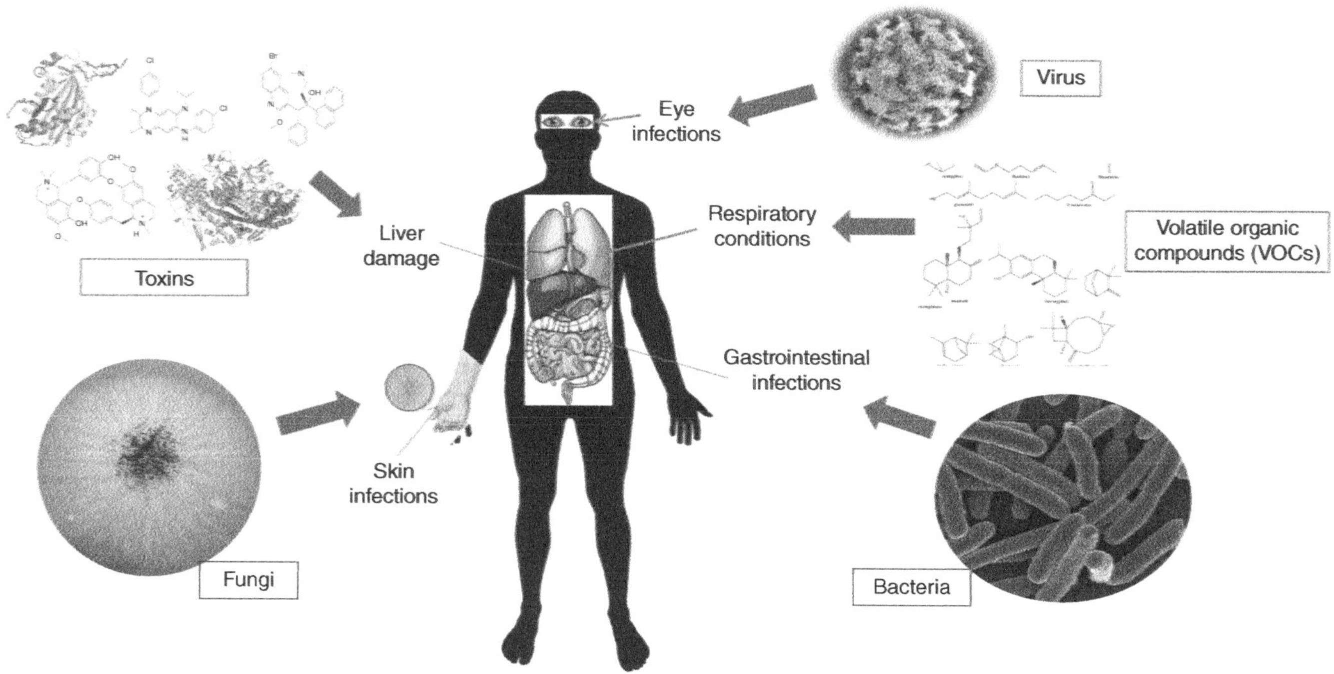

Figure 1.1 Impact of various biocontaminants on human health.

Soldatos et al. (2003) and Tsakas et al. (2011). Aerosol releases indoors, as detailed by Stetzenbach et al. (2004), can lead to health concerns due to organisms like *Francisella tularensis*, *Yersinia pestis*, *Bacillus anthracis*, and the smallpox virus. Animal studies in mice conducted by Rand et al. (2002) revealed that proteases, isosatratoxin-F, and spores of *S. chartarum* induce inflammatory and cytotoxic effects in lungs. Chronic rhinosinusitis, characterized by non-IgE inflammatory responses to fungi with nasal eosinophilia, also releases toxic proteins, as explored by Kern et al. (2007), Ponikau et al. (2005, 2006), and Thrasher and Crawley (2009).

Breathing entails inhalation of 10 m^3 of air exposure to bioaerosols in the form of colloidal suspension, as highlighted by Srikanth et al. (2008). These aerosol contaminants adapt to various environmental conditions, resulting in variations within subareas due to nutrient availability or seasonal changes, influenced by factors like temperature, humidity, and air exchange rates, as studied by Frankel et al. (2012), Kumari et al. (2016), Park et al. (2016), and Kumar et al. (2021).

This chapter offers a comprehensive overview of indoor biocontaminants, encompassing bacteria, molds, algae, and fungi. These contaminants infiltrate various settings under human occupation through avenues like furniture, AC ducts, fans, toilets, and electronic items. Regardless of the setting, be it hospitals or homes/offices, these contaminants persist. The chapter summarizes the role of factors like temperature, humidity, pollen allergens, and air circulation in the spread of biocontaminants, proposing preventive measures such as proper cleaning, the use of disinfectants, and residing in well-ventilated environments to curb their dispersion indoors.

1.2 Indoor Microenvironments That Harbor Biocontaminants

A wide array of components, spanning from minute organisms such as bacteria, viruses, fungi, mites, and insect debris to pollens, is found within various indoor settings (WHO 1990; Moldoveanu 2015; EPA 2020; Kumar et al. 2022). Additionally, pets like dogs, cats, rabbits, birds, and indoor plants also pose the potential to introduce biological contaminants. Even environments under stringent controls, as noted by Chirca (2019), harbor these elements indoors. This interconnection extends beyond confines, as both outdoor and indoor environments remain linked through air exchange, facilitating the dissemination of these pollutants. An essential factor in pinpointing the origin of these biocontaminants lies in the indoor-to-outdoor (I/O) ratio. If the I/O ratio surpasses 1, the source primarily stems from indoors; on the contrary, a value lower than 1 signifies indoor origins of biocontaminants (Karimpour Roshan et al. 2019). Table 1.1 summarizes various biocontaminants occurring in different types of indoor microenvironments.

Table 1.1 Biocontaminants in different indoor microenvironments.

S. no.	Type of indoor microclimate	Type of bio-contaminant	References
1	Home furnishing – bed, carpets, curtains, blankets	House dust mites, molds, bacteria	Sidenius (2002), Nam et al. (2008), Van Den Bemt (2006) and Cho et al. (2006)
2	Humidifiers	*Penicillium*, *Aspergillus*, *Cladosporium*, and *Alternaria*	Lee et al. (2012)
3	AC ducts	*Alternaria*, *Aspergillus*, *Cladosporium*, *Penicillium*, *Rhizopus* and *Mucor*, *S. epidemidis*, *A. globiformis*, *M. luteus*	Liu et al. (2021) and Modebelu and Modebelu (2013)
4	Exhaust/fans	*S. epidemidis*, *A. globiformis*, *M. luteus*	Modebelu and Modebelu (2013)
5	Wooden furniture	*Enterobacteriaceae* family of the genera *Rahnella*, *Pantoea*, *Enterobacter*, and *Klebsiella*. *Aspergillus* spp., *Penicillium* spp., *Absidia* spp. and yeast	Wójcik-Fatla (2022)
6	Books	*Bacillus*, *Clostridium*, *Pseudomonas* and *Micrococcus Penicillium*, *Alternaria*	Dunca (2014)
7	Electronics	*Staphylococci*, *Bacillus*, *Micrococcus*, *Enteric bacteria*	Koscova (2018) and Messina (2011)
8	Refrigerator	*Penicillium*, *Aspergillus*, *Mucor*, *Cladosporium*, *Botyrtis*, *Acremonium*, *Candida*, *Debaryomyces*, *Saccharomyces*, *Torulopsis*	Altunatmaz (2012)
9	Kitchen	*Clostridium botulinum*, *Campylobacter*, *Escherichia coli*, *Salmonella*, and *Shigella*, *E. coli*, *Salmonella*, *Shigella*, *Campylobacter*, *Norovirus* and *Hepatitis A*	Rahimi (2019) and Kusumaningrum (2003)
10	Damp places	*Aspergillus Penicillium*, *Stachybotrys chartarum* and *Memnoniella echinata*, *Firmicutes*, and *Actinobacteria*	Anderson et al. (2011) and Novak Babič (2020)
11	Hospitals	*Mycobacterium tuberculosis*, *Staphylococcus aureus*, *Influenza virus*, *Aspergillus flavus*, *A. fumigates*, *Candida albicans*. Fungal genera, e.g. *Aspergillus* sp., *Penicillium* sp., *Alternaria* sp., *Talaromyces*, *Trichoderma*, etc.	Kausar (2018) and Cho et al. (2018)

Considering that a significant portion of our lives unfolds indoors – encompassing homes, schools, offices, shopping complexes, transportation, and even healthcare facilities (Moldoveanu 2015) – it is crucial to acknowledge that enclosed spaces are invariably teeming with microbes (Chirca 2019).

1.2.1 Bedding/Carpets/Furnishings

Bedding, carpets, and furnishings harbor several biocontaminants, with house dust mites taking the lead. Amid various corners of a household, mattresses stand out as the hotspot for the highest concentration of mite allergens. The composition of mattresses plays a pivotal role in influencing the propagation and growth of these mites (Visitsunthorn et al. 2010). Notably, the rate of mite accumulation proves higher in synthetic and kapok mattresses compared to coir and sponge-like polyurethane variants. These minuscule organisms thrive on partially digested human skin scales. Intriguingly, foam mattresses tend to be more susceptible to mite infestations than their sprung counterparts (Schei et al. 2002; Frost et al. 2010). Even new mattresses transform into reservoirs of mite allergens within three to four months of usage. Allergic reactions to dust mites often stem from the proteins present in their excrement. Astonishingly, a single mite generates approximately 20 of these waste droppings daily, which persist as allergenic triggers even post mortem. The term "house dust mite" collectively denotes two species: *Dermatophagoides pteronyssinus* (*D. pteronyssinus*) and *Dermatophagoides farinae* (*D. farinae*), together constituting a significant portion – 80–90% – of household dust (Arlian et al. 2002; Yahia and Metwally 2019).

Furniture and cushion covers, as well as the stuffing within furniture, serve as breeding grounds for molds and bacteria. Notably, species such as *Penicillium* and *Aspergillus* dominate furniture stuffing, accompanied by Rhizopus and Mucor, barring certain cotton samples where Fusarium takes precedence (Swaebley 1952, pp. 8–10). Egwari (2015) discerned that rubber fabrics tend to be more vulnerable to microbial infestations compared to metallic surfaces.

1.2.2 ACs/Humidifiers

ACs and humidifiers, while extracting humidity from indoor spaces, ironically accumulate higher moisture content within their filters. These filters and ducts create an environment conducive to the proliferation and spore formation of airborne fungi (Hamada and Fujita 2002; Kelkar et al. 2005; Khan et al. 2009; Sánchez Espinosa et al. 2021). Notably, research by Hamada and Fujita (2002) highlighted that AC filters exhibited an average fungal contamination level five times greater than that found in carpets. Predominant fungi included *Cladosporium* and *Penicillium*, as well as *Trichoderma* and *Fusarium* in AC filters. The frequency of AC use correlates with the degree of fungal contamination, with daily use leading to higher levels. Humidifiers, especially those featuring reservoirs, foster microbial growth and have been associated with humidifier fever. Nontuberculous mycobacteria, actinomycetes, and endotoxins have been attributed to respiratory health problems and humidifier fever (Borchers et al. 2017).

1.2.3 AC Ducts/Outdoor Units/Centralized AC Units

Air supply systems are important sources of microbial and fungal proliferation as they provide optimal conditions for growth. Components like filters, heat exchangers, cooling coils, humidifiers, and ducts in AC systems, whether directly or indirectly, contribute to the dissemination of bacteria and fungi (Noris 2009; Meng et al. 2015; Liu et al. 2017, 2018). In a comprehensive study by Liu et al. (2021), four dominant species – *Alternaria_betae–kenyensis*, *Cladosporium_delicatulum*, *Aspergillus_sydowii*, and *Verticillium_dahliae* – were identified across all segments.

1.2.4 Fan/Exhaust

Fans/exhaust, expelling indoor air outward, effectively eliminate moisture and pollutants. In contrast, roof fans recycle enclosed space air and often accumulate black dust laden with molds. *Pseudomonas* sp., *Penicilium* sp., *Aspergillus niger*, *Aspergillus flavus*, *Fusarium* sp., *Candida glabrata*, *Trichophyton mentagrophytes*, and *Paecilomyces variotii* were found in two compromised air supply and exhaust air handling units, as per Piekarska et al. (2018).

1.2.5 Furniture Wooden/Metal Items/Decoratives

Microscopic spores, bacteria, and mold fragments become airborne and settle on indoor surfaces, a phenomenon common to wooden and metal furnishings. The porous nature of wood fosters the growth of bacteria and fungi, transforming it into a potential breeding ground for pathogens. Pathogen-contaminated surfaces can thus contribute to persistent health-related infections (Chen et al. 2020). Furniture also hosts dust mites, as noted by Cunningham (1998). Among the array of fungal strains that infest and degrade wood, molds, white rot fungi, black rot fungi, and blue stain fungus are found to be commonly present. (Embacher et al. 2023).

1.2.6 Books/Stationery

The organic composition of paper creates an environment conducive to the growth of microorganisms. Among the most prominent contaminants reported for paper and paperboard are Bacillus spore-bearing Gram-positive bacteria

(Mohammadzadeh-Vazifeh et al. 2015). Books can inadvertently become sources for transmitting various illnesses, including scarlet fever, typhus fever, tuberculosis, and potentially fatal blood poisoning (Hempel et al. 2014). Notably, fungal species such as *Cladosporium*, *Penicillium*, *Aspergillus*, *Chaetomium*, and *Alternaria* have been identified and isolated (Maggi 2000).

1.2.7 Electronics – TV, Mobile, and Laptop

In today's world of communication, electronic items like TVs, mobiles, and laptops have become indispensable. However, as these devices are handled by multiple users, they become exposed to a variety of pathogenic bacteria. Common microbes reported to thrive on electronic surfaces encompass *Staphylococcus epidermidis*, *Klebsiella* spp., *Staphylococcus aureus*, *Escherichia coli*, *Pseudomonas* spp., *Enterobacter cloacae*, and *Enterobacter* spp. (Olu-Taiwo et al. 2021).

1.2.8 Refrigerator/Deep Freezer

While deep freezers and refrigerators are designed to maintain food freshness and safety, they are not impervious to foodborne infections (Carpentier et al. 2012). Microbes from unwashed fruits and vegetables, spillage, and hand contamination can thrive on internal surfaces even at low temperatures (4 °C). Notable species like *Pseudomonas*, *Aeromonas*, *Listeria monocytogenes*, *Yersinia enterocolitica*, *Penicillium*, and *Cladosporium* have been identified (Ye et al. 2019).

1.2.9 Kitchen Slab and Sink (Wet Surfaces)

Kitchens, among all domestic environments, harbor the highest microbial and insect populations (Flores et al. 2013). Handling raw and cooked food leads to microbial exposure, resulting in diverse microbial communities that are affected by surface conditions, usage patterns, and source dispersal. Bacteria belonging to phyla *Actinobacteria*, *Bacteroidetes*, *Firmicutes*, and *Proteobacteria* are found on wet surfaces. Pathogenic bacteria like *Campylobacter* spp., *Salmonella* spp., *Clostridium perfringens*, and various *E. coli* strains are associated with foodborne illnesses (Flores et al. 2011, 2013). Additionally, wet surfaces and drains in kitchens house spores of Fusarium and Exophiala taxa (Adams et al. 2013b). Cockroaches, ants, fruit flies, drain flies, silverfish, beetles, and weevils are the prevalent biocontaminants in kitchens.

1.2.10 Kitchen Storage (Dry Surfaces)

Pantry pests, including moths, beetles, and weevils, infest dried foods, often entering unopened packages. These pests can be present in various life cycle stages simultaneously – egg, larva, pupa, or adult. Low-moisture stored foods are susceptible to spoilage by xerophilic fungi, with Eurotium species being a common culprit. *Wallemia sebi* is frequently found in cereals and spices, while *Aspergillus* species, particularly *A. flavus*, *A. niger*, *A. candidus*, and *A. ochraceus*, infect nuts. *Penicillium* species frequently contaminate cereals (Pitt and Hocking 2009). Harmful bacteria like *Helicobacter pylori*, *Salmonella* spp., *Campylobacter*, *Yersinia*, *S. epidermidis*, and *Pseudomonas* spp. have been detected on utensils, knives, and cutlery (Moari and De 2010; Erickson et al. 2015; Shayeghi et al. 2020). Various arthropod species also infest stored food in kitchen cabinets (Frost et al. 2010).

1.2.11 Bathroom Floor/Walls

The presence of moisture in buildings facilitates the proliferation of bacteria, fungi, and protozoa. Damp walls have yielded both Gram-positive and Gram-negative bacteria, with numerous species of *Streptomyces*, *Nocardia*, and *Mycobacterium* identified (Thrasher and Crawley 2009). Damp environments, particularly walls and floors, provide an ideal habitat for mold growth. Prominently, the following species have been isolated encompassing *Penicillium*, *Cladosporium*, *Ulocladium*, *Geomyces pannorum*, and *Sistronema brinkmannii*. *Stachybotrys*, a cellulose-loving fungal species, has been identified as a tertiary colonizer on walls (Kuhn and Ghannoum 2003). In modern bathrooms with tiled surfaces, water accumulation in adhesive layers fosters the growth of actinomycetes (filamentous bacteria) and fungi (such as *Aspergillus*, *Aureobasidium*, *Cladosporium*, *Fusarium*, *Paecilomyces*, and *Penicillium*) at the interface between tiles and waterproofing, as evidenced by Bok and Johansson's study (2020).

1.2.12 Toilet and Its Surfaces

Surfaces routinely touched by hands, like doors, faucet handles, soap dispensers, toilet seats, and flush handles exhibit contamination by fecal bacteria like Firmicutes (e.g. *Clostridiales, Ruminococcaceae, Lachnospiraceae*) and Bacteroidetes (e.g. *Prevotellaceae, Bacteroidaceae*). Skin-associated bacteria such as *Staphylococcus aureus* are found on hand-touched surfaces (Flores et al. 2011). Under the rim of toilets, Egert et al. (2010) found α-Proteobacteria and Actinobacteria like *Brevundimonas*. Floor contamination can occur through water splashes. Fouquier et al.'s study (2016) on fungal species in public restrooms revealed diverse isolates including *Cladosporium, Saccharomyces, Alternaria, Cryptococcus, Phoma, Teratosphaeria, Penicillium, Fusarium, Coniosporium, Hormonema, Rhodotorula,* and *Saccharomyces* spp.

1.2.13 Basement/Attic

Basements and attics, often used for storage, are commonly overlooked spaces in homes and offices. Their dark, damp, and poorly ventilated conditions, coupled with wooden materials, create an ideal habitat for mold growth. Notably, *Stachybotrys chartarum* (black mold) thrives on damp walls, with other species like *Penicillium, Cladosporium, Ulocladium, Geomyces pannorum,* and *Sistronema brinkmannii* also isolated from moisture-laden buildings (Kuhn and Ghannoum 2003).

1.2.14 Hospitals

Hospital environments serve as exposure points for bioaerosols and pathogenic microorganisms from various sources. Even asymptomatic healthy staff members can carry pathogens such as *Staphylococcus aureus, Staphylococcus pyogenes, Neisseria meningitidis, Corynebacterium diphtheriae, hepatitis B virus,* and *cytomegalovirus* (Bonadonna et al. 2017). Gram-negative rods like *Escherichia coli, Pseudomonas, Enterobacter,* and *Gram-positive cocci such as Enterococcus* and *S. aureus* are the most concerning pathogens in healthcare-associated infections (Bures et al. 2000). Water cooling systems can release aerosols carrying pathogens like *Legionella* sp., *Mycobacterium* sp., *Aspergillus* sp., *zygomycetes*, and *Fusarium* sp. Fungal species are also present in water distribution systems. These bioaerosols are potential sources of pathogens causing skin and lung diseases. Various opportunistic bacteria and fungi inhabit hospital microenvironments, with Gram-negative bacteria persisting longer compared to Gram-positive counterparts. Increased moisture elevates the persistence of multiple bacteria types, including *Chlamydia trachomatis, Listeria monocytogenes, Salmonella typhimurium, Pseudomonas aeruginosa*, and *Escherichia coli* (Bonadonna et al. 2017).

1.3 Factors Influencing Biocontaminant Accumulation in Microenvironments

1.3.1 Entry of Outdoor Microbes

Over 90% of our time is spent indoors, exposing us to a diverse mix of biological materials and particulate matter from both indoor and outdoor sources (Besis et al. 2021). Elements such as architectural design, interactions with surfaces and objects, rituals, outdoor environment, and occupants collectively impact IAQ (Horve et al. 2020). Outdoor pollutants are carried indoors by humans through clothes and shoes, as well as factors like temperature, air exchange rate, ventilation, and cleaning methods, determine indoor biocontaminant levels (D'ovidio et al. 2021). The indoor environment reflects a complex interplay between outdoor biome and occupants, wherein microbiomes combine with pressures from human behavior and building structures (Adams et al. 2013a). Microbiological contamination becomes problematic when it exceeds certain thresholds for specific environments (Almatawah et al. 2022).

1.3.2 Resident Biocontaminants (in Dust)

Residential dust hosts a community of around 500–1000 microbial species, some beneficial and others potentially harmful (Shan et al. 2019). Indoor dust serves as a source for microbial dispersion, with anthropogenic factors influencing microbial communities. Dust can also contain organic contaminants, contributing to diseases and health issues (Salthammer 2020). Such contaminants stem from various sources, including construction materials and consumer products like polycyclic aromatic hydrocarbons (PAHs), polychlorinated biphenyls (PCBs), and organophosphate flame retardants (OPFRs) (Wang et al. 2018). The pandemic's emphasis on self-care, including disinfecting homes, homeschooling, and remote work, underscores the importance of addressing indoor environments containing dust and microbes (Juel Holst et al. 2020; Thompson et al. 2021).

1.3.3 Human Occupancy

Indoor biocontaminants influence human microbiomes and health outcomes, with airborne contaminants contributing to illnesses like asthma (Shan et al. 2019). Human presence introduces respirable particulate matter, bacteria, and skin-derived microbes to indoor spaces. Human skin sheds about a billion cells daily, with each square centimeter containing 10^2–10^7 bacteria (Hospodsky et al. 2012). These indoor microbes can negatively impact the human nervous, cardiovascular, and developmental systems. Additionally, heavy metals pose exposure risks through dermal contact, inhalation, and ingestion, especially for crawling infants (Krantz and Dorevitch 2004; Grimsley et al. 2012; Arar et al. 2019). High pollutant concentrations can cause acute and chronic health issues, even fatal outcomes (De Juniac et al. 2012). Studies indicate a link between indoor PM2.5 levels and respiratory symptoms in children (Delfino et al. 2004; Jung et al. 2012).

1.3.4 Pets and Aquariums

Around 50% of households in developing countries have pets, primarily dogs, which can release allergens into indoor environments (Ahluwalia and Matsui 2018). These allergens can remain suspended in the air, adhering to clothing or transferring from nostrils to surfaces. Aquariums, however, improve IAQ by increasing humidity and reducing water-soluble pollutants like formaldehyde (Lee et al. 2017). While pets provide companionship, they can also carry allergens and toxins. Cats, for instance, can carry endocrine-disrupting toxins that impact their health and longevity. Different metabolism rates in pets contribute to indoor pollutant accumulation.

1.3.5 Construction Materials and VOCs

Building materials can introduce contaminants like plasticizers, pesticides, and PCBs to indoor spaces (Tran et al. 2020). VOCs, released from sources such as smoking, solvents, renovations, and household products, contribute to IAQ hazards (Vardoulakis et al. 2020). Indoor air contains chemical pollutants, bioaerosols, and VOCs that collectively contribute to health risks, leading to conditions like SBS (Shuang et al. 2014). SBS symptoms include headaches, eye discomfort, nasal issues, fatigue, and concentration difficulties (Schirmer et al. 2011). Adequate ventilation is key to diluting indoor contaminants, which impact visitor and occupant well-being (Ponsoni and Raddi 2010). Parameters like air changes per hour, humidity, filtration, and cleaning practices affect IAQ and subsequent health outcomes (Lutz et al. 2003). Indoor air pollutants from building material emissions cause approximately 5 million deaths annually (Ma et al. 2022).

1.4 Conditions Affecting Spread of Biocontaminants

1.4.1 Light (Sunlight and Artificial Light, Darkness)

Historically, buildings were designed with south-facing orientations to harness sunlight's decontaminating benefits, promoting health and comfort. However, crowding and poor architecture of buildings may result in lower than adequate availability of sunlight. Research indicates that light exposure can control bacterial infections. Studies explore diverse light types, from UV light to blue light, revealing their potential to reduce virus virulence and enhance virion decay (Tang 2009). UV exposure can control microbial growth, though prolonged exposure might lead to the evolution of UV-adapted microbes (Goldman and Travisano 2011). Daylight has proven health benefits, regulating metabolic rhythms in vertebrates and contributing to a healthier indoor environment (Hobday and Dancer 2013).

1.4.2 Temperature

Indoor temperature has varying effects on microbial communities, influenced by factors like AC, building usage, and chemical utilization. While indoor temperature impacts microbial growth patterns, it does not significantly affect human and animal hosts' shedding rates or susceptibility (Lowen et al. 2007). Temperature and air exchange rates correlate with fungi and enzyme activity positively, but with bacteria negatively (Frankel et al. 2012). Steam decontamination effectively inactivates biological agents and toxins, proving useful in biocontaminant control (Bearss et al. 2017).

1.4.3 Humidity/Moisture

Relative humidity significantly affects fungal spore and particle release from infested surfaces. High humidity (>80%) supports microbial survival and activity; it influences bacterial inactivity by targeting proteins and membrane phospholipids. Viruses with structural lipids remain stable at low humidity, while those without lipids are more stable at high humidity (Zhao et al. 2012). Humidity affects particle evaporation, density, diameter, and settling velocity (Dungan 2010).

1.4.4 Air Circulation

IAQ significantly impacts occupants' health and well-being, with bioaerosols containing diverse particle sizes and compositions (Wolkoff 2018). Bioaerosols can range from virus-sized particles (0.003 μm) to bacteria (0.25–20 μm) and pollen (17–58 μm) (Ghosh et al. 2015). Bioaerosols pose health risks, including potential carcinogenicity and respiratory inflammation (Douwes et al. 2000). Poor ventilation in residential buildings exacerbates bioaerosol growth, contributing to health issues like SBS (Wålinder et al. 2001).

1.4.5 Spores, Pollens, and Allergens

Bioaerosols contain pollen and fungal spores, which are potent allergens and inflammogens (Kurup et al. 2002). Indoor fungal concentrations exceed pollen levels in environments such as hospital patient rooms. Dominant indoor fungi include *Aspergillus*, *Botrytis*, and *Cladosporium*. Many molds produce glycoprotein allergens triggering allergies and asthma, impacting a significant portion of the population, and leading to chronic illnesses (Twaroch et al. 2015)

1.5 Preventing Biocontaminants in Microenvironments

1.5.1 Cleaning Habits – Thorough Cleaning, Disinfecting

Preventing the accumulation of dirt and microbial growth is achieved through regular cleaning, effectively removing germs from surfaces using soap or detergent-based cleansers. The efficacy of cleaning alone in eliminating harmful viruses and bacteria from surfaces is evidenced (Le Cann et al. 2011). Ensuring the effectiveness of sanitizers and disinfectants by prior surface cleaning is recommended (Sandle 2018). Routine cleaning suffices to manage germ transmission, through hard and soft surfaces, laundry, and electronics. Surface-specific agents and manufacturer guidelines are followed for cleaning hard and soft surfaces, with proper laundering and drying for fabrics and carpets. Utilizing EPA-registered sprays or milder bleach solutions is suggested for sanitization. Some items can be sanitized in dishwashers with appropriate cycles. The process of disinfection entails the following: (i) Washing surfaces with soap and water. (ii) Verifying suitability of disinfectants for specific surfaces. (iii) Adhering to disinfectant label instructions and precautions. (iv) Allowing adequate contact time for effective bacteria eradication. (v) Maintaining surface moisture throughout the contact period. (vi) Ensuring proper indoor ventilation during disinfection. Enhanced disinfectants, such as those based on hydrogen peroxide, exhibit significant reduction in bacterial levels on surfaces. Usage of such products has been associated with reduced healthcare-associated infections. They demonstrate efficacy against multi-drug-resistant bacteria on soft surfaces. While their potency is not on par with bleach solutions, they carry EPA safety rating category IV and require no personal protective equipment during application.

1.6 Preventing Biocontaminants in Controlled Environments

1.6.1 Cleaning Protocols – Comprehensive Cleaning and Disinfection

The utilization of cleanrooms and clean air technologies is primarily guided by their purpose, determined by air quality assessments. Pharmaceutical cleanrooms, for instance, adhere to EU GUP guidelines, using alphabetic notations, or the ISO 14644 international standard, employing numeric classes for air cleanliness (Sandle 2018). Biosafety cabinets known as Unidirectional Air Flow System (UDAF) create self-contained environments safeguarding against hazardous microorganisms. These cabinets establish protective air barriers and employ exhaust filtration. Some UDAF systems incorporate gloves to minimize human interaction, termed Restricted Access Barrier Systems (RABS). Isolators, gloveboxes, and hatches also provide varying degrees of biocontaminant protection. Specialized cleanroom clothing, like nonshedding gowns or coveralls, headgear, facemasks, gloves, and shoe coverings, is designed to reduce contaminant shedding.

1.6.2 Efficient Ventilation

Ventilation plays a crucial role in ensuring IAQ by introducing and distributing fresh external air while expelling indoor pollutants (Yang et al. 2020). Key components of building ventilation encompass airflow direction, ventilation rate, and airflow pattern. Ideally, airflow should move from clean to contaminated areas, with a sufficient influx of quality outdoor air. Two main methods achieve ventilation: natural and mechanical. Natural ventilation relies on external forces, like wind and thermal buoyancy, while mechanical ventilation employs fans to circulate air through ducts or openings (Elmualim et al. 2003).

1.6.3 HVAC Systems for Enhanced Air Quality

Heating, Ventilation and Air Conditioning (HVAC) systems are essential for IAQ management, maintaining directional airflow and air change rates (Mousavi and Grosskopf 2014). Clean air devices, including hoods, glove boxes, isolators, and mini-environments, are used to reduce turbulence in cleanrooms to prevent dust accumulation (ISO 14644-7).

1.6.4 Optimizing Sunlight in Indoor Environments

As indoor living becomes more prevalent, IAQ is a growing concern. IAQ standards and guidelines exist to regulate exposure to various indoor pollutants such as gases, particles, volatile organic compounds (VOCs), and aldehydes (Lizana et al. 2020). Besides outdoor pollution, indoor sources like combustion, cooking, and materials like furniture and paints impact IAQ. Wallpaper adhesives and polyurethane insulation are also notable indoor VOC sources (Tran et al. 2020).

1.7 Conclusion

Indoor microenvironment harbors biocontaminants in a wide range of objects such as furnishings, humidifiers, centralized AC units, decoratives, and wet surfaces in kitchens and bathrooms; apparently they are found everywhere. As they are widespread in the indoor environments, they directly or indirectly affect the human health. Several ailments and diseases are attributed to the biocontaminants such as infectious diseases, respiratory diseases, and skin allergies. The health threats are serious for susceptible hosts and depend upon the type and amount of biocontaminants, and duration of exposure. Thus it is imperative to identify the source of biocontamination and evaluate the methods of their prevention and decontamination. Appropriate and effective protocols could be designed to curb indoor biocontamination following the strategies based on identification, characterization, and quantification of biocontaminants in microenvironments.

References

Adams, R.I., Miletto, M., Taylor, J.W., and Bruns, T.D. (2013a). Dispersal in microbes: fungi in indoor air are dominated by outdoor air and show dispersal limitation at short distances. *The ISME Journal* 7 (7): 1262–1273.

Adams, R.I., Miletto, M., Taylor, J.W., and Bruns, T.D. (2013b). The diversity and distribution of fungi on residential surfaces. *PLoS One* 8 (11): e78866.

Ahluwalia, S.K. and Matsui, E.C. (2018). Indoor environmental interventions for furry pet allergens, pest allergens, and mold: looking to the future. *The Journal of Allergy and Clinical Immunology: In Practice* 6 (1): 9–19.

Almatawah, Q.A., Al-Rashidi, M.S., Yassin, M.F., and Varghese, J.S. (2022). Microbiological contamination of indoor and outdoor environments in a desert climate. *Environmental Monitoring and Assessment* 194 (5): 355.

Altunatmaz, S.S., Issa, G., and Aydin, A. (2012). Detection of airborne psychrotrophic bacteria and fungi in food storage refrigerators. *Brazilian Journal of Microbiology* 43: 1436–1443.

Andersen, B., Frisvad, J.C., Søndergaard, I. et al. (2011). Associations between fungal species and water-damaged building materials. *Applied and Environmental Microbiology* 77 (12): 4180–4188.

Arar, S., Al-Hunaiti, A., Masad, M.H. et al. (2019). Elemental contamination in indoor floor dust and its correlation with PAHs, fungi, and gram +/− bacteria. *International Journal of Environmental Research and Public Health* 16 (19): 3552.

Arlian, L.G., Morgan, M.S., and Neal, J.S. (2002). Dust mite allergens: ecology and distribution. *Current Allergy and Asthma Reports* 2: 401–411.

Bearss, J.J., Honnold, S.P., Picado, E.S. et al. (2017). Validation and verification of steam sterilization procedures for the decontamination of biological waste in a biocontainment laboratory. *Applied Biosafety* 22 (1): 33–37.

Besis, A., Botsaropoulou, E., Balla, D. et al. (2021). Toxic organic pollutants in Greek house dust: implications for human exposure and health risk. *Chemosphere* 284: 131318.

Bok, G. and Johansson, P., 2020. Microbial growth behind tiles in bathrooms. *E3S Web of Conferences, 12th Nordic Symposium on Building Physics (NSB 2020)*, Tallinn, Estonia (6–9 September 2020), vol. 172, p. 20003. EDP Sciences.

Bonadonna, L., Briancesco, R. and Coccia, A.M., 2017. Analysis of microorganisms in hospital environments and potential risks. *Indoor Air Quality in Healthcare Facilities*, Guest Editor (s): Stefano Capolongo, Gaetano Settimo and Marco Gola, Springer, pp. 53–62.

Borchers, A.T., Chang, C., and Eric Gershwin, M. (2017). Mold and human health: a reality check. *Clinical Reviews in Allergy & Immunology* 52 (3): 305–322.

Bures, S., Fishbain, J.T., Uyehara, C.F. et al. (2000). Computer keyboards and faucet handles as reservoirs of nosocomial pathogens in the intensive care unit. *American Journal of Infection Control* 28 (6): 465–471.

Caillaud, D., Leynaert, B., Keirsbulck, M., and Nadif, R. (2018). Indoor mould exposure, asthma and rhinitis: findings from systematic reviews and recent longitudinal studies. *European Respiratory Review* 27 (148): 170137.

Carpentier, B., Lagendijk, E., Chassaing, D. et al. (2012). Factors impacting microbial load of food refrigeration equipment. *Food Control* 25 (1): 254–259.

Chen, J.C., Munir, M.T., Aviat, F. et al. (2020). Survival of bacterial strains on wood (Quercus petraea) compared to polycarbonate, aluminum and stainless steel. *Antibiotics* 9 (11): 804.

Chew, G.L., Horner, W.E., Kennedy, K. et al. (2016). Procedures to assist health care providers to determine when home assessments for potential mold exposure are warranted. *The Journal of Allergy and Clinical Immunology: In Practice* 4 (3): 417–422.

Chirca, I. (2019). The hospital environment and its microbial burden: challenges and solutions. *Future Microbiology* 14 (12): 1007–1010.

Cho, S.H., Reponen, T., Bernstein, D.I. et al. (2006). The effect of home characteristics on dust antigen concentrations and loads in homes. *Science of the Total Environment* 371 (1–3): 31–43.

Cho, S.Y., Myong, J.P., Kim, W.B. et al. (2018). Profiles of environmental mold: indoor and outdoor air sampling in a hematology hospital in Seoul, South Korea. *International Journal of Environmental Research and Public Health* 15 (11): 2560.

Cunningham, M.J. (1998). Direct measurements of temperature and humidity in dust mite microhabitats. *Clinical & Experimental Allergy* 28 (9): 1104–1112.

D'Ovidio, M.C., Di Renzi, S., Capone, P., and Pelliccioni, A. (2021). Pollen and fungal spores evaluation in relation to occupants and microclimate in indoor workplaces. *Sustainability* 13 (6): 3154.

De Juniac, A., Kreis, I., Ibison, J., and Murray, V. (2012). Epidemiology of unintentional carbon monoxide fatalities in the UK. *International Journal of Environmental Health Research* 22 (3): 210–219.

Delfino, R.J., Quintana, P.J., Floro, J. et al. (2004). Association of FEV1 in asthmatic children with personal and microenvironmental exposure to airborne particulate matter. *Environmental Health Perspectives* 112 (8): 932–941.

Douwes, J., Wouters, I., Dubbeld, H. et al. (2000). Upper airway inflammation assessed by nasal lavage in compost workers: a relation with bio-aerosol exposure. *American Journal of Industrial Medicine 37* (5): 459–468.

Dunca, S.I., Tanase, C., Padurariu, C. et al. (2014). Study of the contaminating microbiota of old paper supports. *European Scientific Journal* 3: 237–251.

Dungan, R.S. (2010). Board-invited review: fate and transport of bioaerosols associated with livestock operations and manures. *Journal of Animal Science* 88 (11): 3693–3706.

Egert, M., Schmidt, I., Bussey, K., and Breves, R. (2010). A glimpse under the rim – the composition of microbial biofilm communities in domestic toilets. *Journal of Applied Microbiology* 108 (4): 1167–1174.

Egwari, L.O., Oniha, M.I., and Ayepola, O.O. (2015). Susceptibility of fabrics in office furniture to microbial attack: microbial burden and health implications. *Antimicrobial Resistance and Infection Control* 4 (1): P36.

Elmualim, A.A., Awbi, H.B., Fullford, D., and Wetterstad, L. (2003). Performance evaluation of a wall mounted convector for pre-heating naturally ventilated spaces. *International Journal of Ventilation* 2 (3): 213–222.

Embacher, J., Zeilinger, S., Kirchmair, M. et al. (2023). Wood decay fungi and their bacterial interaction partners in the built environment – a systematic review on fungal bacteria interactions in dead wood and timber. *Fungal Biology Reviews* 45: 100305.

Erickson, M.C., Liao, J., Cannon, J.L., and Ortega, Y.R. (2015). Contamination of knives and graters by bacterial foodborne pathogens during slicing and grating of produce. *Food Microbiology* 52: 138–145.

Flores, G.E., Bates, S.T., Caporaso, J.G. et al. (2013). Diversity, distribution and sources of bacteria in residential kitchens. *Environmental Microbiology* 15 (2): 588–596.

Flores, G.E., Bates, S.T., Knights, D. et al. (2011). Microbial biogeography of public restroom surfaces. *PLoS One* 6 (11): e28132.

Fouquier, J., Schwartz, T., and Kelley, S.T. (2016). Rapid assemblage of diverse environmental fungal communities on public restroom floors. *Indoor Air* 26 (6): 869–879.

Frankel, M., Bekö, G., Timm, M. et al. (2012). Seasonal variations of indoor microbial exposures and their relation to temperature, relative humidity, and air exchange rate. *Applied and Environmental Microbiology 78* (23): 8289–8297.

Frost, C.L., Braig, H.R., Amendt, J., and Perotti, M.A. (2010). Indoor arthropods of forensic importance: insects associated with indoor decomposition and mites as indoor markers. In: *Current Concepts in Forensic Entomology* (ed. J. Amendt, M.L. Goff, C.P. Campobasso, and M. Grassberger), 93–108. Springer.

Ghosh, B., Lal, H., and Srivastava, A. (2015). Review of bioaerosols in indoor environment with special reference to sampling, analysis and control mechanisms. *Environment International* 85: 254–272.

Goldman, R.P. and Travisano, M. (2011). Experimental evolution of ultraviolet radiation resistance in *Escherichia coli*. *Evolution* 65 (12): 3486–3498.

Grimsley, L.F., Chulada, P.C., Kennedy, S. et al. (2012). Indoor environmental exposures for children with asthma enrolled in the HEAL study, post-Katrina New Orleans. *Environmental Health Perspectives* 120 (11): 1600–1606.

Hamada, N. and Fujita, T. (2002). Effect of air-conditioner on fungal contamination. *Atmospheric Environment* 36 (35): 5443–5448.

Hempel, M., Rakhra, V., Rothwell, A., and Song, D. (2014). Bacterial and fungal contamination in the library setting: a growing concern? *Environmental Health Review* 57 (1): 9–15.

Hobday, R.A. and Dancer, S.J. (2013). Roles of sunlight and natural ventilation for controlling infection: historical and current perspectives. *Journal of Hospital Infection* 84 (4): 271–282.

Horve, P.F., Lloyd, S., Mhuireach, G.A. et al. (2020). Building upon current knowledge and techniques of indoor microbiology to construct the next era of theory into microorganisms, health, and the built environment. *Journal of Exposure Science & Environmental Epidemiology* 30 (2): 219–235.

Hospodsky, D., Qian, J., Nazaroff, W.W. et al. (2012). Human occupancy as a source of indoor airborne bacteria. *PLoS One* 7 (4): e34867.

Juel Holst, G., Pørneki, A., Lindgreen, J. et al. (2020). Household dampness and microbial exposure related to allergy and respiratory health in Danish adults. *European Clinical Respiratory Journal* 7 (1): 1706235.

Jung, K.H., Hsu, S.I., Yan, B. et al. (2012). Childhood exposure to fine particulate matter and black carbon and the development of new wheeze between ages 5 and 7 in an urban prospective cohort. *Environment International* 45: 44–50.

Kader, R., Kennedy, K., and Portnoy, J.M. (2018). Indoor environmental interventions and their effect on asthma outcomes. *Current Allergy and Asthma Reports* 18: 1–10.

Karimpour Roshan, S., Godini, H., Nikmanesh, B. et al. (2019). Study on the relationship between the concentration and type of fungal bio-aerosols at indoor and outdoor air in the Children's Medical Center, Tehran, Iran. *Environmental Monitoring and Assessment* 191: 1–13.

Kausar, M.A. (2018). A review on respiratory allergy caused by insects. *Bioinformation* 14 (9): 540.

Kelkar, U., Bal, A.M., and Kulkarni, S. (2005). Fungal contamination of air conditioning units in operating theatres in India. *Journal of Hospital Infection* 60 (1): 81–84.

Kern, E.B., Sherris, D., Stergiou, A.M. et al. (2007). Diagnosis and treatment of chronic rhinosinusitis: focus on intranasal Amphotericin B. *Therapeutics and Clinical Risk Management* 3 (2): 319–325.

Khan, A.H., Karuppayil, S.M., Manoharachary, C. et al. (2009). Isolation, identification and testing for allergenicity of fungi from air-conditioned indoor environments. *Aerobiologia* 25: 119–123.

Koscova, J., Hurnikova, Z., and Pistl, J. (2018). Degree of bacterial contamination of mobile phone and computer keyboard surfaces and efficacy of disinfection with chlorhexidine digluconate and triclosan to its reduction. *International Journal of Environmental Research and Public Health* 15 (10): 2238.

Krantz, A. and Dorevitch, S. (2004). Metal exposure and common chronic diseases: a guide for the clinician. *Disease-a-Month* 50 (5): 220–262.

Kuhn, D.M. and Ghannoum, M. (2003). Indoor mold, toxigenic fungi, and *Stachybotrys chartarum*: infectious disease perspective. *Clinical Microbiology Reviews* 16 (1): 144–172.

Kumar, P., Kausar, M.A., Singh, A.B., and Singh, R. (2021). Biological contaminants in the indoor air environment and their impacts on human health. *Air Quality, Atmosphere & Health* 14 (11): 1723–1736.

Kumar, P., Singh, A.B., and Singh, R. (2022). Spatial variation and comprehensive health risk assessment of microbial indoor air quality in microenvironments of North Delhi. *Plos one 17* (2): e0264226.

Kumari, P., Woo, C., Yamamoto, N., and Choi, H.L. (2016). Variations in abundance, diversity and community composition of airborne fungi in swine houses across seasons. *Scientific Reports* 6 (1): 37929.

Kurup, V.P., Shen, H.D., and Vijay, H. (2002). Immunobiology of fungal allergens. *International Archives of Allergy and Immunology* 129 (3): 181–188.

Kusumaningrum, H.D. (2003). *Behaviour and Cross-Contamination of Pathogenic Bacteria in Household Kitchens-Relevance to Exposure Assessment.* Wageningen University and Research.

Le Cann, P., Bonvallot, N., Glorennec, P. et al. (2011). Indoor environment and children's health: recent developments in chemical, biological, physical and social aspects. *International Journal of Hygiene and Environmental Health* 215 (1): 1–18.

Lee, J.H., Ahn, K.H., and Yu, I.J. (2012). Outbreak of bioaerosols with continuous use of humidifier in apartment room. *Toxicological Research* 28: 103–106.

Lee, J.M., Ban, H., Lee, Y. et al. (2017). Impact of aquariums on indoor environmental quality. *Journal of Environmental Health Sciences 43* (1): 87–94.

Li, Y., Leung, G.M., Tang, J.W. et al. (2007). Role of ventilation in airborne transmission of infectious agents in the built environment-a multidisciplinary systematic review. *Indoor Air* 17 (1): 2–18.

Liu, G., Xiao, M., Zhang, X. et al. (2017). A review of air filtration technologies for sustainable and healthy building ventilation. *Sustainable Cities and Society* 32: 375–396.

Liu, Z., Deng, Y., Ma, S. et al. (2021). Dust accumulated fungi in air-conditioning system: findings based on field and laboratory experiments. In: *Building Simulation*, vol. 14, 793–811. Tsinghua University Press.

Liu, Z., Ma, S., Cao, G. et al. (2018). Distribution characteristics, growth, reproduction and transmission modes and control strategies for microbial contamination in HVAC systems: a literature review. *Energy and Buildings* 177: 77–95.

Lizana, J., Almeida, S.M., Serrano-Jimenez, A. et al. (2020). Contribution of indoor microenvironments to the daily inhaled dose of air pollutants in children. The importance of bedrooms. *Building and Environment* 183: 107188.

Lowen, A.C., Mubareka, S., Steel, J., and Palese, P. (2007). Influenza virus transmission is dependent on relative humidity and temperature. *PLoS Pathogens* 3 (10): e151.

Lutz, B.D., Jin, J., Rinaldi, M.G. et al. (2003). Outbreak of invasive *Aspergillus* infection in surgical patients, associated with a contaminated air-handling system. *Clinical Infectious Diseases* 37 (6): 786–793.

Ma, T., Wu, P., Ding, Z. et al. (2022). Pet cats, the better sentinels for indoor organic pollutants. *Frontiers in Environmental Science* 10: 2064.

Maggi, O., Persiani, A.M., Gallo, F. et al. (2000). Airborne fungal spores in dust present in archives: proposal for a detection method, new for archival materials. *Aerobiologia* 16: 429–434.

Maori, L. and De, N. (2010). Bacterial contamination of crockery and cutlery within the kiosks' restaurants of the Federal University of Technology, Yola. *African Journal of Microbiology Research* 4 (3): 147–153.

Meng, C., Wang, Q., Song, Y. et al. (2015). Experimental study on both cleaning effect and motion performance of the duct-cleaning robot. *Sustainable Cities and Society* 14: 64–69.

Messina, G., Quercioli, C., Burgassi, S. et al. (2011). How many bacteria live on the keyboard of your computer? *American Journal of Infection Control* 39 (7): 616–618.

Miller, J.D., Dugandzic, R., Frescura, A.M., and Salares, V. (2007). Indoor-and outdoor-derived contaminants in urban and rural homes in Ottawa, Ontario, Canada. *Journal of the Air & Waste Management Association* 57 (3): 297–302.

Modebelu, M.N. and Modebelu, A.F. (2013). Microbial efficiency of air conditioners and other ventilation devices for enhancing future air quality. *Academic Journal of Interdisciplinary Studies* 2 (5): 163.

Mohammadzadeh-Vazifeh, M.M., Hosseini, S.M., Khajeh-Nasiri, S. et al. (2015). Isolation and identification of bacteria from paperboard food packaging. *Iranian Journal of Microbiology* 7 (5): 287.

Moldoveanu, A.M. (2015). Biological contamination of air in indoor spaces. In: *Current Air Quality Issues* (ed. F. Nejadkoorki). IntechOpen.

Moses, L., Morrissey, K., Sharpe, R.A., and Taylor, T. (2019). Exposure to indoor mouldy odour increases the risk of asthma in older adults living in social housing. *International Journal of Environmental Research and Public Health* 16 (14): 2600.

Mousavi, E.S. and Grosskopf, K.R. (2014). Transport of respiratory aerosols in patient corridors subject to a directional and non-directional airflow-a case study. *ASHRAE Transactions* 120 (2).

Murrison, L.B., Brandt, E.B., Myers, J.B., and Hershey, G.K.K. (2019). Environmental exposures and mechanisms in allergy and asthma development. *The Journal of Clinical Investigation* 129 (4): 1504–1515.

Nam, H.S., Siebers, R., Lee, S.H. et al. (2008). House dust mite allergens in domestic homes in Cheonan, Korea. *The Korean Journal of Parasitology* 46 (3): 187.

Noris, F. (2009). Biological and metal contaminants in HVAC filter dust. *Ashrae Transactions* 115: 484.

Novak Babič, M., Gostinčar, C., and Gunde-Cimerman, N. (2020). Microorganisms populating the water-related indoor biome. *Applied Microbiology and Biotechnology* 104: 6443–6462.

Olu-Taiwo, M., Laryea, C.A., Kweku Mykels, D., and Forson, A.O. (2021). Multidrug-resistant bacteria on the mobile phones and computer keyboards of healthcare university students in Ghana. *Canadian Journal of Infectious Diseases and Medical Microbiology* 2021: 6647959.

Park, J., Ichijo, T., Nasu, M., and Yamaguchi, N. (2016). Investigation of bacterial effects of Asian dust events through comparison with seasonal variability in outdoor airborne bacterial community. *Scientific Reports* 6 (1): 35706.

Piekarska, K., Trusz, A., and Szczęśniak, S. (2018). Bacteria and fungi in two air handling units with air recirculating module. *Energy and Buildings* 178: 154–164.

Pitt, J.I. and Hocking, A.D. (2009). *Fungi and Food Spoilage*, vol. 519, 388. New York: Springer.

Ponikau, J.U., Sherris, D.A., Kephart, G.M. et al. (2006). The role of ubiquitous airborne fungi in chronic rhinosinusitis. *Clinical Reviews in Allergy & Immunology* 30: 187–194.

Ponikau, J.U., Sherris, D.A., Kephart, G.M. et al. (2005). Striking deposition of toxic eosinophil major basic protein in mucus: implications for chronic rhinosinusitis. *Journal of Allergy and Clinical Immunology* 116 (2): 362–369.

Ponsoni, K. and Raddi, M.S.G. (2010). Indoor air quality related to occupancy at an air-conditioned public building. *Brazilian Archives of Biology and Technology* 53: 99–103.

Rahimi, S.M., Ebrahimi, M., Barikbin, B., and Zeinali, T. (2019). Evaluation of bacterial and fungal contamination of kitchens of Birjand University of Medical Sciences. *BMC Research Notes* 12 (1): 1–6.

Rand, T.G., Mahoney, M., White, K., and Oulton, M. (2002). Microanatomical changes in alveolar type II cells in juvenile mice intratracheally exposed to *Stachybotrys chartarum* spores and toxin. *Toxicological Sciences* 65 (2): 239–245.

Salthammer, T. (2020). Emerging indoor pollutants. *International Journal of Hygiene and Environmental Health* 224: 113423.

Sánchez Espinosa, K.C., Rojas Flores, T.I., Davydenko, S.R. et al. (2021). Fungal populations in the bedroom dust of children in Havana, Cuba, and its relationship with environmental conditions. *Environmental Science and Pollution Research* 28 (38): 53010–53020.

Sandle, T. (2018). *Biocontamination Control for Pharmaceuticals and Healthcare*. Academic Press.

Schei, M.A., Hessen, J.O., and Lund, E. (2002). House-dust mites and mattresses. *Allergy* 57 (6): 538–542.

Schilling, S., Fusco, F.M., De Iaco, G. et al. (2014). Isolation facilities for highly infectious diseases in Europe – a cross-sectional. *PLoS One* 9 (10): e100401.

Schirmer, W.N., Pian, L.B., Szymanski, M.S.E., and Gauer, M.A. (2011). Air pollution in internal environments and sick building syndrome. *Ciencia & Saude Coletiva* 16 (8): 3583.

Shan, Y., Wu, W., Fan, W. et al. (2019). House dust microbiome and human health risks. *International Microbiology* 22: 297–304.

Shayeghi, F., Matini, E., Rahbar, N.A.S.E.R. et al. (2020). Microbial contamination of kitchen instruments as a minatory to human health. *Pakistan Journal of Medical and Health Sciences 14* (2): 1282–1287.

Shuang, G.W., Kamaruzzaman, S.N., and Zulkifli, N. (2014). Occupant's perception on indoor performance of historical museum: a case of study of national museum and Perak museum, Malaysia. *2nd International Conference on Research in Science, Engineering and Technology* (21–22 March 2014), Dubai (UAE).

Sidenius, K.E., Hallas, T.E., Brygge, T. et al. (2002). House dust mites and their allergens at selected locations in the homes of house dust mite-allergic patients. *Clinical & Experimental Allergy* 32 (9): 1299–1304.

Soldatos, A.P., Bakeas, E.B., and Siskos, P.A. (2003). Occupational exposure to BTEX of workers in car parkings and gasoline service stations in Athens, Greece. *Fresenius Environmental Bulletin* 12 (9): 1064–1070.

Srikanth, P., Sudharsanam, S., and Steinberg, R. (2008). Bio-aerosols in indoor environment: composition, health effects and analysis. *Indian Journal of Medical Microbiology* 26 (4): 302–312.

Stetzenbach, L.D., Buttner, M.P., and Cruz, P. (2004). Detection and enumeration of airborne biocontaminants. *Current Opinion in Biotechnology* 15 (3): 170–174.

Sun, L., Miller, J.D., Van Ryswyk, K. et al. (2022). Household determinants of biocontaminant exposures in Canadian homes. *Indoor Air* 32 (1): e12933.

Swaebly, M.A. (1952). Molds and bacteria in house dust and furniture stuffing. *Journal of the Minnesota Academy of Science* 20 (1): 8–10.

Tang, J.W. (2009). The effect of environmental parameters on the survival of airborne infectious agents. *Journal of the Royal Society Interface* 6 (suppl_6): S737–S746.

Thompson, J.R., Argyraki, A., Bashton, M. et al. (2021). Bacterial diversity in house dust: characterization of a core indoor microbiome. *Frontiers in Environmental Science* 3: 533.

Thrasher, J.D. and Crawley, S. (2009). The biocontaminants and complexity of damp indoor spaces: more than what meets the eyes. *Toxicology and Industrial Health* 25 (9–10): 583–615.

Tran, V.V., Park, D., and Lee, Y.C. (2020). Indoor air pollution, related human diseases, and recent trends in the control and improvement of indoor air quality. *International Journal of Environmental Research and Public Health* 17 (8): 2927.

Tsakas, M.P., Siskos, A.P., and Siskos, P.A. (2011). Indoor air pollutants and the impact on human health. In: *In Chemistry, Emission Control, Radioactive Pollution and Indoor Air Quality* (ed. N. Mazzeo), 447–484. Intechopen.

Twaroch, T.E., Curin, M., Valenta, R., and Swoboda, I. (2015). Mold allergens in respiratory allergy: from structure to therapy. *Allergy, Asthma and Immunology Research* 7 (3): 205–220.

United States Environmental Protection agency (EPA). (2020). Chloramine in drinking water, drinking water requirements for states and public water systems. https://www.epa.gov/dwreginfo/chloramines-drinking-water (accessed 9 June 2020).

Van Den Bemt, L., De Vries, M.P., Van Knapen, L. et al. (2006). Influence of mattress characteristics on house dust mite allergen concentration. *Clinical & Experimental Allergy* 36 (2): 233–237.

Vardoulakis, S., Giagloglou, E., Steinle, S. et al. (2020). Indoor exposure to selected air pollutants in the home environment: a systematic review. *International Journal of Environmental Research and Public Health* 17 (23): 8972.

Visitsunthorn, N., Chirdjirapong, V., Pootong, V. et al. (2010). The accumulation of dust mite allergens on mattresses made of different kinds of materials. *Asian Pacific Journal of Allergy and Immunology* 28 (2–3): 155.

Wålinder, R., Wieslander, G., Norbäck, D. et al. (2001). Nasal lavage biomarkers: effects of water damage and microbial growth in an office building. *Archives of Environmental Health: An International Journal* 56 (1): 30–36.

Wang, Y., Sun, H., Zhu, H. et al. (2018). Occurrence and distribution of organophosphate flame retardants (OPFRs) in soil and outdoor settled dust from a multi-waste recycling area in China. *Science of the Total Environment* 625: 1056–1064.

Watanabe, K., Yanagi, U., Shiraishi, Y. et al. (2022). Bacterial communities in various parts of air-conditioning units in 17 Japanese houses. *Microorganisms* 10 (11): 2246.

Wójcik-Fatla, A., Mackiewicz, B., Sawczyn-Domańska, A. et al. (2022). Timber-colonizing gram-negative bacteria as potential causative agents of respiratory diseases in woodworkers. *International Archives of Occupational and Environmental Health* 95 (6): 1179–1193.

Wolkoff, P. (2018). Indoor air humidity, air quality, and health – an overview. *International Journal of Hygiene and Environmental Health* 221 (3): 376–390.

World Health Organization, 1990. *Indoor Air Quality: Biological Contaminants: Report on a WHO Meeting, Rautavaara* (No. ICP/CEH 073). World Health Organization, Regional Office for Europe. 29 August–2 September 1988.

Yahia, S.H. and Metwally, A.S. (2019). Effect of some housing criteria and seasonal variations on indoor prevalence and distribution of dust mite populations in Sharkia Governorate, Egypt. *Life Sciences Journal* 16 (7): 59–68.

Yang, X., Zhang, Y., Hang, J. et al. (2020). Integrated assessment of indoor and outdoor ventilation in street canyons with naturally-ventilated buildings by various ventilation indexes. *Building and Environment* 169: 106528.

Ye, K., Wang, J., Han, Y. et al. (2019). Investigation on microbial contamination in the cold storage room of domestic refrigerators. *Food Control* 99: 64–67.

Zhao, Y., Aarnink, A.J., Dijkman, R. et al. (2012). Effects of temperature, relative humidity, absolute humidity, and evaporation potential on survival of airborne Gumboro vaccine virus. *Applied and Environmental Microbiology* 78 (4): 1048–1054.

2

Estimation of Indoor Bioaerosols and Occurrence of Sick Building Syndrome Symptoms Within Office Premises in Urban Delhi

Arun Kumar Yadav and Chirashree Ghosh

Environmental Pollution Laboratory, Department of Environmental Studies, University of Delhi, Delhi, New Delhi, India

2.1 Introduction

Bioaerosols are airborne biological particulate matter, size varying from 0.3 to 100 μm and in particularly breathable bioaerosols (size range from 1 to 10 μm) pose a significant primary concern for health study (Lacey and Dutkiewicz 1994; Powitz and Ph 2017; Kim et al. 2018). Bioaerosols are considered one of the major indoor air pollutants carrying pathogens and trigger allergic reactions upon inhalation, contributing to building-related illnesses (Douwes et al. 2003; Hsu et al. 2012; Heo et al. 2017a). Office environments are complex indoor ecosystems that contain various pollutants including trillions of microorganisms. Nowadays, urban dwellers spend 80–90% of their time in the indoor environment. Indoor offices with heating and air conditioning systems offer many benefits to workers but can also expose them to various indoor air pollutants (directly emitting from photocopy machines, printers, papers, files, and other electronic gadgets), which leads to severe health risks. According to the US Environmental Protection Agency (EPA), office workers spend as much as 90% of their time in indoor environments, and insufficient ventilation resulting from an increased workforce in a building can lead to SBS (Apter et al. 1994).

The World Health Organization (WHO) defined it as an excess of work-related irritations of the skin and mucous membranes and other symptoms like fatigue, headache, and attention disorder were also reported by office workers exposed to modern enclosed buildings (World Health Organization 2018). Additionally, certain individuals who are highly sensitive to mold more susceptible to the development of SBS, which manifests through symptoms like nasal allergies, eye

Airborne Biocontaminants and Their Impact on Human Health, First Edition.
Edited by Rajeev Singh and Anamika Singh.

irritation, and chronic lung infection. Moreover, affected individuals might experience tightness in the chest, recurring coughing, wheezing, and difficulty in breathing (Hedge et al. 1995; Cooley et al. 1998; Milton 2000).

2.2 Methodology and Study Design

The study was aimed to determine spatiotemporal variation in bioaerosol distribution and its association with SBS among office workers in two diverse office complexes (Offices I and II).

2.2.1 Sampling Sites

The primary study area was the urban agglomeration of the University of Delhi. The study area is spread over an area of approximately 2 km^2 lying between latitude 28°41′47.6639″N–28°40′44.3708″N and longitude 77°12′33.003″E–77°12′36.8654″E. The satellite image of the study site is shown in Figure 2.1 indicating the GPS location of the study sites, Office I (28.6908361, 77.216768) and Office II (28.690837, 77.216768).

2.2.2 Site Description

Two office complexes were selected based on daily footfalls that were fully air conditioned and equipped with all modern facilities like computers, photostat machines, printers, ovens, etc. Both the offices (in size) covered almost the

Figure 2.1 Sampling site and its GPS location.

same area, with similar building ages. Despite their similar size and activities, these offices had diverse daily footfalls, with Office I having 320 persons per day and Office II having 180 persons per day. A total of 39 staff members (25 from Office I and 14 from Office II) participated in the epidemiological survey.

2.2.3 Bioaerosol Sampling

The study aimed to assess the concentration of bioaerosols, which is one of the major SBS marker pollutants. Bioaerosol sampling was done with a sampling frequency of thrice a month by using the gravitational plate method/settle plate technique (Napoli et al. 2012). The Petri plate of 100 mm sizes using Hi Media blood agar was exposed for 30 minutes at a vertical height of approximately 1.5 m to collect bioaerosols after following sampling guidelines (Omberg and Stetzenbach 2007; Yamamoto et al. 2011; Napoli et al. 2012). The concentration was reported as the number of colony-forming units (CFUs) per unit volume of air. Additionally, onsite measurements of relative humidity and temperature parameters were also done using pocket kestrel 11 A handheld instrument.

2.2.4 Respondent Characteristics and Prevalence of Sick Building Syndrome Symptoms

A structured self-formulated questionnaire was used to find the sociodemographic background, indoor activities, and health status of the respondents. Symptoms related to SBS were assessed using the Central Pollution Control Board (CPCB), Indoor Environmental Quality Survey, and the National Institute for Occupational Safety and Health (NIOSH) 1991 guidelines. As per the study conducted by Ooi et al. in 1998, participants were required to report experiencing SBS symptoms for a minimum of one to three days per week during a consecutive four-week period. Additionally, they have observed that these symptoms improved when the individuals were away from their workplace. For this study, office workers were categorized as subjects when they reported experiencing at least one SBS symptom once a week.

2.2.5 Statistical Analyses

The correlation coefficients and linear regressions for the primary data were computed using SPSS statistical software, specifically version 22.0, and MS Office Excel 2019 version.

2.3 Results and Discussion

2.3.1 Seasonal Variation in Bioaerosol Distribution

The highest concentrations of indoor bioaerosols were observed during post-monsoon season (330.44±29.47 CFU/plate/h in Office I and 246.77±31.92 CFU/plate/h in Office II). The last indoor bioaerosol concentrations were reported during the summer season (269.55±21.59 CFU/plate/h in Office I and 179.44±16.39 CFU/plate/h in Office II). Annual average of bioaerosol counts was reported higher in Office I (297.52±21.59 CFU/plate/h) compared to Office II (205.19±24.86 CFU/plate/h) (Figure 2.2). Among all seasons, indoor bioaerosol concentrations were consistently higher in indoor environment than outdoor environment (Figure 2.2).

2.3.2 Effect of Meteorological Parameters and Footfall Count on Bioaerosol Distribution

In order to understand the relationship between total bioaerosols with meteorological parameters and footfall count, the Pearson Correlation Coefficient was calculated (Table 2.1). In both sites, footfall count shows a moderate relationship

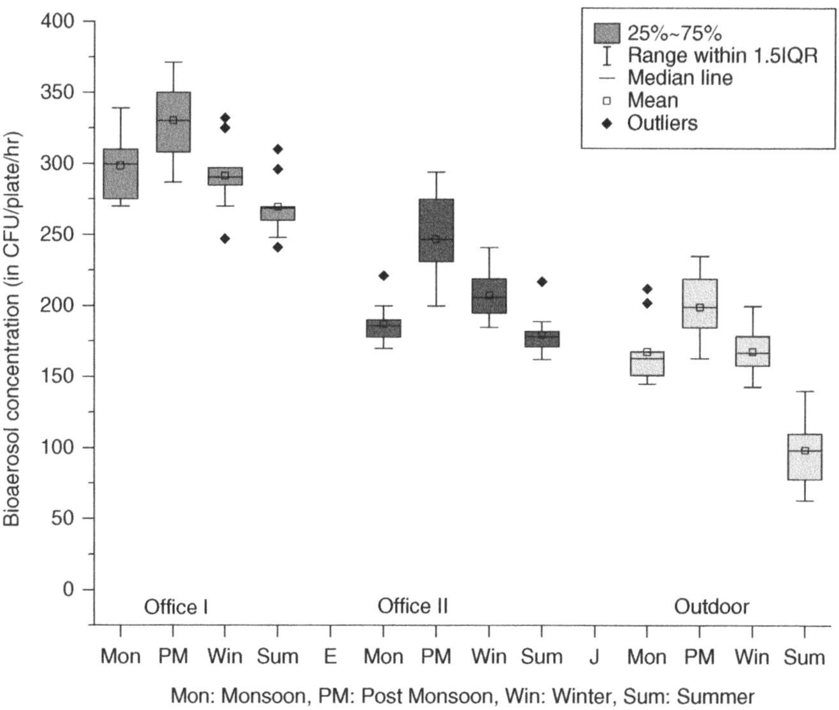

Figure 2.2 Variation of bioaerosol concentration in diverse sites.

Table 2.1 Correlation matrix of bioaerosols with different factors in different zones.

Factors	Bioaerosol count (Office I)	Bioaerosol count (Office II)
Temperature	−0.110	−0.331
Humidity	0.390	0.313
Footfall per day	0.668	0.566
Outdoor bioaerosol count	0.231	0.235

with bioaerosol counts. Temperature shows a negative correlation with bioaerosol growth but not in significant ways.

Indoor environment of Office I showed the highest bioaerosol load concentration (293.82±41.30 CFU/plate/h) compared to Office II (203.78±33.78 CFU/plate/h) and outdoor environment (154.6±47.15 CFU/plate/h). Interestingly, the annual mean temperature of outdoor (32.39±8.41 CFU/plate/h) was higher than Office I (29.66±4.48 CFU/plate/h) and Office II (28.93±4.04 CFU/plate/h) (Table 2.2).

The spatiotemporal variability of bioaerosol concentration in different indoor and outdoor environments was enumerated and correlated with influencing meteorological parameters (temperature and humidity). In the summer season, the indoor/outdoor ratio was maximum followed by winter, and least in the post-monsoon season (Table 2.3). The annual mean average concentration was maximum in Office I (1.98) compared to Office II (1.35), maybe due to higher footfall in Office I compared to Office II.

2.3.3 Association of Bioaerosol Count and Daily Footfall

The analysis indicates a significant positive correlation between bioaerosol count and daily footfall per day ($r = 0.607$, $p < 0.05$), suggesting that as footfall count increases, bioaerosol load also tends to increase in Office I. This indicates that

Table 2.2 Mean and standard deviation of bioaerosols and metrological factors in the indoor and outdoor environment.

	Office I		Office II		Outdoor Office complex	
Factors	Mean	Standard deviation	Mean	Standard deviation	Mean	Standard deviation
Bioaerosol count (in CFU/plate/h)	297.52	25.12	205.19	20.48	158.36	23.49
Temperature (in °C)	29.66	4.47	28.93	4.05	32.39	8.41
Humidity (in %)	51.95	7.21	50.79	8.74	46.86	19.21

Table 2.3 Table I/O (Indoor/Outdoor) ratio in Office complex in different seasons.

Season	Monsoon	Post-monsoon	Winter	Summer	Yearly average
Location					
Office I	1.76	1.66	1.77	2.73	1.98
Office II	1.16	1.14	1.24	1.82	1.35

there is a statistically significant correlation between bioaerosol count and footfall per day. This finding was also supported by an earlier study of underground subway stations (Chen and Hildemann 2009; Heo and Lee 2016; Heo et al. 2017a).

But interestingly in Office II, the correlation coefficient (r) between bioaerosol count and daily footfall was approximately −0.258, indicating a weak negative correlation. However, the p-value was 0.149, which is greater than the significance level of 0.05. Therefore, we fail to reject the null hypothesis, suggesting that there is no statistically significant correlation between bioaerosol count and daily footfall in Office II data set.

2.3.4 Respondent Characteristics

A total of 39 participants were surveyed, with a 100% response rate from both Office Complexes I and II. Sociodemographic data were collected from these questionnaires and compiled in Table 2.4. In Office I, there were 15 male and 10 female respondents, whereas in Office II, there were 8 male and 6 female respondents. The mean age of respondents was 43.4 years in Office I and 42.5 years in Office II. On average, participants from Office I spent 45 hours per week in the buildings, while those from Office II spent 43.5 hours per week.

2.3.5 Occurrence of SBS

Table 2.5 shows the occurrence of symptoms associated with SBS, revealing variations across the selected office complexes. Office complex I showed a

Table 2.4 Sociodemographic information of respondents.

Respondent characteristics	Office 1	Office II
N (Total respondents) = 39	25	14
Gender distribution (Male + female)	15 + 10	8 + 6
Age of respondents (in years)	43.4	42.5
Work time per week (in hours)	45 ± 0.45	43.5 ± 0.39

Table 2.5 Percentage prevalence of SBS symptoms within working staff.

		Office I (25)		Office II (14)	
S. no.	Symptoms (in %)	Men 15	Women 10	Men 8	Women 6
1	Malaise (feeling of discomfort)	**66.67**	**80**	50	66.66
2	Itchy, dry, irritation to eyes	**60**	**60**	50	33.33
3	Unusual tiredness, lethargy, and drowsiness	46.67	**70**	37.5	**66.66**
4	Headache	20	30	**25**	16.66
5	Hoarse, dry throat	**40**	**40**	25	33.33
6	Dizziness or lightheadedness	**20**	10	**25**	16.66
7	Nasal irritation/congestion	26.67	**40**	25	**66.66**
8	Cough, cough with phlegm	26.67	**50**	25	**50.00**
9	Psychological symptoms, tension, irritability, and nervousness	13.33	20	**25**	**33.33**
10	Difficulty concentrating or remembering things	20	20	**25**	**33.33**
11	Feeling depressed	**26.67**	20	25	**33.33**
12	Dermal symptoms, dry or itchy skin	20	**30**	25	**33.33**
13	Do you suffer from diabetes mellitus	13.33	**20**	**25**	16.66
14	Do you suffer from hypertension	40	**50**	37.5	66.66
15	Asthma	13.33	**20**	12.5	**16.66**

Bold values are for highlighting highest values in result.

higher prevalence of SBS symptoms compared to complex II. The most frequently reported symptoms in both offices were "itchy, dry eyes, and eye irritation," along with coughing and sneezing, each accounting for approximately 21% of the reported symptoms. To evaluate the likelihood of SBS symptoms among office workers, office staff were asked to assess their health in relation to SBS over the past three weeks. Employees rated the occurrence levels of the listed SBS symptoms in the questionnaire survey. SBS symptoms were more prevalent in females compared to males in office environment. Our result is also supported by previous studies (Stenberg and Wall 1995; Jafari et al. 2015; Nag 2018) reported around the world. The frequency of these symptoms played a crucial role in determining the likelihood of SBS presence, and the survey results are summarized in Table 2.5.

Table 2.6 Percentage of occupants who experienced the symptoms after leaving the building.

		Office I		Office II	
S. no.	**Type of SBS symptoms (in %)**	**Men**	**Women**	**Men**	**Women**
1	Malaise (feeling of discomfort)	13.33	10	12.5	16.66
2	Itchy, dry, irritation to eyes	13.33	20	nr[a]	16.66
2	Unusual tiredness, lethargy, and drowsiness	20	10	nr	16.66
3	Headache	6.66	10	12.5	16.66
4	Hoarse, dry throat	13.33	20	12.5	nr
5	Nasal irritation/congestion	13.33	10	12.5	33.33
6	Coughing sneeze	20	nr	nr	16.66
7	Dermal symptom: dry or itchy skin	13.33	10	12.5	33.33

[a] nr: not reported.

To identify which symptoms contribute to SBS, (Murphy 2006) proposed that an office building can be classified as a sick building if more than 20% of its occupants exhibit SBS symptoms that disappear after they leave the building. Subsequently, the building's occupants were asked to specify which symptoms disappeared after leaving the office, based on the same list of symptoms. The results of this survey are presented in Table 2.6.

From Table 2.6, it is evident that seven symptoms disappeared in the participants after leaving the office complex building, with the highest being malaise and lethargy, not experienced by 87.5% of the occupants once they left the office complexes. In regard to all 7 symptoms, 20% of occupants reported not experiencing these symptoms after leaving the office room. Consequently, both office buildings can potentially be categorized as sick buildings since 7 out of 15 SBS symptoms were not observed just after the occupants left the premises.

2.3.6 Occupant Satisfaction

The office workers were asked to give their perception regarding satisfaction level and perceived indoor air quality (IAQ) status. The results of office staff's perceptions are shown in Table 2.7.

The results of the study showed that 35.89% of the building's respondents expressed dissatisfaction with the thermal conditions, citing discomfort from both hot and cold temperatures. Furthermore, the IAQ perceived by the office workers was rated as low, leading to a dissatisfaction rate of 46.15% among them.

Table 2.7 Office staff's perception (in %) about indoor air quality.

Perception	Satisfied	Neutral	Dissatisfied
Perception regarding temperature	30.77	33.34	**35.89**
Perception regarding humidity	30.77	**35.89**	33.34
Perception regarding IAQ	23.07	30.76	**46.15**
Perception regarding surrounding cleanliness	33.34	**38.46**	28.21

Bold value are for highlighting highest values in result.

2.4 Conclusion

In this study, we have studied indoor-outdoor bioaerosol distribution in office buildings and the relationship between human presence (footfall) and the surveyed health status of office occupants. The concentration of bioaerosols is closely associated with human indoor activities. Human occupancy and movements contribute to increased bioaerosol concentrations. Interestingly, verbal communication, which involves heavy inhalation and exhalation, has been found to decrease the concentration of bioaerosols in indoor spaces. Further research is needed to fully understand the mechanisms underlying the connection between human activities and bioaerosol levels. Understanding this relationship will serve as a crucial basis for future studies on air infections and the development of air-cleaning technologies. By exploring the intricate interplay between humans and bioaerosols, we can pave the way for advancements in promoting safer IAQ and reducing the risks of airborne infections for occupants.

Acknowledgments

This study is supported by the Ministry of Earth Sciences (MoES) and the University Grants Commission, Delhi.

References

Apter, A., Bracker, A., Hodgson, M. et al. (1994). Epidemiology of the sick building syndrome. *The Journal of Allergy and Clinical Immunology* 94 (18): 277–288. https://doi.org/10.1053/ai.1994.v94.a56006.

Chen, Q. and Hildemann, L.M. (2009). The effects of human activities on exposure to particulate matter and bioaerosols in residential homes. *Environmental Science and Technology* 43 (13): 4641–4646. https://doi.org/10.1021/es802296j.

Cooley, J.D., Wong, W.C., Jumper, C.A., and Straus, D.C. (1998). Correlation between the prevalence of certain fungi and sick building syndrome. *Occupational and Environmental Medicine* 55 (9): 579–584. https://doi.org/10.1136/oem.55.9.579.

Douwes, J., Thorne, P., Pearce, N., and Heederik, D. (2003). Bioaerosol health effects and exposure assessment: progress and prospects. *Annals of Occupational Hygiene* 47 (3): 187–200. https://doi.org/10.1093/annhyg/meg032.

Hedge, A., Erickson, W.A., and Rubin, G. (1995). Individual and occupational correlates of the sick building syndrome. *Indoor Air* 5 (1): 10–21. https://doi.org/10.1111/j.1600-0668.1995.t01-3-00003.x.

Heo, K.J. and Lee, B.U. (2016). Seasonal variation in the concentrations of culturable bacterial and fungal aerosols in underground subway systems. *Journal of Aerosol Science* 92: 122–129.

Heo, K.J., Lim, C.E., Kim, H.B., and Lee, B.U. (2017). Effects of human activities on concentrations of culturable bioaerosols in indoor air environments. *Journal of Aerosol Science* 104 (November 2016): 58–65. https://doi.org/10.1016/j.jaerosci.2016.11.008.

Hsu, Y.C., Kung, P.Y., Wu, T.N., and Shen, Y.H. (2012). Characterization of indoor-air bioaerosols in Southern Taiwan. *Aerosol and Air Quality Research* 12 (4): 651–661. https://doi.org/10.4209/aaqr.2012.03.0070.

Jafari, M.J., Khajevandi, A.A., Najarkola, S.A.M. et al. (2015). Association of sick building syndrome with indoor air parameters. *Tanaffos* 14 (1): 55.

Kim, K.H., Kabir, E., and Jahan, S.A. (2018). Airborne bioaerosols and their impact on human health. *Journal of Environmental Sciences (China)* 67 (January): 23–35. https://doi.org/10.1016/j.jes.2017.08.027.

Lacey, J. and Dutkiewicz, J. (1994). Bioaerosols and occupational lung disease. *Journal of Aerosol Science* 25 (8): 1371–1404. https://doi.org/10.1016/0021-8502(94)90215-1.

Milton, D.K. (2000). Risk of sick leave associated with outdoor air supply rate, humidification, and occupant complaints. *Indoor Air* 10 (4): 212–221. https://doi.org/10.1034/j.1600-0668.2000.010004212.x.

Murphy, M. (2006). *Sick Building Syndrome and the Problem of Uncertainty*. Duke University Press.

Nag, P.K. (2018). Sick building syndrome and other building-related illnesses. In: *Office Buildings: Health, Safety and Environment*, 53–103. Springer. https://doi.org/10.1007/978-981-13-2577-9_3

Napoli, C., Marcotrigiano, V., and Montagna, M.T. (2012). Air sampling procedures to evaluate microbial contamination: a comparison between active and passive methods in operating theatres. *BMC Public Health* 12: 1–6.

Omberg, K. and Stetzenbach, L. (2007). Indoor and outdoor air sampling. *Sampling for Biological Agents in the Environment* 133–164.

Powitz, B.R.W. and Ph, D. (2017). Bioaerosols. November *2008*, pp. 1–7.

Stenberg, B. and Wall, S. (1995). Why do women report 'sick building symptoms' more often than men? *Social Science & Medicine* 40 (4): 491–502.

World Health Organization (2018). WHO Housing and health guidelines. http://www.who.int/phe%0Ahttp://apps.who.int/bookorders (accessed 4 March 2019).

Yamamoto, N., Schmechel, D., Chen, B.T. et al. (2011). Comparison of quantitative airborne fungi measurements by active and passive sampling methods. *Journal of Aerosol Science* 42 (8): 499–507.

3

Sources and Transmission of Bio-Contaminants in Indoor Areas

P. Kavita and Priti Giri

Department of Botany, Maitreyi College, University of Delhi, New Delhi, India

3.1 Introduction

Bio-contaminants are pollutants of biological origin in air, soil, and water.

Microbes (bacteria, viruses, and unicellular organisms), fungi, algae, mites, insect droppings/animal dander, and their by-products are the primary elements in the air (World Health Organization 1990; Moldoveanu 2015). These are caused by animals (dogs, cats, birds, etc.), plants (pollen, smells, allergens), and building materials. Physical circumstances like heating (temperature and humidity), ventilation, carpets, fabrics, and so on can make them worse. According to reports, bio-pollutants can be found in indoor air anywhere, even in tightly regulated settings like the operation rooms of contemporary hospitals (Chirca 2019). 90% or more of people's time is spent indoors such as in homes, classrooms, and colonial structures like stores, vehicles, and businesses and are exposed to these biological pollutants heavily (Moldoveanu 2015).

Even air conditioners, fans, chillers, humidifiers, etc., are major sources for the proliferation and spread of indoor microorganisms. Ventilation equipment, fans, and air conditioners are generally colonized by bacteria, yeasts, and fungi (e.g. *Phialophora, Aspergillus, Penicillium, and Geotrichum*), (Dubey and Maheshwari 2014). The interconnected outdoor and indoor environments allow microorganisms to spread randomly.

An important indoor environment like the hospital is responsible for the spread of airborne pathogenic microorganisms. They are the reservoir for pathogens that are later spread to other people, e.g. doctors, patients, workers, and visitors (Nevalainen and Seuri 2005).

Airborne Biocontaminants and Their Impact on Human Health, First Edition.
Edited by Rajeev Singh and Anamika Singh.

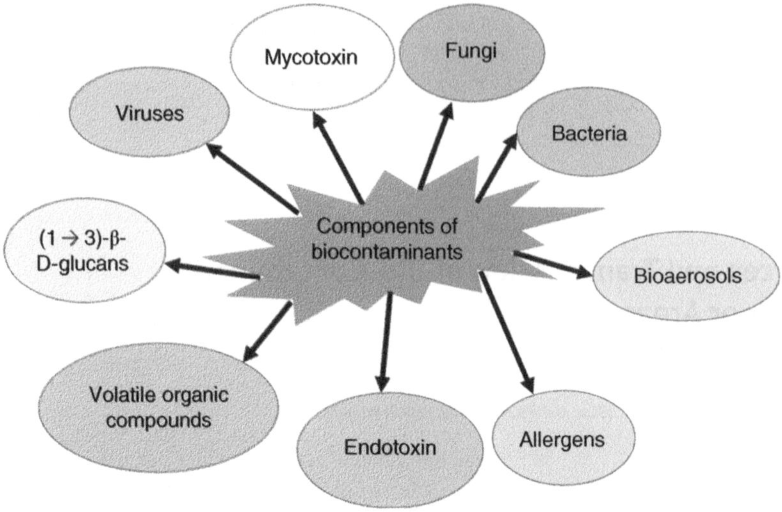

Figure 3.1 Outline of study.

The main airborne pathogens in hospitals are *Aspergillus fumigatus, Mycobacterium tuberculosis (TB), Influenza virus, Aspergillus flavus, Staphylococcus aureus,* and *Candida albicans* (Kausar et al. 2016). Fungal genera, e.g. *Penicillium* sp., *Alternaria* sp., *Aspergillus* sp., *Talaromyces, Trichoderma*, etc., were predominantly present in the hospital hematological environment (Cho et al. 2018). However, *Cladosporium, Penicillium,* and *Aspergillus* can occur in higher concentrations in intensive care units (Gonçalves et al. 2017).

It can be difficult to evaluate the interior environment for the study of biological pollutants. The choice of proper sampling test, knowing where and when to sample, and interpreting results with negative impacts are major issues. If a threat is identified, the source of the contamination must be investigated to reduce the risk of contamination. Thus, the evaluation process most importantly involves correlation of health effects with sampling data. The purpose of this chapter is to provide an overview of indoor airborne bio-contaminants: their sources, physical properties, mode of transmission, associated assessment methods, and possible health consequences (Figure 3.1).

3.2 Sources of Bio-contaminants in Indoor Areas

3.2.1 Bacteria in Indoor Air

Bacteria found in the nose, mouth, ear, etc., and associated with buildings are typically saprophytic and originate from outdoor air (Moldoveanu 2015). Heterotrophic bacteria that grow in wet areas of a building, such as bathroom sinks or in water reservoirs are additional prevalent indoor bacteria. Legionella and nontuberculous environmental mycobacteria develop in biofilms in water pipes or water reservoirs of cooling systems. Actinobacteria (especially Streptomycetes), Bacillus species, and many other bacteria grow together with a number of fungi. Components from bacterial structures that are released into the air include exotoxins, peptidoglycans, microbial volatile organic compounds (VOCs), along with bacterial cells, bacterial spores, and other bacteria metabolites (Nevalainen and Morawaska 2009). Talking, coughing, or sneezing are frequent ways to get rid of upper respiratory tract bacteria. Significant bacterial species including *Micrococcus spp., Streptococcus spp., Staphylococcus spp.,* and *Corynebacterium spp.* were found in the indoor regions (Moldoveanu 2015; Kausar et al. 2016). Pets can affect endotoxin levels in indoor environments and are substantial sources of germs (Ownby et al. 2013). The indoor storage of organic waste further increases the risk of bacterial infiltration (Moldoveanu 2015).

3.2.2 Virus in the Indoor Air

Viruses are biological organisms that can grow on both natural and artificial materials, but they require a host in order to replicate and proliferate. The likelihood that respiratory viruses would survive longer indoors and develop into disease is typically increased (Yu et al. 2004). Human activities including coughing, sneezing, talking, and laughing can spread millions of airborne viruses through indoor droplets (Diglisic et al. 1999). Hantavirus infections are more prevalent among persons who dwell inside due to the prevalence of rats (Aitken and Jeffries 2001).

3.2.3 Fungi in the Indoor Air

Normally, fungal loads indoors are lower than outdoors. If the conditions are present that allow fungal growth, especially in damp houses, the concentrations of fungi in the indoor air can reach very high levels. Measurements of fungal content level are expressed in colony-forming units (CFUs) per cubic meter of air. The fungal loads and types of fungi found in the indoor air depend on the structural conditions, the usage of the building, ventilation, light, and the season. The most common genera of fungi found indoors are yeasts, *Cladosporium, Alternaria, Aspergillus,* and *Penicillium* (Nevalainen and Morawaska 2009). These are the most common fungi found outdoors also and cause serious health problems (World Health Organization 2009; Sharma et al. 2011, 2012). The CFU of fungi depends on the local conditions like the high fungal spore count is very common in damp houses with high water activity (Nevalainen and Seuri 2005). *Penicillium* and *Aspergillus versicolor* grow well at relatively moderate to low moisture content while *Fusarium, Yeast, Phialophora, Stachybotrys, Aspergillus fumigatus, Trichoderma,* etc., are found at the high moisture content.

The various respiratory problems in humans may be caused by mold, mold spores, and their by-products (American Air and Water 2020). This fungus is extremely diverse, and its spore-bearing structures facilitate the process of aerosolization. These fungal spores are ubiquitous and need nutrients (lipids, proteins, and carbohydrates) for growth and development. They can grow well on stored food, cooking oil, books, construction material, wood, paint, etc. (World Health Organization 2009; Rautiainen et al. 2019).

3.2.4 Mycotoxins in the Indoor Air

Secondary metabolites like mycotoxins are toxic in nature and are produced by the fungi. These can cause various diseases in humans and animals leading to their death (Bennett and Klich 2003). These most importantly cause DNA damage and stop RNA synthesis. *Stachybotrys chartarum* produces some mycotoxins, which are associated with acute pulmonary hemorrhage (Sharma et al. 2011, 2012). Carcinogens and mutagens like aflatoxin produced by molds (*Aspergillus favus* and *A. parasiticus*) cause cancer too.

3.2.5 (1, 3)-β–D glucans in the Indoor Air

A kind of β-D-glucose polysaccharides known as (1, 3)-β–D glucans can be found in household dust as a structural component of some bacteria and fungi. The presence of (1,3)-β-D glucans inside rooms may cause respiratory illnesses (Seo et al. 2015). Seo and colleagues examined the concentration of (1,3)-β-D glucans in elementary schools before and after the rainy season and concluded that their concentration and effects change with time (Seo et al. 2015).

3.2.6 Endotoxins in the Indoor Air

Endotoxins are gram-negative bacteria's lipopolysaccharide (LPS)-based cell wall components (Dales et al. 2008; Moldoveanu 2015). When the cell is dying in extreme conditions, bacteria secrete them through microscopic fragments. A high level of endotoxin or gram-negative is linked to pets, food waste storage, humidifiers, and poor ventilation. The level of endotoxin in the indoor area is directly related to the presence of asthma and asthma-like symptoms in the habitants (Michel et al. 1996).

3.2.7 Volatile Organic Compounds (VOCs) and Fungi in the Indoor Air

Low molecular weight molecules with high vapor pressure and limited solubility are known as VOCs. Microbial VOCs (mVOCs) are VOCs that are formed because of microbial metabolism (Rautiainen et al. 2019). Even if quantitative measurements are negative, the detection of mVOC in indoor air signifies the presence of fungi. Some mVOCs such as 2 octen-1-ol serve as markers of biological contamination. These substances have been associated as general indicators of fungal contamination and are strongly emitted in the initial stages of growth of the fungi (*Penicillium* spp., *Aspergillus* spp., *Alternaria alternata*) and therefore they are a nonspecific indicator of fungal contamination (Jeleń and Wasowicz 1998).

3.2.8 Allergens in the Indoor Air

The compounds known as allergens have the potential to be the root of IgE-mediated allergic responses in people. They are glycoproteins or proteins with high molecular weights that range from 10 kDa to more than 100 kDa. Numerous respiratory disorders are brought on by type 1 hypersensitivity reactions, which are caused by allergens such as *Aspergillus, Penicillium, Cladosporium,* and *Alternaria* spp. (World Health Organization 1990; Kumar et al. 2021). Insects and their by-products might trigger severe allergies and act as potent allergens. According to a clinicoimmunological investigation, indoor air may include strong inhalant allergens such as mosquito detritus (Kausar et al. 2007). The two main kinds of mites that produce allergens are *Dermatophagoides farinae* (Der f I) and *Dermatophagoides pteronyssius* (Der p I and Der p II) (Kausar 2018). The major indoor allergen sources included the bakery, poultry, granary, and sugar industry. Common fungi that cause allergic reactions include *Cladosporium, Epicoccum, Stemphylium, Penicillium, Alternaria, Curvularia, Pullularia, Ganoderma, Trichoderma,* and *Aspergillus* (Kausar 2018).

3.2.9 Bioaerosols in the Indoor Air

Bioaerosols are airborne particles that are made by or come from living organisms such as bacteria, viruses, and fungi. Bioaerosols can be either manmade or naturally occurring, and are extremely varied and pervasive. In recent years, the sampling and analysis of airborne microorganisms has received attention and the concerns about mold contamination in indoor environments, the risk of bioterrorism, and the occurrence of related health effects, such as infectious diseases, acute toxic effects, allergies have drawn attention to the sampling and analysis of airborne microorganisms (Douwes et al. 2003; O'Riordan and Smaldone 2004). Bioaerosols may contain bacterial cells and cellular fragments, fungal spores, and by-products of microbial metabolism that are present as particulate, liquid, or volatile organic molecules (Stetzenbach 2005).

3.3 Mode of Transmission of Bio-Contaminants in Indoor Air

Biological pollutant sources in indoor air may alter according to various circumstances and activities, such as inhabitants, building systems, furniture and food in various locations, the presence and concentration of microbial agents (Figure 3.2) (Kausar et al. 2007; Yassin and Almouqatea 2010; Sheik et al. 2015; Kumar et al. 2021).

3.3.1 Humans

In the indoor environment, humans are the main sources for transportation and survival of biological pollutants. A normal person may carry 1012–1014 microorganisms in various body parts (Ananthanaryan and Paniker 2009). Indoor air quality (IAQ) may also be in danger from unsuitable outdoor air, poor air distribution, improper air filtering, and inadequate ventilation system maintenance in enclosed indoor spaces (Mandal and Brandl 2011). Both community structure and airborne microorganisms are impacted by human inhabitation. In human-occupied areas compared to uninhabited outdoor areas, there are two times as many germ bacteria that are connected to humans (Prussin and Marr 2015). *Micrococcus, Staphylococcus, Bacillus,* and *Streptococcus* are common bacterial

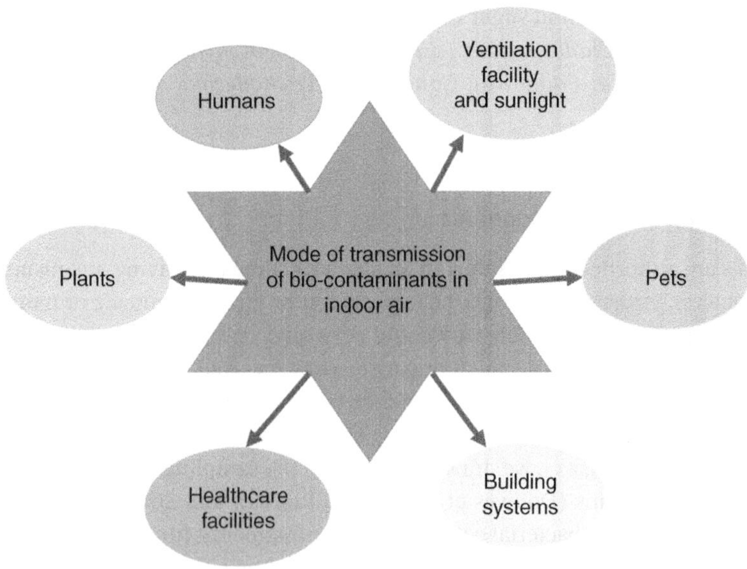

Figure 3.2 Bubble map showing mode of transmission of indoor bio-contaminants.

species associated with construction (Charlson et al. 2011). The microflora of the male and female inhabited areas varies, even though most bacterial species are harmful to humans and a natural component of microflora. *Lactobacillus* genera are more prevalent in females than *Roseburia* spp., *Dermabacter* spp., and *Corynebacterium* spp., genera that are most common in males (Charlson et al. 2011; Prussin and Marr 2015). Also, human skin is found connected to a few fungal genera including *Cryptococcus, Rhodotorula, Trichosporon,* and *Candida*. These spread across occupied buildings as bioaerosols. Pathogenic viruses are also linked to humans; however, it is unknown how they affect the human population due to the knowledge vacuum in this area (Barberán et al. 1814).

3.3.2 Building Systems

Possible biological contamination may be present in building materials and other components. Through the formation of bioaerosols in toilets, bathtubs, sinks, drain traps, aquariums, etc., biological contaminants can spread throughout a building (World Health Organization 1990; Kumar et al. 2021). Degradation of construction materials can result in organic chemicals that are ideal for the development of microbes. High temperature and high relative humidity both encourage the development and spread of indoor bacteria. A significant source of potential contamination in the indoor environment includes some natural materials used for curtains, cushions, ornamental items, straw baskets, plush toys, carpeting and furnishings. Sick building syndrome (SBS) and other issues like building-related illness may be brought on by undesirable biological pollution in a building (World Health Organization 1990; Mandal and Brandl 2011).

3.3.3 Pets

According to certain studies, children's and newborn's health may benefit from exposure to bioaerosols and dust from dogs (Fujimura et al. 2010, 2014; Prussin and Marr 2015). In the indoor environment, there are 56 and 24 bacterial genera present in cats and dogs, respectively. In dogs, *Arthrobacter* spp., *Bacteroides* spp., *Blautia* spp., *Neisseria* spp., and *Moraxella* spp. are commonly found and cats are predominately infected with *Sporosarcina* spp., *Porphyromonas* spp., *Prevotella* spp., *Bifidobacterium* spp., *Sporosarcina* spp., *Jeotgalicoccus* spp. and *Moraxella* spp. (Prussin and Marr 2015; Kumar et al. 2021).

3.3.4 Healthcare Facilities

One of the main places where different bacteria propagate is spreading various microorganisms, in the hospital area. Through the formation of bioaerosols, patients with various diseases, ventilation systems, sweeping, dressing beds, and other activities spread airborne microorganisms (Knights et al. 2011; Kumar et al. 2021).

3.3.5 Ventilation Facility and Sunlight

Microorganisms from the outside air enter indoor buildings through numerous ports of entry and exit. Poor air distribution and a lack of outdoor air are caused by inadequate ventilation, which lowers the quality of the air inside a building (World Health Organization 1990). Particle sizes in bioaerosol are crucial because dense, large particles sink more quickly than less dense, larger particles. The ultraviolet (UV) light rays dimerize the purine nucleotides in bacterial DNA, which is bactericidal and controls the number of bacteria in the air outside (World Health Organization 1990). According to various reports, the correct use of a ventilation system can significantly improve IAQ (Ghanizadeh and Godini 2018).

3.3.6 Plants

Additionally, significant sources of indoor microbes are plants. Food supplies in kitchens also spread airborne microorganisms, as per the studies in closed artificial ecosystems conducted in China (Sun et al. 2016).

3.4 Health Consequences of Bio-Contaminants in Indoor Air

There are several ways that indoor air pollution might affect people's health. The respiratory system is particularly affected by the impacts. Exposure to many biological contaminants that have been found in indoor air can cause allergic responses. Additionally, there is evidence to support the notion that breathing in biological pollutants in indoor air may be linked to neurological symptoms, such as headaches, fatigue, and forgetfulness. The inflammation of the airways is a contributing factor in some of the consequences of the biological contamination of indoor air on the respiratory system. This can appear in both nonallergic and allergic (involving IgE and IgG antibodies) respiratory diseases. In the case of asthma and rhinitis, it is crucial to distinguish between those two entities.

3.4.1 Asthma

In addition to eosinophils, allergic asthma requires IgE-mediated sensitivity and underlying atopic reaction of the respiratory airways (Nevalainen and Morawaska 2009; Annesi-Maesano 2013). Asthma is becoming more and more common among adult people. Due to nonspecific bronchial hyper-responsiveness, asthmatic patients will react adversely toward biological and other contaminants in the indoor environment. This is a bronchoconstrictor response to numerous stimuli that have been present indoors. The amount of endotoxin in household dust is closely connected to the severity of asthma and requirement for medicine in those with home dust mite allergies, in a dose-dependent manner (Nevalainen and Morawaska 2009). Asthma attacks are brought on by the concentration of pollen allergens in the air due to interactions between the allergens and other sources of small particles. Airborne particles like diesel exhaust particles have an adjuvant function for generation of cytokines and IgE antibodies (Nevalainen and Morawaska 2009). Compared to allergic individuals implanted with ragweed alone, the intranasal administration of ragweed allergens with diesel exhaust particles dramatically raised allergen-specific IgE production and the Th2-type cytokine mRNA expression. According to a UK study, viral infections and allergens together can exacerbate asthma symptoms in adults (Nevalainen and Morawaska 2009).

3.4.2 Rhinitis

The inflammation of the upper respiratory tract mucosa is known as rhinitis. Sneezing, conjunctivitis, nasal congestion, sore throat, and rhinorrhea are all signs of rhinitis. There are allergic and nonallergic causes of rhinitis (Nevalainen and Morawaska 2009; Haleem Khan and Mohan 2012).

3.4.3 Tuberculosis

One of the leading bacterial infections killing people worldwide is TB. It is a typical community-acquired illness brought on by the inhalation of aerosolized bacilli seen in the droplet nuclei of expectorating TB patients with positive sputum. *Mycobacterium TB* is the causal agent of TB, which is spread by infected individuals to others by bioaerosols produced during coughing, sneezing, and talking activities (Srikanth et al. 2008). The potential for transmission inside the hospital environment has been highlighted in the United Kingdom, by several outbreaks of multidrug-resistant (MDR) TB (Srikanth et al. 2008).

3.4.4 Leprosy

Mycobacterium laprae is a slow-growing, intracellular acid-fast bacteria that causes the infectious disease leprosy, which is spread through the air. Respiratory droplets from the infected person's nasal mucosa may be used to transmit the causative agent. The symptoms may take 2–10 years to manifest; it colonizes and typically infects at a lower temperature. The principal organs affected include the skin, superficial nerves, nails, lymph nodes, and upper respiratory mucosa. Leprosy is being given increased attention since it cannot be cultured in a lab setting (Ananthanaryan and Paniker 2009).

3.4.5 Legionellosis

People who live in commercial or industrial establishments can contract *Legionellosis pneumophila*, a pneumonia that can be contracted in hospitals and the public. Legionella appears because of the active aerosolizing processes (aeration of polluted water). Legionella is frequently colonized in artificial water systems, cooling tower biofilms, and air conditioning systems. However, using monochloramine to disinfect community residual drinking water may help ward off Legionnaires' disease. For reaching far areas and penetrating biofilm in water sources, monochloramine is superior to chlorine, although it requires a higher pH for effective disinfection (Kool et al. 1999; United States Environmental Protection Agency (USEPA) n.d.).

3.4.6 Mycosis

Both humans and animals frequently contract the fungus known as mycosis. Depending on the host's reaction to the pathogen and various tissues infected, it might be classified as cutaneous, subcutaneous, systemic, or superficial (Fung and Hughson 2003; World Health Organization 2009). Bird droppings and cons dust may be involved in systemic mycosis such as coccidioidomycosis, cryptococcosis, and histoplasmosis (Fung and Hughson 2003; World Health Organization 2009). Mycosis is a more serious infection in immunocompromised people, such as those with AIDS, cancer patients receiving chemotherapy and leukemia patients (World Health Organization 1990; Fung and Hughson 2003; Prussin and Marr 2015).

3.4.7 Mycotoxicosis

Exposure to secondary metabolites made by fungus results in mycotoxicosis. Mycotoxins cause cytotoxicity, interfere with DNA/RNA synthesis, and protein synthesis and cause cancer, among other harmful effects (Seltzer 1994; Fung and Hughson 2003). Only a small number of molds can produce mycotoxins, which are influenced by seasonal changes and environmental factors such as relative humidity, temperature, oxygen, and carbon dioxide levels (Moldoveanu 2015; American Air and Water 2020). Mycotoxin is a powerful carcinogen that has been identified from a few species of fungi. The mycotoxin aflatoxin produced by *Aspergillus favus* and *A. parasiticus* is capable and can cause liver cancer (cirrhosis). Ochratoxin A is a mycotoxin, which is produced by *Aspergillus ochraceus, A. carbonarius, A. niger,* and *Penicillium verrucosum* and is a human carcinogen (Moldoveanu 2015; American Air and Water 2020). Exposure to these carcinogens by ingestion and inhalation has discovered a connection between wood dust and other malignancies including sinus cancer (Seltzer 1994; Fung and Hughson 2003; American Air and Water 2020).

3.4.8 Influenza Virus (H1N1)

Numerous viruses, such as the Influenza virus (H1N1), the severe acute respiratory syndrome (SARS) virus, respiratory syncytial virus (RSV), intestinal enteric viruses, the measles, mumps, and rubella viruses, and hantavirus from rodent feces, are easily transmitted through air (Diglisic et al. 1999; Yu et al. 2004).

3.4.9 Coronavirus Disease (COVID-19)

A highly contagious respiratory disease known as SARS is spread by a new coronavirus that causes illness, death, and very serious atypical pneumonia (Srikanth et al. 2008). Aerosol-generating procedures including bronchoscopy, treatment with aerosolized medication, and endotracheal intubation, that are performed in hospitals may increase the spread of SARS (Yu et al. 2004). The coronavirus disease (COVID-19), which is brought on by the coronavirus (SARS-CoV-2), is currently affecting the entire world. According to the most recent WHO reports, there have been around 77 million verified cases of COVID-19, including 6.9 million fatalities.

3.4.10 Chronic Bronchitis

Coughing up phlegm (a productive cough) for at least three months per year for two years or longer is chronic bronchitis. Bronchitis shows an increase in activity of mucus-secreting glands and changes to the properties of the mucus itself. It is possible to have chronic bronchitis with or without airway blockage. There is inflammation of the airways, and more neutrophils are present. There is therefore a significant overlap in symptoms between bronchitis, nonallergic asthma and allergic asthma. People with chronic bronchitis may also experience symptoms including dyspnea, chest tightness, and wheezing. The pathophysiology of bronchitis can be attributed to biological contamination caused by the presence of endotoxin and β-glucan (Nevalainen and Morawaska 2009).

3.4.11 Organic Dust Toxic Syndrome

The illness has symptoms that are like influenza, including an intense fever, a dry cough, dyspnea, tightness in the chest, headache, chills, malaise, exhaustion, and aches and pains in the muscles and joints. ODTS is a nonallergic-mediated airway inflammation (Nevalainen and Morawaska 2009).

3.4.12 Hypersensitivity Pneumonitis or Extrinsic Allergic Alveolitis

The illness is one of the most serious ones brought on by exposure to fungus and thermophilic actinomycetes (spore-forming bacteria). It is also known as "pigeon breeder's lung" or "farmer's lung." The severe pulmonary disease with pneumonia-like fever, inflammation of pulmonary parenchyma, cough, shortness of breath, tightness of chest, and development of pulmonary granulomas is the clinical manifestation of HP. HP and organic dust toxic syndrome are extremely similar (Fung and Hughson 2003; Nevalainen and Morawaska 2009; World Health Organization 2009).

3.4.13 Sick Building Syndrome

The WHO established the term "sick building syndrome" (SBS) in the 1980s. The term "sick building syndrome (SBS)" can be used to describe situations in which people occupying various interior environments experience sudden health issues or discomfort. The symptoms experienced by those who have been exposed include dryness and irritation of the eyes, nose, and throat; itching of skin and eyes; dry cough; nausea; vomiting; headach; dizziness; and exhaustion. The symptoms go away after exposure is stopped (Shoemaker and House 2006; Joshi 2008; World Health Organization 2009).

3.4.14 Chronic Airflow Obstruction

Numerous studies have linked exposure to organic dust in industrial and agricultural occupational populations with chronic airflow blockages as determined by spirometry (particularly FEV1). Exposure to endotoxins may be a significant factor too (Simpson et al. 1998; Nevalainen and Morawaska 2009).

3.5 Method of Detection and Control of Bio-Contaminants

3.5.1 Detection of Bio-Contaminants

Both culturable and nonculturable microorganisms are viable. Only culturable microorganisms are counted and identified during sampling, which causes the concentration of bioaerosol to be underestimated. Therefore, it is crucial to estimate both culturable and nonculturable organisms using appropriate microscopy to identify bacteria and fungi (using gram staining for bacteria, and lactophenol cotton blue and calcofluor white for fungi, respectively) as well as traditional microbiology techniques like observation of growth trait, cellular or spore morphology, and biochemical tests for identification (Eduarda and Heederik 1998). Following sample collection, colonies of bacteria and fungi are grown on culture media over a three to seven day period at a defined temperature before being identified. For even better identification, molecular biology methods such as restriction fragment length polymorphism (RFLP) analysis are performed (Hensel and Petzoldt 1995; Madelin and Madelin 2020). Polymerase chain reaction (PCR) and enzyme-linked immunosorbent assay (ELISA) are analytical techniques used for nonviable and viable microorganisms; these are typically qualitative, while semi-quantitative and quantitative methods are developing (Srikanth et al. 2008). The use of PCR was initially proven as a method to identify bacteria (Alvarez et al. 1994) and viruses (Sawyer et al. 1994) in air samples in 1994. *Pneumocystis jiroveci* (Olsson et al. 1998) and *Aspergillus* spp. (Leenders et al. 1999) were later identified to be airborne by PCR. Additionally, it has been applied to the improved identification of actual bioterrorism agents. The use of quantitative PCR for bioaerosol monitoring is being examined.

Endotoxin Assay: Polycarbonate capillary pore membrane filters are used in the filter method to gather samples from the air. After sampling, it is extracted using sonication and the limulus amoebocyte lysate (LAL) test, also known as the chromogenic kinetical limulus test, is used to determine whether endotoxins are present. The LAL test is a comparative bioassay method rather than an analytical one since changes in parameters other than LPS concentrations affect the observed endotoxin activity levels (Finney 1978). A modified spectrophotometer is used to measure optical density (Milton et al. 1992).

Fungal Biomass Assay: Ergosterol evaluated using gas chromatography-mass spectrometry (Pasanen et al. 1999) and fungal extracellular polysaccharides measured using immunoassays are two examples of indicators for the evaluation of fungal biomass (Douwes et al. 1999).

3.5.2 Control of Biological Contaminants

Primarily, the prevention of conditions that serve as substrate for the growth of viable particles and then the closure and removal of such growth are common but important methods for controlling biological pollution in indoor air (World Health Organization 1990). It could be possible to regulate bioaerosols by using air filters and optimizing ventilation performances (Vlad et al. 2013). Most control methods are used inside center to lower the dew point, which inhibits the growth of a variety of bacteria. Aerosolization from contaminated water can cause dangerous bioaerosols to form and spread in both indoor and outdoor spaces (World Health Organization 1990). The risk of airborne disease transmission inside a building increases with occupant density, e.g. the higher the density, the greater the risk. Allergy-causing nonviable aerosols may spread due to the presence of vectors like dogs and insects (World Health Organization 1990); so, a check on these vectors helps in control.

3.5.3 Control Measures for Reducing Bioaerosols

Certain management strategies can be used to lower bioaerosol loading in indoor spaces. These include accurate source microbial identification and eradication in residential and commercial settings, equipment maintenance, humidity control, natural ventilation, filter use in ventilation, and air purification using disinfectants and biocides. One way to ensure controlled bioaerosol concentrations is the regular application of disinfectants and biocides. By fumigating with different microbicidal agents, the air in operating rooms and other crucial spaces like isolation rooms can be disinfected. The most widely used commercial surface and environmental disinfectant, Bacillocid has excellent cleaning properties in addition to having bactericidal, virucidal, sporicidal, and fungicidal activities (Srikanth et al. 2008). It is either liberally sprayed or mopped, providing 30 minutes of allowing contact time, and offers complete asepsis in 30–60 minutes. It does not require cleaning with detergent, carbolic acid, or formalin fumigation. It does not call for shutting down the cleaned spaces such as operating rooms, for a full day (Ghosh et al. 2004).

3.6 Summary and Conclusion

IAQ can harm people's health and cause potentially fatal diseases since indoor contains bio-pollutants such as pathogenic agents like bacteria, fungi, viruses, toxins, and other infectious materials. Bio-pollutants may cause a variety of conditions that affect the lungs, intestines, kidneys, and central nervous system. Currently, the COVID-19 outbreak is creating a serious risk to human health by degrading IAQ. The indoor environment needs to be better ventilated, and the population needs to be made aware of difficulties with IAQ and their effects in order to improve quality of life and reduce health issues brought on by bio-pollutants. In healthcare settings, bioaerosols can cause occupational hazards and nosocomial infections. Modern constructed environments may produce bioaerosols. Bioaerosol monitoring in hospitals can be used to track nosocomial infections and identify the source and spread of airborne germs in order to reduce hospital-associated infections (HAI). This bioaerosol monitoring will be useful for gauging biosafety when handling biohazardous products. Interdisciplinary research is essential because of how complicated these bioaerosols. The study of bioaerosols and their impacts on IAQ, environmental quality, and human health are better understood in the West. In order to solve the health issues facing emerging nations, more knowledge of focused surveillance for infection control is required.

References

Aitken, C. and Jeffries, D.J. (2001). Nosocomial spread of viral disease. *Clinical Microbiology Reviews* 14 (3): 528–546.

Alvarez, A.J., Buttner, M.P., Toranzos, G.A. et al. (1994). Use of solid-phase PCR for enhanced detection of airborne microorganisms. *Applied and Environmental Microbiology* 60 (1): 374–376.

American Air and Water (2020). Mold, mold spores and indoor air quality. American Air and Water. https://www. americanairandwater.com/mold (accessed 8 June 2020).

Ananthanaryan, R. and Paniker, C.K.J. (2009). *Text of Microbiology*, 8e. Universities press.

Annesi-Maesano, I. (2013). Indoor exposition to molds and health outcome. *Review of Allergy and Clinical Immunology* 23: 21–25.

Barberán, A., Dunn, R.R., Reich, B.J. et al. (1814). The ecology of microscopic life in household dust. *Proceedings of the Royal Society B: Biological Sciences* 2015 (282): 20151139.

Bennett, J.W. and Klich, M. (2003). Mycotoxins. *Clinical Microbiology Reviews* 16: 497–516. CrossRef| PubMed| CAS| Web of Science® Times Cited. 2003;428.

Charlson, E.S., Bittinger, K., Haas, A.R. et al. (2011). Topographical continuity of bacterial populations in the healthy human respiratory tract. *American Journal of Respiratory and Critical Care Medicine* 184 (8): 957–963.

Chirca, I. (2019). The hospital environment and its microbial burden: challenges and solutions. *Future Microbiology* 14 (12): 1007–1010.

Cho, S.Y., Myong, J.P., Kim, W.B. et al. (2018). Profiles of environmental mold: indoor and outdoor air sampling in a hematology hospital in Seoul, South Korea. *International Journal of Environmental Research and Public Health* 15 (11): 2560.

Dales, R., Liu, L., Wheeler, A.J., and Gilbert, N.L. (2008). Quality of indoor residential air and health. *CMAJ* 179 (2): 147–152.

Diglisic, G., Rossi, C.A., Doti, A., and Walshe, D.K. (1999). Seroprevalence study of Hantavirus infection in the community based population. *Maryland Medical Journal (Baltimore, Md.: 1985)* 48 (6): 303–306.

Douwes, J., van der Sluis, B., Doekes, G. et al. (1999). Fungal extracellular polysaccharides in house dust as a marker for exposure to fungi: relations with culturable fungi, reported home dampness, and respiratory symptoms. *Journal of Allergy and Clinical Immunology* 103 (3): 494–500.

Douwes, J., Thorne, P., Pearce, N., and Heederik, D. (2003). Bioaerosol health effects and exposure assessment: progress and prospects. *The Annals of Occupational Hygiene* 47 (3): 187–200.

Dubey, R.C. and Maheshwari, D.K. (2014). *Textbook of Microbiology*, 5e. Ram Nagar, New Delhi 110055 (India): S. Chand & Co., Limited.

Eduarda, W. and Heederik, D. (1998). Methods for quantitative assessment of airborne levels of noninfectious microorganisms in highly contaminated work environments. *American Industrial Hygiene Association Journal* 59 (2): 113–127.

Finney, D.J. (1978). *Statistical Method in Biological Assay*. Charles Griffin & Company.

Fujimura, K.E., Johnson, C.C., Ownby, D.R. et al. (2010). Man's best friend? The effect of pet ownership on house dust microbial communities. *Journal of Allergy and Clinical Immunology* 126 (2): 410–412.

Fujimura, K.E., Demoor, T., Rauch, M. et al. (2014). House dust exposure mediates gut microbiome Lactobacillus enrichment and airway immune defense against allergens and virus infection. *National Academy of Sciences of the United States of America* 111 (2): 805–810.

Fung, F. and Hughson, W.G. (2003). Health effects of indoor fungal bioaerosol exposure. *Applied Occupational and Environmental Hygiene* 18 (7): 535–544.

Ghanizadeh, F. and Godini, H. (2018). A review of the chemical and biological pollutants in indoor air in hospitals and assessing their effects on the health of patients, staff and visitors. *Reviews on Environmental Health* 33 (3): 231–245.

Ghosh, A., Hazra, A., and Mandal, S.C. (2004). New drugs in India over the past 15 years: analysis of trends. *National Medical Journal of India* 17: 10–16.

Gonçalves, C.L., Mota, F.V., Ferreira, G.F. et al. (2017). Airborne fungi in an intensive care unit. *Brazilian Journal of Biology* 78: 265–270.

Haleem Khan, A.A. and Mohan, K.S. (2012). Fungal pollution of indoor environments and its management. *Saudi Journal of Biological Sciences* 19 (4): 405–426.

Hensel, A. and Petzoldt, K. (1995). Biological and biochemical analysis of bacteria and viruses, chapter 13. In: *Bio-Aerosols Handbook* (ed. C.S. Cox and C.M. Wathes), 335–360. New York: Lewis Publishers.

Jeleń, H. and Wasowicz, E. (1998). Volatile fungal metabolites and their relation to the spoilage of agricultural commodities. *Food Reviews International* 14 (4): 391–426.

Joshi, S.M. (2008). The sick building syndrome. *Indian Journal of Occupational and Environmental Medicine* 12 (2): 61.

Kausar, M.A. (2018). A review on respiratory allergy caused by insects. *Bioinformation* 14 (9): 540.

Kausar, M.A., Vijayan, V.K., Bansal, S.K. et al. (2007). Mosquitoes as sources of inhalant allergens: clinicoimmunologic and biochemical studies. *Journal of Allergy and Clinical Immunology* 120 (5): 1219–1221.

Kausar, M.A., Arif, J.M., Alanazi, S.M. et al. (2016). Assessment of microbial load in indoor environment of university and hospitals of hail, ksas. *Biochemical and Cellular Archives* 16 (1): 177–183.

Knights, D., Kuczynski, J., Charlson, E.S. et al. (2011). Bayesian community-wide culture-independent microbial source tracking. *Nature Methods* 8 (9): 761–763.

Kool, J.L., Bergmire-Sweat, D., Butler, J.C. et al. (1999). Hospital characteristics associated with colonization of water systems by Legionella and risk of nosocomial legionnaires' disease: a cohort study of 15 hospitals. *Infection Control and Hospital Epidemiology* 20 (12): 798–805.

Kumar, P., Kausar, M.A., Singh, A.B., and Singh, R. (2021). Biological contaminants in the indoor air environment and their impacts on human health. *Air Quality, Atmosphere and Health* 14 (11): 1723–1736. https://doi.org/10.1007/s11869-021-00978-z.

Leenders, A.C., van Belkum, A., Behrendt, M. et al. (1999). Density and molecular epidemiology of Aspergillus in air and relationship to outbreaks of Aspergillus infection. *Journal of Clinical Microbiology* 37 (6): 1752–1757.

Madelin, T.M. and Madelin, M.F. (2020). Biological analysis of fungi and associated molds. In: *Bioaerosols Handbook* (ed. Christopher S. Cox and Christopher M. Wathes), 361–386. CRC Press.

Mandal, J. and Brandl, H. (2011). Bioaerosols in indoor environment-a review with special reference to residential and occupational locations. *The Open Environmental & Biological Monitoring Journal* 4 (1): 83.

Michel, O., Kips, J., Duchateau, J. et al. (1996). Severity of asthma is related to endotoxin in house dust. *American Journal of Respiratory and Critical Care Medicine* 154 (6): 1641–1646.

Milton, D.K., Feldman, H.A., Neuberg, D.S. et al. (1992). Environmental endotoxin measurement: the kinetic Limulus assay with resistant-parallel-line estimation. *Environmental Research* 57 (2): 212–230.

Moldoveanu, A.M. (2015). Biological contamination of air in indoor spaces. In: *Current Air Quality Issues* (ed. F. Nejadkoorki). IntechOpen.

Nevalainen, A. and Morawaska, L. (2009). Biological agents in indoor environments. Assessment of health risks. Work conducted by a WHO Expert Group between 2000-2003. World Health Organization, Copenhagen, Denmark.

Nevalainen, A. and Seuri, M. (2005). Of microbes and men. *Indoor Air* 15 (Suppl 9): 58–64. https://doi.org/10.1111/j.1600-0668.2005.00344.x. PMID: 15910530.

Olsson, M., Lidman, C., Latouche, S. et al. (1998). Identification of Pneumocystis carinii f. sp. hominis gene sequences in filtered air in hospital environments. *Journal of Clinical Microbiology* 36 (6): 1737–1740.

O'Riordan, T. and Smaldone, G. (2004). Respiratory medical societies and the threat of bioterrorism. *Thorax* 59 (3): 265.

Ownby, D.R., Peterson, E.L., Wegienka, G. et al. (2013). Are cats and dogs the major source of endotoxin in homes? *Indoor Air* 23 (3): 219–226.

Pasanen, A.L., Yli-Pietilä, K., Pasanen, P. et al. (1999). Ergosterol content in various fungal species and biocontaminated building materials. *Applied and Environmental Microbiology* 65 (1): 138–142.

Prussin, A.J. and Marr, L.C. (2015). Sources of airborne microorganisms in the built environment. *Microbiome* 3: 1.

Rautiainen, P., Hyttinen, M., Ruokolainen, J. et al. (2019). Indoor air-related symptoms and volatile organic compounds in materials and air in the hospital environment. *International Journal of Environmental Health Research* 29 (5): 479–488.

Sawyer, M.H., Chamberlin, C.J., Wu, Y.N. et al. (1994). Detection of varicella-zoster virus DNA in air samples from hospital rooms. *Journal of Infectious Diseases* 169 (1): 91–94.

Seltzer, J.M. (1994). Biological contaminants. *Journal of Allergy and Clinical Immunology* 94 (2): 318–326.

Seo, S., Ji, Y.G., Yoo, Y. et al. (2015). Submicron fungal fragments as another indoor biocontaminant in elementary schools. *Environmental Science: Processes & Impacts* 17 (6): 1164–1172.

Sharma, R., Deval, R., Priyadarshi, V. et al. (2011). Indoor fungal concentration in the homes of allergic/asthmatic children in Delhi, India. *Allergy & Rhinology* 2 (1): ar-2011.

Sharma, R., Gaur, S.N., Singh, V.P., and Singh, A.B. (2012). Association between indoor fungi in Delhi homes and sensitization in children with respiratory allergy. *Medical Mycology* 50 (3): 281–290.

Sheik, G.B., Al, A.A., and Al Shehri, Z.S. (2015). Assessment of bacteria and fungi in air from college of applied medical sciences (Male) at AD-Dawadmi, Saudi Arabia. *International Research of Journal of Biological Sciences* 4 (9): 48–53.

Shoemaker, R.C. and House, D.E. (2006). Sick building syndrome (SBS) and exposure to water-damaged buildings: time series study, clinical trial and mechanisms. *Neurotoxicology and Teratology* 28 (5): 573–588.

Simpson, J.C., Niven, R.M., Pickering, C.A. et al. (1998). Prevalence and predictors of work-related respiratory symptoms in workers exposed to organic dusts. *Occupational and Environmental Medicine* 55 (10): 668–672.

Srikanth, P., Sudharsanam, S., and Steinberg, R. (2008). Bio-aerosols in indoor environment: composition, health effects and analysis. *Indian Journal of Medical Microbiology* 26 (4): 302–312.

Stetzenbach, L.D. (2005). Airborne bacteria, chapter 7. In: *Topley and Wilsonís Microbiology and Microbial Infections: Bacteriology-I*, vol. 10 (ed. P.S. Borriello, P.R. Murray, and G. Funke), 185–194. Washington DC: ASM Press.

Sun, Y., Xie, B., Wang, M. et al. (2016). Microbial community structure and succession of airborne microbes in closed artificial ecosystem. *Ecological Engineering* 88: 165–176.

United States Environmental Protection agency (USEPA) (n.d.). Chloramine in drinking water, drinking water requirements for states and public water systems. https://www.epa.gov/dwreginfo/chloramines-drinking-water (accessed 9 June 2020).

Vlad, D.C., Popescu, R., Filimon, M.N. et al. (2013). Assessment of microbiological indoor air quality in public buildings: a case study (Timisoara, Romania). *African Journal of Microbiology Research* 7 (19): 1957–1963.

World Health Organization (1990). Indoor air quality: biological contaminants: report on a WHO meeting, Rautavaara, 29 August–2 September 1988. World Health Organization. Regional Office for Europe.

World Health Organization (2009). *Guidelines for Indoor Air Quality: Dampness and Mould*. Copenhagen: World Health Organization.

Yassin, M.F. and Almouqatea, S. (2010). Assessment of airborne bacteria and fungi in an indoor and outdoor environment. *International Journal of Environmental Science and Technology* 7: 535–544.

Yu, I.T., Li, Y., Wong, T.W. et al. (2004). Evidence of airborne transmission of the severe acute respiratory syndrome virus. *New England Journal of Medicine* 350 (17): 1731–1739.

4

Climatic/Meteorological Conditions and Their Role in Biological Contamination

A Comprehensive Review

Pankaj Bhatt[1,2], *Vipin Kumar*[2], *Suruchi Singh*[3], *and Kunal Kanojia*[4]

[1] *Lloyd Institute of Management and Technology, Greater Noida, Uttar Pradesh, India*
[2] *Department of Pharmaceutical Sciences, Gurukul Kangri (Deemed to be University), Haridwar, Uttarakhand, India*
[3] *Department of Pharmacology, Accurate College of Pharmacy, Greater Noida, Uttar Pradesh, India*
[4] *Department of Pharmaceutics, KIET Group of Institutions (KIET School of Pharmacy), Muradnagar, Ghaziabad, Uttar Pradesh, India*

4.1 Introduction

Biological contamination represents a substantial risk to diverse ecosystems, public well-being, and the safety of our food supply. This term refers to the existence of harmful microorganisms such as bacteria, viruses, fungi, and parasites in our surroundings that have negative consequences for human, animal, and plant life. A multitude of factors, including weather and atmospheric conditions, shape the dispersion and proliferation of these harmful agents (Eliana Andrea et al. 2019). Comprehending the origins and effects of biological contamination is imperative for protecting the health of both individuals and the environment. Microorganisms have the potential to pollute bodies of water, soil, the atmosphere, and our food sources, which can lead to the outbreak of infectious diseases, a decline in agricultural yields, and the disruption of ecological equilibrium (Anderson et al. 2003). Biological contamination can arise from multiple origins, such as untreated sewage, runoff from agricultural activities, discharge of industrial waste, and improper disposal of waste materials. The issue of biological contamination may be further aggravated by climate change, which brings about alterations in temperature, patterns of precipitation, and the occurrence of extreme weather

Airborne Biocontaminants and Their Impact on Human Health, First Edition.
Edited by Rajeev Singh and Anamika Singh.

events (Al-Ghazali and Al-Azawi 1990). It is crucial to investigate the connection between climate and meteorological factors and how they impact the presence, longevity, and spread of biological contaminants. These climatic and meteorological elements significantly influence the behavior and endurance of microorganisms, affecting their capacity to pollute various environmental domains (Kumar et al. 2021). Temperature, humidity, precipitation, wind patterns, and other climatic factors can directly or indirectly influence the growth, proliferation, and transmission of microorganisms (Abreu et al. 2010). Gaining insights into the repercussions of these conditions holds immense importance in forecasting and effectively addressing the risks linked to biological contamination. Analyzing the complex interplay between climatic factors and contamination allows us to formulate efficient strategies for prevention, control, and reduction (Alam and Zurek 2004).

We dedicate this chapter to a thorough exploration of the diverse climatic and meteorological factors that contribute to biological contamination. Utilizing insights from contemporary scientific literature, case studies, and illustrative examples, we will delve into the mechanisms through which these conditions influence microbial ecosystems and the dissemination of contaminants. Furthermore, we will examine the effects of climate change on biological contamination and propose future research and recommendations to address this urgent concern.

4.2 Microbial Ecology and Climatic Factors

4.2.1 Overview of Microbial Communities

Microbial communities are complex and diverse assemblages of microorganisms that inhabit various environments, including soil, water, air, and the human body (Aruscavage et al. 2010). These communities comprise bacteria, viruses, fungi, and parasites, engaging in complex interactions among themselves and with their environment. Gaining insight into the composition and behavior of microbial communities is crucial for a comprehensive grasp of the factors that impact biological contamination (Pope et al. 1993). Microbial communities are shaped by numerous factors, including environmental conditions, resource availability, and interactions among different microbial species (Balshem et al. 2011). The makeup and variety of these communities can significantly differ among various ecosystems, as each environment hosts its own set of microorganisms and demonstrates distinct patterns of microbial interactions (Aitken and Jeffries 2001). Progress in molecular methods, including DNA sequencing and metagenomics, has transformed our capacity to examine microbial communities. These techniques offer valuable insights into the genetic capabilities and functional roles of individual microorganisms within their ecological settings (Barberán et al. 2015).

4.2.2 Influence of Temperature on Microbial Growth and Survival

Temperature plays a pivotal role in shaping the growth and viability of microorganisms. These microorganisms have distinct temperature preferences for their best growth, and temperature variations can significantly impact their activity (Bragoszewska et al. 2018). Mesophilic microorganisms flourish in moderate temperature ranges, usually spanning from 20 to 40 °C. This category encompasses numerous common bacteria and fungi that inhabit diverse environments. In contrast, psychrophilic microorganisms have a preference for cold temperatures and can thrive in polar regions or high-altitude settings where temperatures consistently remain low (Bragoszewska et al. 2018; Cho et al. 2018). Thermophilic microorganisms have evolved to flourish in elevated temperature settings, like hot springs or compost heaps. Variations in temperature can exert substantial effects on microbial communities, influencing their capacity to thrive, propagate, and endure (Curtis et al. 2009). Higher temperatures can expedite microbial growth and metabolic activities, elevating the potential for contamination in specific settings. For instance, in food processing or storage, insufficient temperature regulation can facilitate the proliferation of pathogenic bacteria, thereby posing health hazards (Dey et al. 2019). On the contrary, severe cold can impede microbial growth, although specific psychrophilic microorganisms can endure and sustain metabolic activity even at subzero temperatures. Furthermore, the process of freezing and thawing can disrupt microbial habitats, leading to alterations in their population and ecological functions (Fang et al. 2019). Figure 4.1 shows the Effect of Temperature on Microbial Growth.

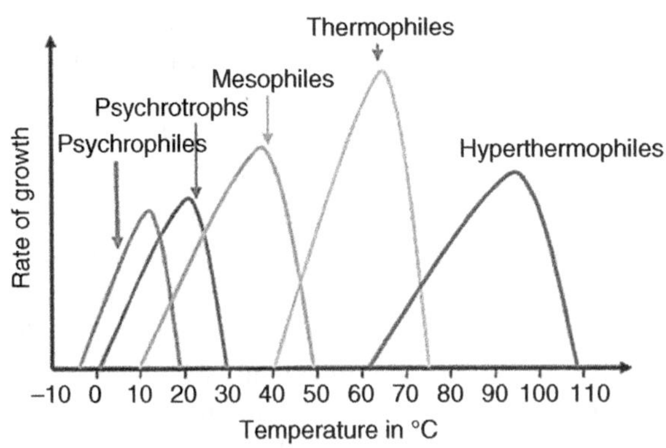

Figure 4.1 Study the effect of temperature on microbial growth.

4.2.3 Effect of Humidity on Microbial Proliferation

Humidity, which signifies the level of moisture in the air, holds a pivotal role in facilitating microbial propagation. Microorganisms rely on water for conducting fundamental biological activities, rendering moisture availability a pivotal factor influencing their development and persistence (Fung and Hughson 2010). High humidity levels create favorable conditions for microbial growth and reproduction, particularly for moisture-loving microorganisms (Barak and Liang 2008). Elevated humidity levels result in greater water accessibility, providing a conducive environment for the proliferation of diverse microorganisms. In tropical rainforests or regions characterized by high humidity, for instance, fungi and bacteria can flourish, participating in the decomposition of organic matter and the cycling of nutrients (Godish and Godish 2007). On the flip side, decreased humidity levels can hinder microbial proliferation and diminish the chances of contamination. Microbial activity and population densities typically decrease in arid regions with limited water resources. Nevertheless, certain microorganisms have developed adaptive mechanisms to endure low humidity conditions, such as the formation of protective spores or transitioning into dormant states until moisture becomes accessible (Jeleń and Wąsowicz 2009). On the flip side, decreased humidity levels can hinder microbial proliferation and diminish the chances of contamination. In arid regions characterized by limited water resources, microbial activity and population densities are typically reduced. Nevertheless, certain microorganisms have developed adaptive mechanisms to endure low humidity conditions, such as the formation of protective spores or transitioning into dormant states until moisture becomes accessible (Kang and Frank 1989).

4.2.4 Precipitation and Its Role in Microbial Dissemination

Precipitation, encompassing rain, snow, and hail, has a substantial impact on the dispersal of microbial contaminants. Rainfall occurrences can exert profound influences on the movement and destiny of microorganisms across diverse settings (Tornevi et al. 2014). When rain descends upon surfaces like soil, vegetation, or infrastructure, it possesses the capability to displace any microorganisms that might be present, thereby playing a role in their redistribution. Rainfall has the potential to carry microorganisms into bodies of water, which could lead to the contamination of drinking water sources or recreational waters (Barak et al. 2002). The runoff stemming from agricultural fields or urban locales during rainstorms has the potential to transport microbial contaminants, pesticides, and various pollutants, thereby exacerbating water contamination concerns (Tryland et al. 2011). Heavy rainfall and subsequent flooding events pose additional risks. Floodwaters can overwhelm sewage systems, leading to the release of untreated wastewater into the environment. This can introduce a high concentration of microbial contaminants, including pathogens, into water sources, amplifying the risk of waterborne diseases (Tryland et al. 2011). Moreover, precipitation patterns can influence the availability of water for microbial growth. Regions with high and consistent rainfall may create favorable conditions for microbial proliferation, whereas drought conditions can limit water availability, affecting microbial activity and abundance (Levy et al. 2018; Lai et al. 2020).

4.2.5 Wind Patterns and Their Impact on the Spread of Contaminants

Wind patterns can significantly influence the spread of biological contaminants. Wind currents can transport microorganisms over long distances by attaching themselves to dust particles, aerosols, or other airborne particles (Rendana 2020). Aerial transportation can result in the spread of contaminants from their point of origin to different regions, posing potential consequences for both human and environmental well-being. Microorganisms have the capacity to become airborne through a range of mechanisms, including soil erosion by the wind, agricultural procedures, or industrial operations (Bernstein et al. 2007). Once suspended in the air, these microorganisms can maintain their viability and be transported by winds to remote areas. A prime example is the airborne transmission of fungal spores, which can play a role in the extensive dissemination of plant diseases. Furthermore, wind patterns can impact the deposition of contaminants onto various surfaces (Beuchat 2006). Dust storms, for example, have the ability to transport microbial contaminants over long distances and deposit them on the ground, vegetation, and man-made structures. Such events elevate the potential for exposure and contamination, posing risks to both human populations and ecosystems (Kumar et al. 2021). Comprehending the correlation between wind patterns and the dispersion of contaminants is essential in evaluating and effectively addressing the potential hazards linked to biological contamination (Incineration 2000). Through the amalgamation of insights into wind patterns and microbial ecology, scientists and policymakers can formulate efficient approaches for the prevention, management, and reduction of potential risks (Kumar et al. 2021).

4.3 Climatic Conditions and Waterborne Contamination

4.3.1 Rainfall and Surface Water Contamination

Rainfall stands as a pivotal climatic element, exerting a substantial influence on waterborne contamination. When rainwater interacts with diverse surfaces like rooftops, roads, agricultural fields, or polluted regions, it has the capacity to accumulate and convey a range of contaminants, encompassing microorganisms, chemicals, and pollutants, into surface water reservoirs (Levy et al. 2018). Rainfall can wash off microorganisms from vegetation, animal waste, or urban areas, introducing them into rivers, lakes, and streams. Pathogenic bacteria, viruses, and parasites can contaminate surface water sources, posing significant risks to public health (Levy et al. 2016). Moreover, rainfall can exacerbate other sources of water pollution. It can mobilize and transport pollutants from agricultural runoff, industrial discharges, and improperly managed waste sites, further contaminating surface water bodies (Levy et al. 2018). The intensity and duration of rainfall events also influence the magnitude of contamination. Heavy or prolonged rainfall can lead to higher runoff volumes, increasing the likelihood of surface water contamination (Cann et al. 2013). Additionally, combined sewer overflows can occur during intense rainfall, where the excess stormwater overwhelms sewage systems, leading to the release of untreated wastewater into surface water bodies (Hayes et al. 2020).

4.3.2 Floods and Their Role in Waterborne Pathogen Transmission

Floods, caused by excessive rainfall or other natural events, can have severe implications for waterborne pathogen transmission. When rivers, lakes, or coastal areas exceed their capacity, floodwaters can inundate surrounding regions, bringing contaminants and pathogens into contact with water sources used for drinking, sanitation, and recreation (Brandl and Amundson 2008). Floodwaters can carry pathogenic microorganisms such as bacteria, viruses, and parasites during floods, thereby contaminating both surface water and groundwater sources. This can lead to outbreaks of waterborne diseases, such as cholera, dysentery, or hepatitis A (Caldwell et al. 2003). Floods can also disrupt infrastructure and sanitation systems, compromising the treatment and distribution of drinking water. Contaminated floodwaters can infiltrate wells or seep into underground aquifers, jeopardizing the safety of groundwater resources (Cooley et al. 2006). Beyond direct contamination, floods can trigger population displacement, overcrowding in makeshift shelters, and compromised hygiene measures, thereby amplifying the potential for the transmission of waterborne diseases (Delaquis et al. 2007).

4.3.3 Drought and Its Influence on Water Quality

Drought conditions, characterized by prolonged periods of minimal rainfall, can exert significant impacts on water quality and the potential for waterborne contamination (Dreux et al. 2007). Several factors come into play during droughts, increasing the risk of contamination (Mosley 2015). First, as drought reduces water levels in rivers, lakes, and reservoirs, it can lead to the concentration of contaminants, rendering the remaining water more vulnerable to pollution.

Elevated levels of pollutants, encompassing microorganisms, chemicals, and heavy metals, can pose adverse effects on water quality and public health (Baldwin et al. 2008; Mosley 2015). Furthermore, drought conditions can cause water treatment systems to deteriorate. Reduced water volumes and decreased flow rates can impair the effectiveness of water treatment processes, potentially compromising the removal of contaminants and elevating the risk of waterborne diseases (Barros et al. 1995). During droughts, water scarcity may force communities to resort to alternative, untreated water sources like shallow wells or stagnant ponds. These alternative sources may harbor microbial contaminants, thereby increasing the likelihood of waterborne illnesses (Barros et al. 1995). Furthermore, drought conditions can influence agricultural practices. Water scarcity may necessitate changes to irrigation methods, including the use of untreated or recycled wastewater for crop irrigation. If these water sources contain contaminants such as pathogenic microorganisms or agricultural chemicals, they can contribute to waterborne contamination (Delpla et al. 2009).

4.4 Climatic Conditions and Airborne Contamination

4.4.1 Aerosolized Microorganisms and Atmospheric Conditions

Airborne contamination is defined as the presence and distribution of microorganisms, pollutants, and allergens in the atmosphere. Aerosolized microorganisms, including bacteria, fungi, viruses, and their spores, have the capacity to be transported by air currents, impacting both human health and environmental processes (Cruz-Paredes et al. 2021). The survival and movement of aerosolized microorganisms in the atmosphere are significantly influenced by atmospheric conditions. Elements such as temperature, humidity, wind speed, and levels of air pollution can impact the viability, dispersion, and fate of these microorganisms in the air (Alster et al. 2020). Temperature plays a pivotal role in microbial survival, growth, and reproduction rates. Different microorganisms have distinct optimal temperature ranges, and extreme temperatures can hinder their viability. High temperatures can accelerate the inactivation of specific microorganisms, while low temperatures can preserve them for extended periods (Bååth 2001). Humidity is also a key factor in the survival and dissemination of aerosolized microorganisms. Elevated humidity levels provide the necessary moisture for microbial survival and can prevent desiccation. Conversely, low humidity levels can lead to the desiccation of microorganisms, reducing their viability. Nonetheless, certain microorganisms have adapted mechanisms to endure both high and low humidity conditions (Islam et al. 2019). Wind speed and direction influence the dispersion and distribution of airborne contaminants.

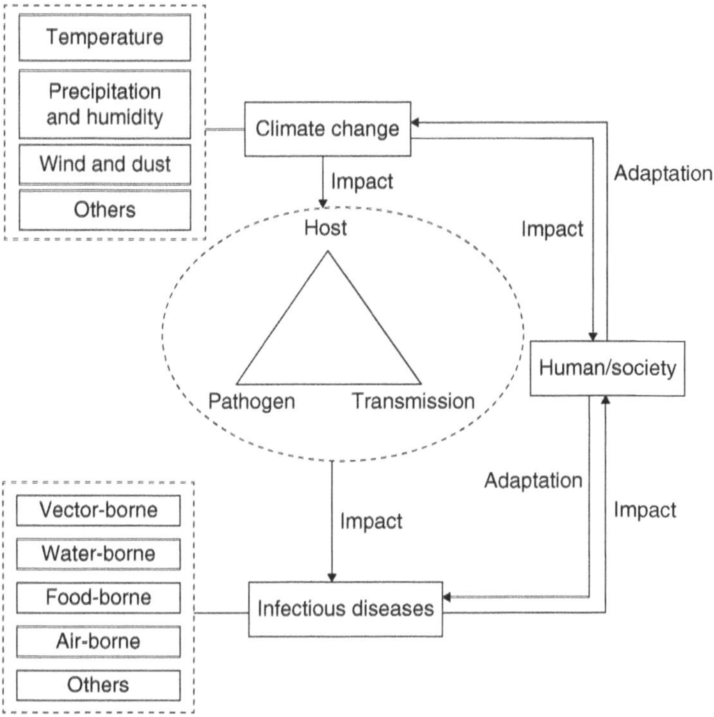

Figure 4.2 The interplay between climate change, infectious diseases in humans, and human society.

Microorganisms can attach themselves to dust particles, pollen, or other airborne particles, allowing them to be transported over significant distances by wind currents. Wind patterns can determine the direction and extent of microbial dispersal, which can potentially impact the spread of pathogens and allergens (Banhazi et al. 2010). Air pollution, including particulate matter and various pollutants, can interact with aerosolized microorganisms, affecting their viability and behavior. Pollutants can modify the structure and function of microbial cells, influencing their survival and potential health consequences when inhaled (Cambra-López et al. 2010). Figure 4.2 depicts the interplay between climate change, infectious diseases in humans, and human society.

4.4.2 Dust Storms and Their Role in Microbial Dissemination

Dust storms are meteorological phenomena characterized by robust winds that pick up and carry substantial quantities of dust particles across extensive regions (Erickson et al., 2010a). These storms play a pivotal role in the dispersion of aerosolized microorganisms, pollutants, and various particles within the atmosphere (Nourmoradi et al. 2015). Dust storms typically originate from arid and semi-arid areas, where loose soil and dry surface conditions facilitate the easy mobilization of dust. During these events, microorganisms such as bacteria, fungi, and their spores can adhere to dust particles and become suspended in the air (Nourmoradi et al. 2015). As dust storms traverse landscapes, they have the capacity to transport aerosolized microorganisms over considerable distances (Agranovski et al. 2007). Microbes attached to dust particles can retain their viability throughout this journey, potentially facilitating the dissemination of pathogens to new regions or even across continents (Alexandersen et al. 2001). Moreover, dust storms can deposit microbial contaminants onto various surfaces, including soil, vegetation, and infrastructure. This can lead to localized contamination and potential exposure pathways for humans and other organisms (Angenent et al. 2005).

4.4.3 Pollen and Allergenic Contaminants Influenced by Weather

Weather conditions significantly influence the release, dispersal, and concentration of pollen and other allergenic contaminants in the atmosphere. Pollen grains from plants are major contributors to allergic reactions in humans, causing seasonal allergies such as hay fever (Oh 2022). Temperature, humidity, wind patterns, and precipitation influence the phenology and abundance of plants, consequently affecting pollen production and dispersion. Warmer temperatures can lead to earlier flowering and extended pollen seasons, while humidity levels can affect pollen grain hydration and dispersal (Traidl-Hoffmann et al. 2009). Wind patterns play a crucial role in the transport and distribution of pollen. Wind currents can carry pollen grains over long distances, increasing the likelihood of exposure and allergic reactions in susceptible individuals (Galli et al. 2008). Precipitation can affect pollen concentrations in the air by causing wet deposition. Rainfall can wash pollen grains out of the atmosphere, temporarily reducing allergenic pollen levels. Conversely, following rainfall, there can be a surge in pollen release due to increased humidity and favorable conditions for plant growth (Tanno et al. 2016). Weather conditions can also interact with air pollution, exacerbating the impact of allergenic contaminants. Pollen grains can adsorb air pollutants, such as ozone or particulate matter, increasing their allergenicity and potentially intensifying allergic responses (Reinmuth-Selzle et al. 2017). Understanding the influence of weather on pollen dispersal and allergenic contaminants is essential for forecasting and managing allergic diseases (Erickson et al. 2010b). By monitoring weather patterns and pollen counts, individuals can take proactive measures to mitigate their exposure and alleviate symptoms during high-risk periods (Traidl-Hoffmann et al. 2009).

4.5 Climatic Conditions and Soil Contamination

4.5.1 Temperature and Its Impact on Soil Microbial Communities

Temperature plays a crucial role in shaping soil microbial communities and influencing soil contamination dynamics. Soil microorganisms, including bacteria, fungi, archaea, and viruses, exhibit varying responses to temperature, which in turn affects their activity, growth, and diversity (Fu et al. 2023). Different microorganisms have distinct temperature preferences, and soil temperature fluctuations can favor the growth of certain microbial groups over others (Ferens and Hovde 2011). Warmer temperatures generally enhance microbial activity and accelerate biochemical processes, such as nutrient cycling and organic matter decomposition. This can influence the breakdown of contaminants in the soil and potentially promote their degradation (Barberán et al. 2011). Conversely, extreme temperatures, either excessively high or low, can stress microbial populations and reduce their activity (Nottingham et al. 2019). Extreme heat can lead to the decline of sensitive microorganisms and alter the overall composition of soil microbial communities. Cold temperatures can slow down microbial metabolic rates, limiting the degradation of contaminants and potentially allowing them to persist in the soil (Rinnan et al. 2009). Understanding the relationship between temperature and soil microbial communities is crucial for managing soil contamination. By comprehending the temperature ranges favorable for the growth of specific microorganisms involved in bioremediation processes, scientists can optimize strategies to mitigate soil pollution effectively (Canarini et al. 2016).

4.5.2 Moisture and Soilborne Pathogen Survival

Moisture availability is another critical climatic factor that influences soil contamination, particularly in relation to soilborne pathogens. Soil moisture content affects the survival, proliferation, and dispersal of various pathogens, including bacteria, viruses, nematodes, and fungi (Franz and Van Bruggen 2008; Panth et al. 2020). Inadequate moisture levels can limit microbial growth and survival, thus reducing the potential for pathogen persistence in the soil. Dry conditions can desiccate pathogens, causing their inactivation or dormancy. However, some soilborne pathogens have mechanisms to withstand low moisture levels and can remain viable in a dormant state until favorable conditions arise (Alegbeleye and Sant'Ana 2020). On the other hand, excessive moisture can create favorable conditions for pathogen survival and dissemination. High soil moisture levels promote the growth and activity of certain pathogens, increasing the risk of contamination. Pathogens can move through the soil more easily in moist conditions, facilitating their spread to new areas or infecting plant roots (Singh et al. 2023). Moreover, precipitation events can affect soilborne pathogen dissemination. Rainfall or irrigation water can transport pathogens through surface runoff or infiltration into the soil, potentially contaminating agricultural fields or water sources used for irrigation (Rahman et al. 2021). Understanding the relationship between moisture and soilborne pathogens is crucial for developing strategies to mitigate contamination risks. Proper soil moisture management, irrigation practices, and water treatment can help reduce the survival and dissemination of pathogens in soil environments (Jeger 2022).

4.5.3 Erosion and Its Connection to Soil Contamination

Erosion, both wind and water-driven, is a significant process that can contribute to soil contamination (Zeng et al. 2021). Climatic conditions, particularly rainfall patterns, wind intensity, and surface runoff, influence erosion rates and the transport of contaminants from one location to another. Water erosion occurs when rainfall intensity exceeds the soil's infiltration capacity, leading to the detachment and transport of soil particles (Cheng and Jin 2013). This can lead to the loss of topsoil, which contains essential nutrients and organic matter, while carrying contaminants downstream, including pesticides, heavy metals, and pathogens (Gagliardi and Karns 2000; Fones et al. 2020). These contaminants can contaminate surface water bodies, affect aquatic ecosystems, and potentially pose risks to human health (Ahamad et al. 2019). On the other hand, wind currents lift dry, loose soil particles in arid and semi-arid regions, causing wind erosion. Contaminated soil particles, such as those carrying pesticides or industrial pollutants, can become airborne during wind erosion events (McIvor et al. 2017). This can lead to the dispersal of contaminants over long distances, affecting air quality and potentially contaminating water bodies, vegetation, and human settlements. Erosion control measures, such as contour plowing, terracing, and the use of cover crops, can help reduce soil erosion and subsequent contamination. By implementing these practices, soil loss can be minimized, and the transport of contaminants can be significantly mitigated (Gilley 2005).

4.6 Climatic Conditions and Food Contamination

4.6.1 Temperature and Food Spoilage

Temperature is a critical climatic factor that influences the spoilage of food products. The growth and activity of microorganisms responsible for food spoilage, such as bacteria, molds, and yeasts, are strongly influenced by temperature conditions (André et al. 2013). Higher temperatures accelerate microbial growth and enzymatic activity, leading to faster food spoilage. The rapid multiplication of pathogenic bacteria in perishable food items significantly escalates the risk of foodborne illnesses (Misiou and Koutsoumanis 2022). Moreover, elevated temperatures can facilitate chemical reactions that deteriorate the quality of food, such as lipid oxidation or protein denaturation. In contrast, lower temperatures can decelerate microbial growth and enzymatic reactions, thereby prolonging the shelf life of food products (Sun et al. 2022). Cold storage or refrigeration effectively hinders the proliferation of spoilage microorganisms and aids in preserving the quality and safety of perishable foods. Maintaining proper temperature control throughout the entire food supply chain, spanning from production to storage and transportation, emerges as a critical measure in preventing food spoilage and upholding food safety (Copetti 2023).

4.6.2 Humidity and Microbial Growth in Food Products

Humidity levels also wield significant influence over microbial growth and the potential for food contamination. Elevated humidity fosters ideal conditions for the proliferation of moisture-dependent microorganisms, encompassing bacteria, molds, and yeasts (Bai et al. 2017). Excessive moisture can result in water accumulation on food surfaces, establishing an environment conducive to microbial development. Consequently, this can lead to the growth of molds on items like bread, fruits, or vegetables, as well as the emergence of slimy or odorous textures in various food products (Tsiraki et al. 2017). In addition to microbial growth, heightened humidity levels can contribute to the deterioration of food quality through processes like caking, clumping, or crispness loss. They can also encourage chemical reactions that lead to flavor degradation, alteration of color, and a decline in nutritional value (Li et al. 2016). On the other hand, lower humidity levels can result in food desiccation, which may cause changes in texture, moisture loss, and potential quality reduction. This aspect is particularly relevant for dry food items like cereals, nuts, or spices, which can lose their freshness and become stale when exposed to low humidity conditions. Therefore, maintaining appropriate humidity levels, both during food processing and storage, emerges as a critical practice in preventing microbial growth, preserving food quality, and extending the shelf life of various food products (Chen et al. 2019).

4.6.3 Impacts of Extreme Weather Events on Food Safety

Extreme weather events, including hurricanes, floods, droughts, and heatwaves, can exert significant impacts on food safety (Duchenne-Moutien and Neetoo 2021). These events have the potential to disrupt various stages of the food supply chain, from production and processing to transportation and storage, thereby increasing the likelihood of contamination and compromising food safety. Floods and heavy rainfall, for instance, can lead to the contamination of agricultural fields with pathogens from polluted water sources or sewage overflows (Chakraborty et al. 2000). As a result, crops such as fruits, vegetables, and grains can be contaminated with harmful bacteria or parasites. Moreover, extreme heatwaves or prolonged drought conditions can affect food production by reducing crop yields, impacting the health of livestock, and elevating the risk of foodborne illnesses (Campos et al. 2013). Droughts can lead to water scarcity, influencing irrigation practices and potentially resulting in the utilization of untreated or contaminated water for crop irrigation. Furthermore, extreme weather events can disrupt the cold chain, which is critical for preserving the safety of perishable food products (Bawuro et al. 2018). Storm-induced power outages or other weather-related incidents can lead to refrigeration failures, jeopardizing the quality of refrigerated or frozen foods (Ávila-Quezada et al. 2010). In response to extreme weather events, proactive measures such as the implementation of early warning systems, enhancing infrastructure resilience, and developing contingency plans for food safety are essential steps in mitigating the risks associated with food contamination (Agriopoulou et al. 2020).

4.7 Impact of Climate Change on Biological Contamination

4.7.1 Long-Term Effects of Climate Change on Microbial Ecosystems

Climate change is altering environmental conditions worldwide, which has significant implications for microbial ecosystems and biological contamination (A'Bear et al. 2014). Changes in temperature, precipitation patterns, and other climatic factors can disrupt the delicate balance of microbial communities, potentially leading to shifts in their composition, diversity, and activity. Rising temperatures associated with climate change can directly impact microbial populations (Abramoff and Finzi 2015). Some microorganisms may thrive in warmer conditions, leading to increased growth rates and altered metabolic activities (Ajwa et al. 1999). This can have consequences for biological contamination as certain pathogenic microorganisms may experience enhanced survival and proliferation, posing a higher risk to human and ecosystem health (Flemming and Wuertz 2019). Furthermore, climate change-induced shifts in precipitation patterns can alter soil moisture levels, affecting microbial communities in terrestrial ecosystems. Changes in rainfall distribution and intensity can impact the survival and dispersal of soilborne pathogens, potentially leading to changes in contamination patterns (Classen et al. 2015).

4.7.2 Altered Climatic Patterns and Their Consequences for Contamination

Globally, climate change is causing significant shifts in climatic patterns. These changes can have far-reaching consequences for biological contamination across various domains, including water, air, and food (Noyes et al. 2009). In water systems, increased temperatures can lead to changes in the distribution and abundance of waterborne pathogens. Warmer waters may create more favorable conditions for the growth of harmful microorganisms, such as Vibrio species or harmful algal blooms (Bhandari et al. 2023). This can increase the risk of waterborne diseases and toxin accumulation in aquatic organisms, impacting both human health and the health of aquatic ecosystems (Kittel 2013). Changes in precipitation patterns can also influence waterborne contamination. More intense rainfall events can result in increased runoff, leading to higher levels of contaminants, such as pesticides or fecal matter, entering water bodies. Conversely, prolonged droughts can lead to water scarcity, potentially compromising water quality and increasing the concentration of contaminants in limited water sources (Chaturvedi et al. 2021). Alternate climatic patterns in the atmosphere can influence the dispersal of airborne contaminants. Changes in wind patterns can influence the transport and distribution of aerosolized microorganisms, pollens, and other allergenic particles. This can impact respiratory health and increase the prevalence of respiratory allergies and infections. Additionally, climate change can impact food safety and foodborne contamination (Ghosh et al. 2023). Temperature and humidity changes can affect the growth and survival of foodborne pathogens, altering contamination risks along the food supply chain. Furthermore, shifts in climatic conditions can impact crop yields, agricultural practices, and pest distribution, potentially affecting food security and the prevalence of agricultural contaminants (Croley and Luukkonen 2003).

4.7.3 Strategies for Mitigating Climate Change-Related Contamination Risks

To mitigate the risks of biological contamination associated with climate change, various strategies can be implemented:

1) Strengthening Surveillance and Monitoring Systems: Robust surveillance and monitoring systems are crucial for detecting changes in contamination patterns and identifying emerging risks. This includes monitoring microbial ecosystems, water quality, air pollutants, and food safety parameters (Mosadeghrad et al. 2023).
2) Improving Adaptive Water Management: Given the influence of climate change on waterborne contamination, effective water management strategies are essential. This involves sustainable water use practices, implementing

water treatment technologies, and ensuring adequate sanitation systems to prevent the spread of waterborne pathogens (Malhi et al. 2021).

3) Enhancing Food Safety Measures: Climate change can impact food safety along the entire food production and distribution chain. Strengthening food safety regulations, implementing good agricultural and manufacturing practices, and enhancing traceability systems can help mitigate contamination risks associated with changing climatic conditions (Sitch et al. 2003).
4) Promoting Sustainable Agricultural Practices: Sustainable agriculture practices, such as precision farming, organic farming, and integrated pest management, can reduce the reliance on chemical inputs and minimize contamination risks. These practices can enhance soil health, reduce water pollution, and promote resilient agricultural systems (Deryng et al. 2016).
5) Investing in Research and Innovation: Continued research and innovation are vital for understanding the complex interactions between climate change and biological contamination. This includes developing predictive models, innovative technologies for contamination detection, and sustainable solutions for managing contamination risks in a changing climate (Myers et al. 2014).

4.8 Case Studies and Examples

4.8.1 Documented Instances of Climatic/Meteorological Conditions and Biological Contamination

Several documented case studies provide evidence of the relationship between climatic or meteorological conditions and biological contamination. These instances highlight the diverse ways in which changes in climate can impact the occurrence and spread of contaminants, posing risks to various ecosystems and human health (Xing et al. 2023).

1) Hurricane-Induced Waterborne Contamination: Major hurricanes, such as Hurricane Katrina in 2005 or Hurricane Harvey in 2017, have resulted in significant flooding, leading to waterborne contamination. These events have been associated with the spread of microbial pathogens, pollutants, and toxins in affected areas, affecting drinking water sources, recreational waters, and aquatic ecosystems (Singh and Kumar 2022).
2) Heatwaves and Foodborne Illness Outbreaks: Heatwaves can create conditions favorable for the growth of foodborne pathogens. For example, in 1996, a heatwave in the United Kingdom led to a significant increase in Salmonella infections associated with the consumption of contaminated poultry products. The elevated temperatures during the heatwave facilitated bacterial growth and survival, contributing to the outbreak (Levy et al. 2018).

3) Pollen Allergens and Climate Change: Climate change influences the timing, quantity, and allergenicity of pollen. As temperatures rise, plants may produce more pollen, leading to increased exposure and higher rates of allergic reactions. Changes in rainfall patterns and humidity levels can also affect pollen production and dispersal, influencing the prevalence and severity of pollen-related allergies (Noyes et al. 2009).

4.8.2 Lessons Learned and Potential Interventions

1) Integrated Monitoring and Early Warning Systems: The case studies highlight the importance of establishing integrated monitoring and early warning systems. These systems should encompass various environmental parameters, including climatic conditions, microbial indicators, and contaminant levels. By detecting early signs of contamination, authorities can implement timely interventions and preventive measures to mitigate risks (Malhi et al. 2021).
2) Resilient Infrastructure and Emergency Preparedness: Building resilient infrastructure and enhancing emergency preparedness are crucial to minimize the impacts of climatic events on biological contamination. This includes improved sanitation systems, water treatment facilities, and disaster response plans that account for the potential increase in contamination risks under changing climatic conditions (Raza et al. 2019).
3) Public Education and Awareness: Educating the public about the relationship between climate change, meteorological conditions, and biological contamination is essential. Enhancing public awareness about the risks, preventive measures, and proper hygiene practices can empower individuals to make informed decisions, reduce exposure to contaminants, and contribute to contamination prevention efforts (Oh 2022).
4) Climate-Resilient Agricultural Practices: Agricultural sectors can adopt climate-resilient practices that reduce the risk of biological contamination. This includes implementing precision irrigation systems, crop rotation, soil conservation techniques, and integrated pest management strategies. These practices promote sustainable agriculture, enhance soil health, and minimize the use of agrochemicals that contribute to contamination (Deryng et al. 2016).
5) Collaboration and Knowledge Exchange: Sharing experiences, best practices, and research findings among different stakeholders is crucial for addressing the challenges posed by biological contamination and climate change. Collaboration between scientists, policymakers, industries, and communities can foster innovation, develop effective interventions, and improve overall resilience to climatic conditions and contamination risks (Pacheco et al. 2021).

4.9 Detection and Monitoring of Climatic-Related Contamination

4.9.1 Methods for Identifying and Quantifying Biological Contaminants

Accurate and reliable methods for identifying and quantifying biological contaminants are crucial for effective detection and monitoring. Various techniques are employed to assess the presence and abundance of contaminants in different environmental matrices (Kumar et al. 2021).

1) Microbiological Analysis: Microbiological analysis involves the culturing and identification of microorganisms from samples, such as water, soil, or food. This method allows for the detection of specific pathogens or indicators of contamination, providing valuable information on the presence and viability of biological contaminants (Gonçalves et al. 2017).
2) Molecular Methods: Molecular methods, such as polymerase chain reaction (PCR) and DNA sequencing, enable the detection and identification of microorganisms based on their genetic material. These techniques offer high sensitivity and specificity and can detect low levels of contaminants, including pathogens, in environmental samples (Diglisic et al. 1999).
3) Immunoassays: To detect and quantify target microorganisms or their antigens, immunoassays use specific antibodies. Immunoassay methods like enzyme-linked immunosorbent assays (ELISA) and immunofluorescence assays are often used to quickly find contaminants in water, food, or environmental samples (Di Caprio 2020).
4) Biosensors: To detect and measure the presence of contaminants, biological components, such as antibodies or enzymes, are combined with transducers. These portable devices offer real-time monitoring capabilities and can be used in the field for quick and on-site detection of biological contaminants (Janghorban et al. 2023).

4.9.2 Remote Sensing and Technological Advancements

Remote sensing technologies and advancements in data analysis play a significant role in the detection and monitoring of climatic-related contamination. These tools provide valuable information on environmental conditions, contamination patterns, and their relationship with climatic factors (Murray et al. 2022).

1) Satellite Imagery: Satellite-based remote sensing enables the monitoring of large-scale environmental parameters, such as land cover changes, water quality, and algal blooms. This information can help identify areas prone to

contamination, track the spread of contaminants, and assess the impacts of climatic events on contamination dynamics (Yang et al. 2022).

2) Geographic Information Systems (GIS): GIS technology combines spatial data with analytical tools to visualize, analyze, and model contamination patterns. By integrating data on climatic conditions, contamination sources, and vulnerable populations, GIS can aid in identifying high-risk areas, developing targeted interventions, and optimizing resource allocation for monitoring and response efforts (Mashala et al. 2023).
3) Sensor Networks: Sensor networks consist of distributed sensors that collect data on various environmental parameters in real-time. These networks can be deployed in water bodies, air quality monitoring stations, or agricultural fields to continuously monitor climatic conditions and contamination levels. The data collected from these networks can support early warning systems, prompt response actions, and the assessment of contamination risks (Noyes et al. 2009).
4) Big Data Analytics and Machine Learning: The analysis of large datasets, coupled with machine learning algorithms, can help identify patterns, predict contamination events, and provide insights for decision-making. By processing vast amounts of climatic, environmental, and contamination data, these analytical approaches contribute to the development of proactive monitoring systems and targeted interventions (Anselmo et al. 2018; Chen et al. 2022).

4.9.3 Surveillance Systems for Early Detection and Response

Surveillance systems are essential for early detection and timely response to climatic-related contamination. These systems involve the continuous monitoring of environmental parameters, contamination sources, and health indicators to identify potential risks and take appropriate action (Noyes et al. 2009).

1) Water Quality Surveillance: Water quality surveillance programs involve routine monitoring of water sources, including surface water bodies, groundwater, and drinking water systems. This monitoring assesses microbial indicators, chemical contaminants, and physical parameters to detect changes in water quality and potential contamination events (Zinsstag et al. 2018).
2) Syndromic Surveillance: Syndromic surveillance focuses on monitoring disease symptoms and health indicators in real-time. By analyzing data from healthcare facilities, emergency departments, or community health centers, syndromic surveillance systems can detect early signs of contamination-related illnesses and facilitate rapid response and intervention (dos Ribeiro et al. 2019).
3) One Health Approaches: One Health approaches integrate data from human health, animal health, and environmental monitoring systems to understand

and respond to contamination risks comprehensively. By considering the interconnectedness of humans, animals, and the environment, these surveillance systems can detect potential contamination sources and emerging risks more effectively (Prata et al. 2022).

Effective detection and monitoring of climatic-related contamination require the integration of multiple methods, technological advancements, and surveillance systems. By employing these tools and approaches, stakeholders can enhance their understanding of the complex interactions between climatic conditions and contamination dynamics, enabling proactive measures to mitigate risks and protect public health (Humboldt-Dachroeden and Mantovani 2021).

4.10 Future Research Directions and Recommendations

4.10.1 Identifying Research Gaps and Unanswered Questions

As our understanding of the complex relationship between climatic conditions and biological contamination grows, it is essential to identify research gaps and address unanswered questions (Turkoglu Sasmazel et al. 2021; Samut et al. 2023). By focusing on these areas, future research can contribute to a more comprehensive understanding of the impacts of climatic conditions on contamination dynamics and develop effective strategies for prevention and mitigation (Duchenne-Moutien and Neetoo 2021).

1) Long-term Effects: Further investigation is needed to understand the long-term effects of climatic conditions on microbial ecosystems and contamination risks. Longitudinal studies can provide valuable insights into how prolonged shifts in temperature, precipitation patterns, or extreme weather events influence the persistence, distribution, and virulence of contaminants (Dublan et al. 2014).
2) Interaction of Multiple Factors: Research should explore the interactions between climatic conditions and other environmental factors that contribute to contamination, such as land use, pollution levels, or human activities. Understanding these complex interactions can help develop integrated risk assessments and management strategies (Bielaszewska et al. 2011).
3) Emerging Contaminants: The study of emerging contaminants, including pathogens, pollutants, and toxins, is critical for anticipating and addressing future contamination risks. Research efforts should focus on monitoring, detection, and characterization of these emerging contaminants and their response to climatic conditions (Purse et al. 2017).

4.10.2 Integration of Climatic Data Into Contamination Risk Assessments

To effectively manage and mitigate biological contamination, it is important to integrate climatic data into contamination risk assessments. This integration can enhance the accuracy and predictive capabilities of risk assessments, enabling targeted interventions and proactive measures (Carlton et al. 2016).

1) Climate-Informed Modeling: Developing climate-informed models can help predict the potential impacts of climatic conditions on contamination risks. By incorporating historical and projected climate data into modeling frameworks, researchers can assess the vulnerabilities of different ecosystems and populations to contamination under different climatic scenarios (Effler et al. 2001).
2) Data Sharing and Integration: Enhancing data sharing and integration across different disciplines and sectors is crucial for comprehensive contamination risk assessments. This includes integrating climatic data with environmental monitoring data, health surveillance systems, and agricultural data to gain a holistic understanding of contamination risks and inform evidence-based decision-making (Duchenne-Moutien and Neetoo 2021).
3) Stakeholder Engagement: Engaging stakeholders, including policymakers, scientists, communities, and industry representatives, in the development and implementation of contamination risk assessments is vital. By incorporating diverse perspectives and expertise, risk assessments can capture the complexities of climatic conditions and contamination dynamics more effectively (Rizou et al. 2020).

4.10.3 Strategies for Adapting to Changing Climatic Conditions

As climatic conditions continue to change, it is necessary to develop strategies for adapting to these shifts and minimizing the associated contamination risks. Adapting to changing climatic conditions requires a multidimensional approach that encompasses various sectors and stakeholders (Orozco et al. 2008).

1) Climate-Smart Agriculture: Promoting climate-smart agricultural practices can reduce contamination risks associated with changes in temperature and precipitation patterns. These practices include crop diversification, efficient water management, soil conservation, and integrated pest management, all aimed at enhancing resilience and minimizing contamination risks (Pohoroo and Ranghoo-Sanmukhiya 2017).

2) Water Management and Infrastructure: Investing in resilient water management systems and infrastructure is essential for adapting to changing climatic conditions. This includes improving water storage capacities, implementing efficient water treatment technologies, and enhancing wastewater management to ensure the availability of clean and safe water resources (Gitahi et al. 2002).
3) Health Systems Preparedness: Strengthening health systems' preparedness and response capabilities is crucial for addressing contamination risks under changing climatic conditions. This involves enhancing disease surveillance, improving laboratory capacity, and developing early warning systems to detect and respond to contamination-related health threats promptly (Wright and Holden 2018).
4) Community Engagement and Education: Engaging communities and promoting public awareness about the impacts of climatic conditions on contamination can empower individuals to take preventive measures and adopt resilient practices. Educational campaigns, community-based initiatives, and capacity-building programs can help communities understand the risks and develop adaptive strategies (Xiao et al. 2014).

By focusing future research on identified gaps, integrating climatic data into contamination risk assessments, and implementing strategies for adaptation, we can enhance our ability to address the challenges posed by biological contamination under changing climatic conditions. These efforts will contribute to safeguarding ecosystems, protecting public health, and fostering resilience in the face of climatic uncertainties (Pastor et al. 1996).

4.11 Conclusion

This chapter highlights the significant influence of climatic conditions on biological contamination. The findings demonstrate that temperature, humidity, precipitation, wind patterns, and extreme weather events play a crucial role in the proliferation, dissemination, and survival of contaminants in water, air, soil, and food. The implications for public health and environmental management are substantial, as understanding these dynamics is essential for developing targeted interventions, enhancing surveillance systems, and promoting adaptive strategies. Moving forward, future research should address remaining gaps, and integrate climatic data into risk assessments, and policymakers should prioritize the incorporation of climatic considerations into policies and plans. By taking action based on these insights, we can effectively address climatic-related contamination, safeguard public health, and foster resilient communities.

References

A'Bear, A.D., Jones, T.H., and Boddy, L. (2014). Potential impacts of climate change on interactions among saprotrophic cord-forming fungal mycelia and grazing soil invertebrates. *Fungal Ecology* 10 (1): 34–43. https://doi.org/10.1016/J.FUNECO.2013.01.009.

Abramoff, R.Z. and Finzi, A.C. (2015). Are above- and below-ground phenology in sync? *New Phytologist* 205 (3): 1054–1061. https://doi.org/10.1111/NPH.13111.

Abreu, I.M.d.O., Junqueira, A.M.R., Peixoto, J.R., and de Oliveira, S.A. (2010). Microbiological quality and productivity of lettuce under chemical and organic fertilization. *Food Science and Technology* 30 (Suppl. 1): 108–118. https://doi.org/10.1590/S0101-20612010000500018.

Agranovski, I.E., Safatov, A.S., Pyankov, O.V. et al. (2007). Long-term sampling of viable airborne viruses. 39 (9): 912–918. https://doi.org/10.1080/02786820500297012.

Agriopoulou, S., Stamatelopoulou, E., and Varzakas, T. (2020). Advances in occurrence, importance, and mycotoxin control strategies: prevention and detoxification in foods. *Foods* 9 (2): 137. https://doi.org/10.3390/FOODS9020137.

Ahamad, A., Raju, N.J., Madhav, S. et al. (2019). Impact of non-engineered Bhalswa landfill on groundwater from Quaternary alluvium in Yamuna flood plain and potential human health risk, New Delhi, India. *Quaternary International* 507: 352–369. https://doi.org/10.1016/J.QUAINT.2018.06.011.

Aitken, C. and Jeffries, D.J. (2001). Nosocomial spread of viral disease. *Clinical Microbiology Reviews* 14 (3): 528–546. https://doi.org/10.1128/CMR.14.3.528-546.2001/ASSET/3C99D659-F5CF-4C57-BACA-F108222C285C/ASSETS/GRAPHIC/CM0310027002.JPEG.

Ajwa, H.A., Dell, C.J., and Rice, C.W. (1999). Changes in enzyme activities and microbial biomass of tallgrass prairie soil as related to burning and nitrogen fertilization. *Soil Biology and Biochemistry* 31 (5): 769–777. https://doi.org/10.1016/S0038-0717(98)00177-1.

Alam, M.J. and Zurek, L. (2004). Association of *Escherichia coli* O157:H7 with houseflies on a cattle farm. *Applied and Environmental Microbiology* 70 (12): 7578–7580. https://doi.org/10.1128/AEM.70.12.7578-7580.2004.

Alegbeleye, O.O. and Sant'Ana, A.S. (2020). Manure-borne pathogens as an important source of water contamination: an update on the dynamics of pathogen survival/transport as well as practical risk mitigation strategies. *International Journal of Hygiene and Environmental Health* 227: 113524. https://doi.org/10.1016/J.IJHEH.2020.113524.

Alexandersen, S., Oleksiewicz, M.B., and Donaldson, A.I. (2001). The early pathogenesis of foot-and-mouth disease in pigs infected by contact: a quantitative time-course study using TaqMan RT-PCR. *Journal of General Virology* 82 (4): 747–755. https://doi.org/10.1099/0022-1317-82-4-747/CITE/REFWORKS.

Al-Ghazali, M.R. and Al-Azawi, S.K. (1990). Listeria monocytogenes contamination of crops grown on soil treated with sewage sludge cake. *Journal of Applied Bacteriology* 69 (5): 642–647. https://doi.org/10.1111/J.1365-2672.1990.TB01557.X.

Alster, C.J., von Fischer, J.C., Allison, S.D., and Treseder, K.K. (2020). Embracing a new paradigm for temperature sensitivity of soil microbes. *Global Change Biology* 26 (6): 3221–3229. https://doi.org/10.1111/GCB.15053.

Anderson, G.L., Caldwell, K.N., Beuchat, L.R., and Williams, P.L. (2003). Interaction of a free-living soil nematode, *Caenorhabditis elegans*, with surrogates of foodborne pathogenic bacteria. *Journal of Food Protection* 66 (9): 1543–1549. https://doi.org/10.4315/0362-028X-66.9.1543.

André, S., Zuber, F., and Remize, F. (2013). Thermophilic spore-forming bacteria isolated from spoiled canned food and their heat resistance. Results of a French ten-year survey. *International Journal of Food Microbiology* 165 (2): 134–143. https://doi.org/10.1016/J.IJFOODMICRO.2013.04.019.

Angenent, L.T., Kelley, S.T., St. Amand, A. et al. (2005). Molecular identification of potential pathogens in water and air of a hospital therapy pool. *Proceedings of the National Academy of Sciences* 102 (13): 4860–4865. https://doi.org/10.1073/PNAS.0501235102.

Anselmo, A.C., Gokarn, Y., and Mitragotri, S. (2018). Non-invasive delivery strategies for biologics. *Nature Reviews Drug Discovery* 18 (1): 19–40. https://doi.org/10.1038/nrd.2018.183.

Aruscavage, D., Phelan, P.L., Lee, K., and LeJeune, J.T. (2010). Impact of changes in sugar exudate created by biological damage to tomato plants on the persistence of *Escherichia coli* O157:H7. *Journal of Food Science* 75 (4): M187–M192. https://doi.org/10.1111/J.1750-3841.2010.01593.X.

Ávila-Quezada, G., Sánchez, E., Gardea-Béjar, A.A., and Acedo-Félix, E. (2010). *Salmonella* spp. and *Escherichia coli*: survival and growth in plant tissue. *New Zealand Journal of Crop and Horticultural Science* 38 (2): 47–55. https://doi.org/10.1080/01140671003767834.

Bååth, E. (2001). Estimation of fungal growth rates in soil using 14C-acetate incorporation into ergosterol. *Soil Biology and Biochemistry* 33 (14): 2011–2018. https://doi.org/10.1016/S0038-0717(01)00137-7.

Bai, Y.P., Guo, X.N., Zhu, K.X., and Zhou, H.M. (2017). Shelf-life extension of semi-dried buckwheat noodles by the combination of aqueous ozone treatment and modified atmosphere packaging. *Food Chemistry* 237: 553–560. https://doi.org/10.1016/J.FOODCHEM.2017.05.156.

Baldwin, D.S., Gigney, H., Wilson, J.S. et al. (2008). Drivers of water quality in a large water storage reservoir during a period of extreme drawdown. *Water Research* 42 (19): 4711–4724. https://doi.org/10.1016/J.WATRES.2008.08.020.

Balshem, H., Helfand, M., Schünemann, H.J. et al. (2011). GRADE guidelines: 3. Rating the quality of evidence. *Journal of Clinical Epidemiology* 64 (4): 401–406. https://doi.org/10.1016/J.JCLINEPI.2010.07.015.

Banhazi, T.M., Rutley, D.L., and Pitchford, W.S. (2010). Validation and fine-tuning of a predictive model for air quality in livestock buildings. *Biosystems Engineering* 105 (3): 395–401. https://doi.org/10.1016/J.BIOSYSTEMSENG.2009.12.011.

Barak, J.D. and Liang, A.S. (2008). Role of soil, crop debris, and a plant pathogen in *Salmonella enterica* contamination of tomato plants. *PLoS One* 3 (2): e1657. https://doi.org/10.1371/JOURNAL.PONE.0001657.

Barak, J.D., Whitehand, L.C., and Charkowski, A.O. (2002). Differences in attachment of *Salmonella enterica* serovars and *Escherichia coli* O157:H7 to alfalfa sprouts. *Applied and Environmental Microbiology* 68 (10): 4758–4763. https://doi.org/10.1128/AEM.68.10.4758-4763.2002/ASSET/DF25E7DE-3894-4F41-8857-91230EAA9AF0/ASSETS/GRAPHIC/AM1020677007.JPEG.

Barberán, A., Bates, S.T., Casamayor, E.O., and Fierer, N. (2011). Using network analysis to explore co-occurrence patterns in soil microbial communities. *The ISME Journal* 6 (2): 343–351. https://doi.org/10.1038/ismej.2011.119.

Barberán, A., Dunn, R.R., Reich, B.J. et al. (2015). The ecology of microscopic life in household dust. *Proceedings of the Royal Society B: Biological Sciences* 282 (1814): https://doi.org/10.1098/RSPB.2015.1139.

Barros, M.C., Mendo, M.J.M., and Negrão, F.C.R. (1995). Surface water quality in Portugal during a drought period. *Science of The Total Environment* 171 (1–3): 69–76. https://doi.org/10.1016/0048-9697(95)04671-7.

Bawuro, A.A., Voegborlo, R.B., and Adimado, A.A. (2018). Bioaccumulation of heavy metals in some tissues of fish in Lake Geriyo, Adamawa State, Nigeria. *Journal of Environmental and Public Health* 2018: https://doi.org/10.1155/2018/1854892.

Bernstein, N., Sela, S., and Neder-Lavon, S. (2007). Assessment of contamination potential of lettuce by *Salmonella enterica* serovar newport added to the plant growing medium. *Journal of Food Protection* 70 (7): 1717–1722. https://doi.org/10.4315/0362-028X-70.7.1717.

Beuchat, L.R. (2006). Vectors and conditions for preharvest contamination of fruits and vegetables with pathogens capable of causing enteric diseases. *British Food Journal* 108 (1): 38–53. https://doi.org/10.1108/00070700610637625/FULL/XML.

Bhandari, S., Kukreja, S., Singh, B. et al. (2023). Role of beneficial microbes in biotic and abiotic stress. *Relationship between Microbes and the Environment for Sustainable Ecosystem Services: Microbial Tools for Sustainable Ecosystem Services* 3 (3): 243–259. https://doi.org/10.1016/B978-0-323-89936-9.00003-5.

Bielaszewska, M., Mellmann, A., Zhang, W. et al. (2011). Characterisation of the *Escherichia coli* strain associated with an outbreak of haemolytic uraemic syndrome in Germany, 2011: a microbiological study. *The Lancet Infectious Diseases* 11 (9): 671–676. https://doi.org/10.1016/S1473-3099(11)70165-7.

Bragoszewska, E., Mainka, A., Pastuszka, J.S. et al. (2018). Assessment of bacterial aerosol in a preschool, primary school and high school in Poland. *Atmosphere* 9 (3): 87. https://doi.org/10.3390/ATMOS9030087.

Brandl, M.T. and Amundson, R. (2008). Leaf age as a risk factor in contamination of lettuce with *Escherichia coli* O157:H7 and *Salmonella enterica*. *Applied and Environmental Microbiology* 74 (8): 2298–2306. https://doi.org/10.1128/AEM.02459-07/ASSET/0B821053-600A-452F-97E5-6C8BAB5D64E9/ASSETS/GRAPHIC/ZAM0080887440007.JPEG.

Caldwell, K.N., Anderson, G.L., Williams, P.L., and Beuchat, L.R. (2003). Attraction of a free-living nematode, *Caenorhabditis elegans*, to foodborne pathogenic bacteria and its potential as a vector of Salmonella Poona for preharvest contamination of Cantaloupe. *Journal of Food Protection* 66 (11): 1964–1971. https://doi.org/10.4315/0362-028X-66.11.1964.

Cambra-López, M., Aarnink, A.J.A., Zhao, Y. et al. (2010). Airborne particulate matter from livestock production systems: a review of an air pollution problem. *Environmental Pollution* 158 (1): 1–17. https://doi.org/10.1016/J.ENVPOL.2009.07.011.

Campos, C.J.A., Kershaw, S.R., and Lee, R.J. (2013). Environmental influences on faecal indicator organisms in coastal waters and their accumulation in bivalve shellfish. *Estuaries and Coasts* 36 (4): 834–853. https://doi.org/10.1007/S12237-013-9599-Y.

Canarini, A., Carrillo, Y., Mariotte, P. et al. (2016). Soil microbial community resistance to drought and links to C stabilization in an Australian grassland. *Soil Biology and Biochemistry* 103: 171–180. https://doi.org/10.1016/J.SOILBIO.2016.08.024.

Cann, K.F., Thomas, D.R., Salmon, R.L. et al. (2013). Extreme water-related weather events and waterborne disease. *Epidemiology and Infection* 141 (4): 671. https://doi.org/10.1017/S0950268812001653.

Carlton, E.J., Woster, A.P., DeWitt, P. et al. (2016). A systematic review and meta-analysis of ambient temperature and diarrhoeal diseases. *International Journal of Epidemiology* 45 (1): 117–130. https://doi.org/10.1093/IJE/DYV296.

Chakraborty, S., Tiedemann, A.V., and Teng, P.S. (2000). Climate change: potential impact on plant diseases. *Environmental Pollution* 108 (3): 317–326. https://doi.org/10.1016/S0269-7491(99)00210-9.

Chaturvedi, A., Pandey, B., Yadav, A.K., and Saroj, S. (2021). An overview of the potential impacts of global climate change on water resources. *Water Conservation in the Era of Global Climate Change* 99–120. https://doi.org/10.1016/B978-0-12-820200-5.00012-9.

Chen, Y., Chen, G., Wei, R. et al. (2019). Quality characteristics of fresh wet noodles treated with nonthermal plasma sterilization. *Food Chemistry* 297: 124900. https://doi.org/10.1016/J.FOODCHEM.2019.05.174.

Chen, J., Chen, S., Fu, R. et al. (2022). Remote sensing big data for water environment monitoring: current status, challenges, and future prospects. *Earth's Future* 10 (2): e2021EF002289. https://doi.org/10.1029/2021EF002289.

Cheng, G. and Jin, H. (2013). Permafrost and groundwater on the Qinghai-Tibet Plateau and in northeast China. *Hydrogeology Journal* 21: 5–23. https://doi.org/10.1007/s10040-012-0927-2.

Cho, S.Y., Myong, J.P., Kim, W.B. et al. (2018). Profiles of environmental mold: indoor and outdoor air sampling in a Hematology Hospital in Seoul, South Korea. *International Journal of Environmental Research and Public Health* 15 (11): 2560. https://doi.org/10.3390/IJERPH15112560.

Classen, A.E., Sundqvist, M.K., Henning, J.A. et al. (2015). Direct and indirect effects of climate change on soil microbial and soil microbial-plant interactions: what lies ahead? *Ecosphere* 6 (8): 1–21. https://doi.org/10.1890/ES15-00217.1.

Cooley, M.B., Chao, D., and Mandrell, R.E. (2006). *Escherichia coli* O157:H7 survival and growth on lettuce is altered by the presence of epiphytic bacteria. *Journal of Food Protection* 69 (10): 2329–2335. https://doi.org/10.4315/0362-028X-69.10.2329.

Copetti, M.V. (2023). Sanitizers for controlling fungal spoilage in some food industries. *Current Opinion in Food Science* 52: 101072. https://doi.org/10.1016/J.COFS.2023.101072.

Croley, T.E. and Luukkonen, C.L. (2003). Potential effects of climate change on ground water in Lansing, Michigan. *JAWRA Journal of the American Water Resources Association* 39 (1): 149–163. https://doi.org/10.1111/J.1752-1688.2003.TB01568.X.

Cruz-Paredes, C., Tájmel, D., and Rousk, J. (2021). Can moisture affect temperature dependences of microbial growth and respiration? *Soil Biology and Biochemistry* 156: 108223. https://doi.org/10.1016/J.SOILBIO.2021.108223.

Curtis, L., Lieberman, A., Stark, M. et al. (2009). Adverse health effects of indoor molds. *Journal of Nutritional & Environmental Medicine* 14 (3): 261–274. https://doi.org/10.1080/13590840400010318.

Delaquis, P., Bach, S., and Dinu, L.D. (2007). Behavior of *Escherichia coli* O157:H7 in leafy vegetables. *Journal of Food Protection* 70 (8): 1966–1974. https://doi.org/10.4315/0362-028X-70.8.1966.

Delpla, I., Jung, A.V., Baures, E. et al. (2009). Impacts of climate change on surface water quality in relation to drinking water production. *Environment International* 35 (8): 1225–1233. https://doi.org/10.1016/J.ENVINT.2009.07.001.

Deryng, D., Elliott, J., Folberth, C. et al. (2016). Regional disparities in the beneficial effects of rising CO_2 concentrations on crop water productivity. *Nature Climate Change* 6 (8): 786–790. https://doi.org/10.1038/nclimate2995.

Dey, D., Ghosal, K., and Bhattacharya, S.G. (2019). Aerial fungal spectrum of Kolkata, India, along with their allergenic impact on the public health: a quantitative and qualitative evaluation. *Aerobiologia* 35 (1): 15–25. https://doi.org/10.1007/S10453-018-9534-6/METRICS.

Di Caprio, F. (2020). Methods to quantify biological contaminants in microalgae cultures. *Algal Research* 49: 101943. https://doi.org/10.1016/J.ALGAL.2020.101943.

Diglisic, G., Rossi, C.A., Doti, A., and Walshe, D.K. (1999). Seroprevalence study of Hantavirus infection in the community based population. *Maryland Medical Journal (Baltimore, Md.: 1985)* 48 (6): 303–306. https://europepmc.org/article/med/10628132.

Dreux, N., Albagnac, C., Federighi, M. et al. (2007). Viable but non-culturable Listeria monocytogenes on parsley leaves and absence of recovery to a culturable state. *Journal of Applied Microbiology* 103 (4): 1272–1281. https://doi.org/10.1111/J.1365-2672.2007.03351.X.

Dublan, M.D.L.A., Ortiz-Marquez, J.C.F., Lett, L., and Curatti, L. (2014). Plant-adapted *Escherichia coli* show increased lettuce colonizing ability, resistance to oxidative stress and chemotactic response. *PLoS One* 9 (10): e110416. https://doi.org/10.1371/JOURNAL.PONE.0110416.

Duchenne-Moutien, R.A. and Neetoo, H. (2021). Climate change and emerging food safety issues: a review. *Journal of Food Protection* 84 (11): 1884–1897. https://doi.org/10.4315/JFP-21-141.

Effler, P., Isaäcson, M., Arntzen, L. et al. (2001). Factors contributing to the emergence of *Escherichia coli* O157 in Africa. *Emerging Infectious Diseases* 7 (5): 812. https://doi.org/10.3201/EID0705.017507.

Eliana Andrea, M.M., Ana Carolina, T.E., Tito José, C.B. et al. (2019). Evaluation of contaminants in agricultural soils in an Irrigation District in Colombia. *Heliyon* 5 (8): https://doi.org/10.1016/J.HELIYON.2019.E02217.

Erickson, M.C., Liao, J., Payton, A.S. et al. (2010a). Preharvest internalization of *Escherichia coli* O157:H7 into lettuce leaves, as affected by insect and physical damage. *Journal of Food Protection* 73 (10): 1809–1816. https://doi.org/10.4315/0362-028X-73.10.1809.

Erickson, M.C., Webb, C.C., Diaz-Perez, J.C. et al. (2010b). Surface and Internalized *Escherichia coli* O157: H7 on field-grown spinach and lettuce treated with spray-contaminated irrigation water. *Journal of Food Protection* 73 (6): 1023–1029. https://doi.org/10.4315/0362-028X-73.6.1023.

Fang, Z., Zhang, J., Guo, W., and Lou, X. (2019). Assemblages of culturable airborne fungi in a typical urban, Tourism-driven Center of Southeast China. *Aerosol and Air Quality Research* 19 (4): 820–831. https://doi.org/10.4209/AAQR.2018.02.0042.

Ferens, W.A. and Hovde, C.J. (2011). *Escherichia coli* O157:H7: animal reservoir and sources of human infection. *Foodborne Pathogens and Disease* 8 (4): 465–487. https://doi.org/10.1089/FPD.2010.0673.

Flemming, H.C. and Wuertz, S. (2019). Bacteria and archaea on Earth and their abundance in biofilms. *Nature Reviews Microbiology* 17 (4): 247–260. https://doi.org/10.1038/s41579-019-0158-9.

Fones, H.N., Bebber, D.P., Chaloner, T.M. et al. (2020). Threats to global food security from emerging fungal and oomycete crop pathogens. *Nature Food* 1 (6): 332–342. https://doi.org/10.1038/s43016-020-0075-0.

Franz, E. and Van Bruggen, A.H.C. (2008). Ecology of *E. coli* O157:H7 and *Salmonella enterica* in the primary vegetable production chain. *Critical Reviews in Microbiology* 34 (3–4): 143–161. https://doi.org/10.1080/10408410802357432.

Fu, F., Li, J., Li, Y. et al. (2023). Simulating the effect of climate change on soil microbial community in an *Abies georgei* var. smithii forest. *Frontiers in Microbiology* 14: 1189859. https://doi.org/10.3389/FMICB.2023.1189859/BIBTEX.

Fung, F. and Hughson, W.G. (2010). Health effects of indoor fungal bioaerosol exposure. 18 (7): 535–544. https://doi.org/10.1080/10473220301451.

Gagliardi, J.V. and Karns, J.S. (2000). Leaching of *Escherichia coli* O157:H7 in diverse soils under various agricultural management practices. *Applied and Environmental Microbiology* 66 (3): 877–883. https://doi.org/10.1128/AEM.66.3.877-883.2000/ASSET/8DE254FF-95C4-415A-BF34-7AE1A31714BC/ASSETS/GRAPHIC/AM0301293002.JPEG.

Galli, S.J., Tsai, M., and Piliponsky, A.M. (2008). The development of allergic inflammation. *Nature* 454 (7203): 445–454. https://doi.org/10.1038/nature07204.

Ghosh, D., Ghorai, P., Debnath, S. et al. (2023). Impression of climatic variation on flora, fauna, and human being: a present state of art. *Visualization Techniques for Climate Change with Machine Learning and Artificial Intelligence* 101–122. https://doi.org/10.1016/B978-0-323-99714-0.00004-2.

Gilley, J.E. (2005). Erosion|water-induced. *Encyclopedia of Soils in the Environment* 4: 463–469. https://doi.org/10.1016/B0-12-348530-4/00262-9.

Gitahi, S.M., Harper, D.M., Muchiri, S.M. et al. (2002). Organochlorine and organophosphorus pesticide concentrations in water, sediment, and selected organisms in Lake Naivasha (Kenya). *Hydrobiologia* 488 (1): 123–128. https://doi.org/10.1023/A:1023386732731/METRICS.

Godish, D.R. and Godish, T.J. (2007). Relationship between sampling duration and concentration of culturable airborne mould and bacteria on selected culture media. *Journal of Applied Microbiology* 102 (6): 1479–1484. https://doi.org/10.1111/J.1365-2672.2006.03200.X.

Gonçalves, C.L., Mota, F.V., Ferreira, G.F. et al. (2017). Airborne fungi in an intensive care unit. *Brazilian Journal of Biology* 78 (2): 265–270. https://doi.org/10.1590/1519-6984.06016.

Hayes, N.M., Haig, H.A., Simpson, G.L., and Leavitt, P.R. (2020). Effects of lake warming on the seasonal risk of toxic cyanobacteria exposure. *Limnology And Oceanography Letters* 5 (6): 393–402. https://doi.org/10.1002/LOL2.10164.

Humboldt-Dachroeden, S. and Mantovani, A. (2021). Assessing environmental factors within the one health approach. *Medicina* 57 (3): 240. https://doi.org/10.3390/MEDICINA57030240.

Islam, M.A., Ikeguchi, A., and Naide, T. (2019). Concentrations of aerosol numbers and airborne bacteria, and temperature and relative humidity, and their interrelationships in a tie-stall dairy barn. *Animals: An Open Access Journal from MDPI* 9 (12): https://doi.org/10.3390/ANI9121023.

Janghorban, M., Kazemi, S., Tormon, R. et al. (2023). Methods and analysis of biological contaminants in the biomanufacturing industry. *Chemosensors* 11 (5): 298. https://doi.org/10.3390/CHEMOSENSORS11050298.

Jeger, M.J. (2022). The impact of climate change on disease in wild plant populations and communities. *Plant Pathology* 71 (1): 111–130. https://doi.org/10.1111/PPA.13434.

Jeleń, H. and Wąsowicz, E. (2009). Volatile fungal metabolites and their relation to the spoilage of agricultural commodities. *Food Reviews International* 14 (4): 391–426. https://doi.org/10.1080/87559129809541170.

Kang, Y.-J. and Frank, J.F. (1989). Biological aerosols: a review of airborne contamination and its measurement in dairy processing plants. *Journal of Food Protection* 52 (7): 512–524. https://doi.org/10.4315/0362-028X-52.7.512.

Kittel, T.G.F. (2013). The vulnerability of biodiversity to rapid climate change. *Climate Vulnerability: Understanding and Addressing Threats to Essential Resources* 4: 185–201. https://doi.org/10.1016/B978-0-12-384703-4.00437-8.

Kumar, P., Kausar, M.A., Singh, A.B., and Singh, R. (2021). Biological contaminants in the indoor air environment and their impacts on human health. *Air Quality, Atmosphere, & Health* 14 (11): 1723. https://doi.org/10.1007/S11869-021-00978-Z.

Lai, H., Hales, S., Woodward, A. et al. (2020). Effects of heavy rainfall on waterborne disease hospitalizations among young children in wet and dry areas of New Zealand. *Environment International* 145: 106136. https://doi.org/10.1016/J.ENVINT.2020.106136.

Levy, K., Woster, A.P., Goldstein, R.S., and Carlton, E.J. (2016). Untangling the impacts of climate change on waterborne diseases: a systematic review of relationships between diarrheal diseases and temperature, rainfall, flooding, and drought. *Environmental Science & Technology* 50 (10): 4905. https://doi.org/10.1021/ACS.EST.5B06186.

Levy, K., Smith, S.M., and Carlton, E.J. (2018). Climate change impacts on waterborne diseases: moving toward designing interventions. *Current Environmental Health Reports* 5 (2): 272–282. https://doi.org/10.1007/S40572-018-0199-7.

Li, M., Zhu, K.X., Sun, Q.J. et al. (2016). Quality characteristics, structural changes, and storage stability of semi-dried noodles induced by moderate dehydration:

understanding the quality changes in semi-dried noodles. *Food Chemistry* 194: 797–804. https://doi.org/10.1016/J.FOODCHEM.2015.08.079.

Malhi, G.S., Kaur, M., and Kaushik, P. (2021). Impact of climate change on agriculture and its mitigation strategies: a review. *Sustainability* 13 (3): 1318. https://doi.org/10.3390/SU13031318.

Mashala, M.J., Dube, T., Mudereri, B.T. et al. (2023). A systematic review on advancements in remote sensing for assessing and monitoring land use and land cover changes impacts on surface water resources in semi-arid tropical environments. *Remote Sensing* 15 (16): 3926. https://doi.org/10.3390/RS15163926.

McIvor, I., Youjun, H., Daoping, L. et al. (2017). Agroforestry: conservation trees and erosion prevention. *Reference Module in Food Science* https://doi.org/10.1016/B978-0-08-100596-5.22382-2.

Misiou, O. and Koutsoumanis, K. (2022). Climate change and its implications for food safety and spoilage. *Trends in Food Science & Technology* 126: 142–152. https://doi.org/10.1016/J.TIFS.2021.03.031.

Mosadeghrad, A.M., Isfahani, P., Eslambolchi, L. et al. (2023). Strategies to strengthen a climate-resilient health system: a scoping review. *Globalization and Health* 19 (1): 1–11. https://doi.org/10.1186/S12992-023-00965-2.

Mosley, L.M. (2015). Drought impacts on the water quality of freshwater systems; review and integration. *Earth-Science Reviews* 140: 203–214. https://doi.org/10.1016/J.EARSCIREV.2014.11.010.

Murray, C., Larson, A., Goodwill, J. et al. (2022). Water quality observations from space: a review of critical issues and challenges. *Environments* 9 (10): 125. https://doi.org/10.3390/ENVIRONMENTS9100125.

Myers, S.S., Zanobetti, A., Kloog, I. et al. (2014). Increasing CO_2 threatens human nutrition. *Nature* 510 (7503): 139–142. https://doi.org/10.1038/nature13179.

National Research Council (US) Committee on Health Effects of Waste Incineration (2000). *Environmental Transport and Exposure Pathways of Substances Emitted from Incineration Facilities*. Washington, D.C.: National Academies Press.

Nottingham, A.T., Bååth, E., Reischke, S. et al. (2019). Adaptation of soil microbial growth to temperature: using a tropical elevation gradient to predict future changes. *Global Change Biology* 25 (3): 827–838. https://doi.org/10.1111/GCB.14502.

Nourmoradi, H., Moradnejadi, K., Moghadam, F.M. et al. (2015). The effect of dust storm on the microbial quality of ambient air in Sanandaj: a city located in the west of Iran. *Global Journal of Health Science* 7 (7): 114. https://doi.org/10.5539/GJHS.V7N7P114.

Noyes, P.D., McElwee, M.K., Miller, H.D. et al. (2009). The toxicology of climate change: environmental contaminants in a warming world. *Environment International* 35 (6): 971–986. https://doi.org/10.1016/J.ENVINT.2009.02.006.

Oh, J.W. (2022). Pollen allergy in a changing planetary environment. *Allergy, Asthma & Immunology Research* 14 (2): 168. https://doi.org/10.4168/AAIR.2022.14.2.168.

Orozco, R.L., Iturriaga, M.H., Tamplin, M.L. et al. (2008). Animal and environmental impact on the presence and distribution of *Salmonella* and *Escherichia coli* in hydroponic tomato greenhouses. *Journal of Food Protection* 71 (4): 676–683. https://doi.org/10.4315/0362-028X-71.4.676.

Pacheco, S.E., Guidos-Fogelbach, G., Annesi-Maesano, I. et al. (2021). Climate change and global issues in allergy and immunology. *Journal of Allergy and Clinical Immunology* 148 (6): 1366–1377. https://doi.org/10.1016/J.JACI.2021.10.011.

Panth, M., Hassler, S.C., and Baysal-Gurel, F. (2020). Methods for management of soilborne diseases in crop production. *Agriculture* 10 (1): 16. https://doi.org/10.3390/AGRICULTURE10010016.

Pastor, D., Boix, J., Fernández, V., and Albaigés, J. (1996). Bioaccumulation of organochlorinated contaminants in three estuarine fish species (*Mullus barbatus, Mugil cephalus* and *Dicentrarcus labrax*). *Marine Pollution Bulletin* 32 (3): 257–262. https://doi.org/10.1016/0025-326X(95)00166-K.

Pohoroo, A. and Ranghoo-Sanmukhiya, V.M. (2017). Food-borne bacterial load in fresh and frozen fish sold in Mauritius. *International Food Research Journal* 24 (5): 2193–2200.

Pope, A.M., Patterson, R., Burge, H., and Institute of Medicine (U.S.), Committee on the Health Effects of Indoor Allergens (1993). *Indoor Allergens: Assessing and Controlling Adverse Health Effects*, National Academies Press (US), Washington (DC). 308.

Prata, J.C., Rocha-Santos, T., and Ribeiro, A.I. (2022). An introduction to the concept of One Health. *One Health: Integrated Approach to 21st Century Challenges to Health* 1–31. https://doi.org/10.1016/B978-0-12-822794-7.00004-6.

Purse, B.V., Masante, D., Golding, N. et al. (2017). How will climate change pathways and mitigation options alter incidence of vector-borne diseases? A framework for leishmaniasis in South and Meso-America. *PLoS One* 12 (10): e0183583. https://doi.org/10.1371/JOURNAL.PONE.0183583.

Rahman, N.S.N.A., Hamid, N.W.A., and Nadarajah, K. (2021). Effects of abiotic stress on soil microbiome. *International Journal of Molecular Sciences* 22 (16): 22. https://doi.org/10.3390/IJMS22169036.

Raza, A., Razzaq, A., Mehmood, S.S. et al. (2019). Impact of climate change on crops adaptation and strategies to tackle its outcome: a review. *Plants* 8 (2): 34. https://doi.org/10.3390/PLANTS8020034.

Reinmuth-Selzle, K., Kampf, C.J., Lucas, K. et al. (2017). Air pollution and climate change effects on allergies in the anthropocene: abundance, interaction, and modification of allergens and adjuvants. *Environmental Science and Technology* 51 (8): 4119–4141. https://doi.org/10.1021/ACS.EST.6B04908/ASSET/IMAGES/LARGE/ES-2016-04908F_0006.JPEG.

Rendana, M. (2020). Impact of the wind conditions on COVID-19 pandemic: a new insight for direction of the spread of the virus. *Urban Climate* 34: 100680. https://doi.org/10.1016/J.UCLIM.2020.100680.

dos Ribeiro, C., S., van de Burgwal, L.H.M., and Regeer, B.J. (2019). Overcoming challenges for designing and implementing the One Health approach: a systematic review of the literature. *One Health* 7: 100085. https://doi.org/10.1016/J.ONEHLT.2019.100085.

Rinnan, R., Rousk, J., Yergeau, E. et al. (2009). Temperature adaptation of soil bacterial communities along an Antarctic climate gradient: predicting responses to climate warming. *Global Change Biology* 15 (11): 2615–2625. https://doi.org/10.1111/J.1365-2486.2009.01959.X.

Rizou, M., Galanakis, I.M., Aldawoud, T.M.S., and Galanakis, C.M. (2020). Safety of foods, food supply chain and environment within the COVID-19 pandemic. *Trends in Food Science & Technology* 102: 293–299. https://doi.org/10.1016/J.TIFS.2020.06.008.

Samut, H., Namli, Ş., Ozdemir, F.N. et al. (2023). Climate change and food safety: temperature impact on the attachment of *Escherichia coli* pathogroups on cress leaf. *Journal of Food Safety* e13059. https://doi.org/10.1111/JFS.13059.

Singh, A.B. and Kumar, P. (2022). Climate change and allergic diseases: an overview. *Frontiers in Allergy* 3: 964987. https://doi.org/10.3389/FALGY.2022.964987.

Singh, B.K., Delgado-Baquerizo, M., Egidi, E. et al. (2023). Climate change impacts on plant pathogens, food security and paths forward. *Nature Reviews Microbiology* 2023: 1–17. https://doi.org/10.1038/s41579-023-00900-7.

Sitch, S., Smith, B., Prentice, I.C. et al. (2003). Evaluation of ecosystem dynamics, plant geography and terrestrial carbon cycling in the LPJ dynamic global vegetation model. *Global Change Biology* 9 (2): 161–185. https://doi.org/10.1046/J.1365-2486.2003.00569.X.

Sun, X., Wang, J., Dong, M. et al. (2022). Food spoilage, bioactive food fresh-keeping films and functional edible coatings: research status, existing problems and development trend. *Trends in Food Science & Technology* 119: 122–132. https://doi.org/10.1016/J.TIFS.2021.12.004.

Tanno, L.K., Calderon, M.A., Smith, H.E. et al. (2016). Dissemination of definitions and concepts of allergic and hypersensitivity conditions. *World Allergy Organization Journal* 9 (1): 24. https://doi.org/10.1186/S40413-016-0115-2.

Tornevi, A., Bergstedt, O., and Forsberg, B. (2014). Precipitation effects on microbial pollution in a river: lag structures and seasonal effect modification. *PLoS One* 9 (5): https://doi.org/10.1371/JOURNAL.PONE.0098546.

Traidl-Hoffmann, C., Jakob, T., and Behrendt, H. (2009). Determinants of allergenicity. *Journal of Allergy and Clinical Immunology* 123 (3): 558–566. https://doi.org/10.1016/J.JACI.2008.12.003.

Tryland, I., Robertson, L., Blankenberg, A.G.B. et al. (2011). Impact of rainfall on microbial contamination of surface water. *International Journal of Climate Change Strategies and Management* 3 (4): 361–373. https://doi.org/10.1108/17568691111175650/FULL/XML.

Tsiraki, M.I., Karam, L., Abiad, M.G. et al. (2017). Use of natural antimicrobials to improve the quality characteristics of fresh "Phyllo" – a dough-based wheat product – shelf life assessment. *Food Microbiology* 62: 153–159. https://doi.org/10.1016/J.FM.2016.10.001.

Turkoglu Sasmazel, H., Alazzawi, M., and Kadim Abid Alsahib, N. (2021). Atmospheric pressure plasma surface treatment of polymers and influence on cell cultivation. *Molecules* 26 (6): 1665. https://doi.org/10.3390/molecules26061665.

Wright, K.M. and Holden, N.J. (2018). Quantification and colonisation dynamics of *Escherichia coli* O157:H7 inoculation of microgreens species and plant growth substrates. *International Journal of Food Microbiology* 273: 1–10. https://doi.org/10.1016/J.IJFOODMICRO.2018.02.025.

Xiao, Z., Nou, X., Luo, Y., and Wang, Q. (2014). Comparison of the growth of *Escherichia coli* O157: H7 and O104: H4 during sprouting and microgreen production from contaminated radish seeds. *Food Microbiology* 44: 60–63. https://doi.org/10.1016/J.FM.2014.05.015.

Xing, K., Zhang, S.M., Jia, M.Q., and Zhao, F. (2023). Response of wheat aphid to insecticides is influenced by the interaction between temperature amplitudes and insecticide characteristics. *Frontiers in Physiology* 14: https://doi.org/10.3389/FPHYS.2023.1188917.

Yang, H., Kong, J., Hu, H. et al. (2022). A review of remote sensing for water quality retrieval: progress and challenges. *Remote Sensing* 14 (8): 1770. https://doi.org/10.3390/RS14081770.

Zeng, D., Chen, G., Zhou, P. et al. (2021). Factors influencing groundwater contamination near municipal solid waste landfill sites in the Qinghai-Tibetan plateau. *Ecotoxicology and Environmental Safety* 211: 111913. https://doi.org/10.1016/J.ECOENV.2021.111913.

Zinsstag, J., Crump, L., Schelling, E. et al. (2018). Climate change and One Health. *FEMS Microbiology Letters* 365 (11): 85. https://doi.org/10.1093/FEMSLE/FNY085.

5

Role of Gases, VOC, PM2.5, and PM10 in Biological Contamination in Indoor Areas

Pradeep Kumar[1,2], *Sevaram Singh*[3,4], *Ekta Tyagi*[5], *Vinay Mohan Pathak*[6,7], *Shiwangi Gupta*[1], *and Rajeev Singh*[8]

[1] *Department of Environmental Studies, Satyawati College, University of Delhi, Delhi, New Delhi, India*
[2] *Department of Agricultural and Biosystems Engineering, South Dakota State University, Brookings, SD, USA*
[3] *Translational Health Science and Technology Institute, NCR Biotech Science Cluster, Faridabad, Haryana, India*
[4] *Jawaharlal Nehru University, Delhi, New Delhi, India*
[5] *Department of Bioscience and Biotechnology, Banasthali University, Banasthali, Rajasthan, India*
[6] *Department of Microbiology, University of Delhi (South Campus), Delhi, New Delhi, India*
[7] *Department of Botany and Microbiology, Gurukul Kangri Deemed to be University, Haridwar, Uttarakhand, India*
[8] *Department of Environmental Science, Jamia Millia Islamia, Delhi, New Delhi, India*

5.1 Introduction

IAQ is influenced by a variety of factors, including gases, VOCs, and PM (Lee et al. 2020). Gases, VOCs, PM2.5, and PM10 can all contribute to biological contamination in indoor areas, albeit in different ways. Gases such as carbon dioxide (CO_2) and carbon monoxide (CO) are not directly involved in biological contamination (Kumar et al. 2021a). Understanding the role of gases, VOCs, PM2.5 (PM with a diameter of 2.5 μm or smaller), and PM10 (PM with a diameter of 10 μm or smaller) in biological contamination is crucial for developing effective strategies to maintain a healthy indoor environment (WHO 2013). However, elevated levels of CO_2, typically due to inadequate ventilation, can indicate poor IAQ, which may indirectly affect biological contamination. Inadequate ventilation can lead to a build-up of other contaminants, including those that contribute to biological contamination (Kumar et al. 2023) (Figure 5.1). Furthermore, gases emitted by combustion sources such as gas stoves, fireplaces, and tobacco smoke can release pollutants like nitrogen dioxide (NO_2) and CO. These gases can have adverse effects on the respiratory system and, in some cases, provide favorable conditions for the growth of bacteria and other microorganisms.

Airborne Biocontaminants and Their Impact on Human Health, First Edition.
Edited by Rajeev Singh and Anamika Singh.

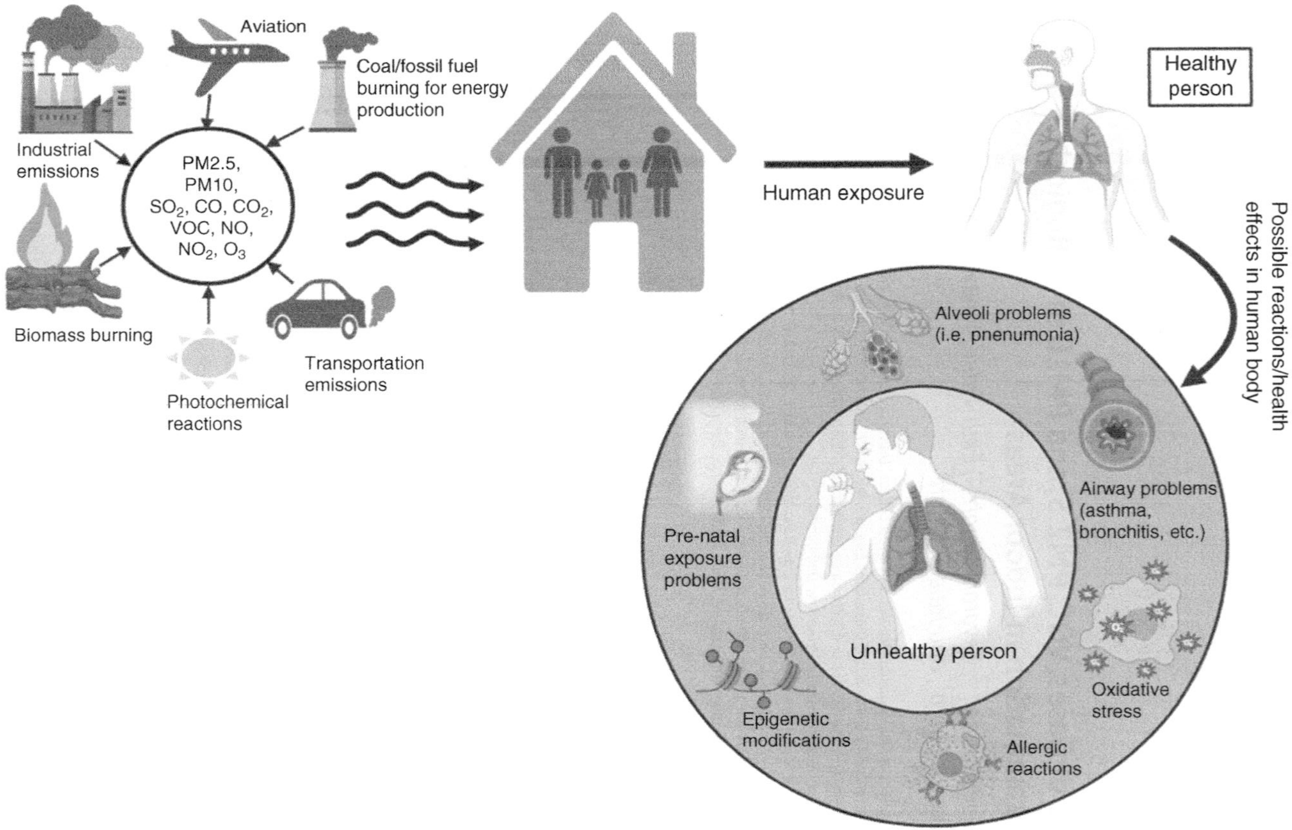

Figure 5.1 Emission of the major air pollutants (PM, VOC, and gases) from various sources and their associated health effects in human body through direct and indirect mechanisms. *Source:* Created with Biorender.com).

VOCs are emitted as gases from various sources, including building materials, furniture, cleaning products, and personal care items. Some VOCs, such as formaldehyde, benzene, and toluene, can have detrimental effects on IAQ and human health (Harčárová et al. 2020). While VOCs themselves may not directly cause biological contamination, they can interact with other factors and contribute to the growth and spread of microorganisms. Moreover, some VOCs emitted by fungi and molds can directly contribute to biological contamination (Kumar et al. 2021a).

PM refers to tiny solid or liquid particles suspended in the air. PM2.5 and PM10 are two categories based on their size. PM2.5 and PM10 can carry and serve as a medium for biological contaminants such as bacteria, viruses, fungi, and allergens (Kumar et al. 2021a). These particles can harbor microorganisms and facilitate their transport and dispersal within indoor environments. PM can also provide surfaces for microorganisms to attach and grow (Kumar et al. 2022). Dust particles, for example, can accumulate in ducts, filters, and ventilation systems, becoming reservoirs for bacteria and fungi (Batterman and Burge 1995). These microorganisms can then be released back into the indoor air, leading to biological contamination. Particles can also impact the respiratory system of individuals, making them more susceptible to respiratory infections caused by biological contaminants (Kumar et al. 2021b, 2023). PM2.5 can penetrate deep into the lungs, potentially exacerbating respiratory conditions and compromising the immune system's ability to defend against microbial threats (Yang et al. 2020).

This chapter aims to shed light on the individual and synergistic impacts of gases, VOCs, PM2.5, and PM10 on IAQ and the potential implications for human health. Furthermore, it also explores the toxicity mechanisms involved in the health effects on humans (Table 5.1).

5.2 Gases and Biological Contamination

Gases present in indoor environments can directly or indirectly contribute to biological contamination. CO_2 levels are often used as a proxy for IAQ and ventilation (WHO 2010). High levels of CO_2 can indicate poor ventilation, which can lead to increased moisture levels, promoting the growth of mold and fungi. Gases play a significant role in indoor air pollution and can have adverse effects on human health and the environment (Tran et al. 2020). Here are some key gases commonly associated with indoor air pollution and their impacts are discussed in the subsequent sections.

5.2.1 Carbon Dioxide (CO_2)

CO_2 is one of the key gases that contribute to indoor air pollution. CO_2 is a natural component of air and is produced by humans during respiration. In indoor environments, high levels of CO_2 can indicate poor ventilation, which can lead to

Table 5.1 Overview of the toxicity mechanisms of the various gases in indoor air and their human health risks.

Pollutants	Toxicity mechanisms	Possible health effects
CO	Carboxy-hemoglobin formation and reduced oxygen (O_2) supply	Vomiting, nausea, fatigue, fainting, headaches, and unconsciousness are all possible effects
SO_2	Enhanced bronchial functioning	Skin-related problems and lung disorders
NO_2	Significant increase in broncho movement, as well as a higher possibility of infection in the lungs	Wheezing, coughing, fever, pain in the chest, breathing difficulties headache, edema of the lungs, irritation of the throat, breathlessness, and diaphoresis
CO_2	Acetylcholinesterase inhibition prevents acetylcholine hydrolysis	Headaches and tiredness; increased focus causes nausea, vomiting, and drowsiness
VOCs	May induce cancer, mutation, neurotoxicity and genotoxicity	Eye, mouth, skin, and nasal irritation Headaches, respiratory difficulties, weariness, and sick building syndrome (SBS) kidney, central nervous system, and liver damage
Radon	Production of reactive oxygen species (ROS), inflammatory effects	Respiratory disorder, pulmonary cancer
Ozone	Effects on the pulmonary system at the biochemical and morphological levels	Bronchitis, pneumonia, asthma exacerbation, and permanent pulmonary destruction
PM	Systemic inflammation and oxidative stress	Cardiac or pulmonary disease-related premature mortality, heartbeat fluctuations
PAHs	May act as carcinogen	Cancer of the lungs, laryngeal cancer, nasopharyngeal cancer

Source: Kumar et al. (2023)/Elsevier.

discomfort, drowsiness, and decreased cognitive function (WHO 2010). While it is a natural component of the air we breathe, elevated levels of CO_2 indoors can have several negative effects on human health and comfort. Prolonged exposure to elevated CO_2 levels may cause headaches, fatigue, and difficulty concentrating. The primary source of indoor CO_2 is the exhalation of breath by humans and animals. Other sources may include combustion processes and infiltration from outdoor air (WHO 2010).

CO_2 is often used as an indicator of IAQ and ventilation effectiveness. As people breathe in O_2 and exhale CO_2, the concentration of CO_2 in a closed or poorly ventilated space will increase over time (WHO 2010). High levels of CO_2 indicate inadequate ventilation, which can lead to a buildup of other pollutants and affect the comfort and well-being of occupants. Elevated levels of CO_2 can cause discomfort and reduce cognitive function (WHO 2010). When CO_2 levels rise, people may experience feelings of stuffiness, drowsiness, and difficulty concentrating. In poorly ventilated spaces, such as crowded meeting rooms or poorly designed workspaces, increased CO_2 concentrations may lead to reduced productivity and performance. CO_2 can also be produced indoors through combustion processes, such as using gas stoves, fireplaces, or other fuel-burning appliances. Incomplete combustion can release not only CO_2 but also other harmful pollutants like CO, NO_2, and PM (WHO 2010).

While moderate levels of CO_2 are not directly toxic to humans, prolonged exposure to elevated CO_2 concentrations can have health effects. Studies have shown that chronic exposure to high CO_2 levels may cause headaches, fatigue, dizziness, and increased heart rate (Liu et al. 2017; Azuma et al. 2018; Kumar et al. 2023). Additionally, it may exacerbate respiratory conditions in individuals with pre-existing lung problems.

To maintain a healthy indoor environment and reduce the impact of CO_2 on indoor air pollution, proper ventilation is essential. Adequate ventilation helps to dilute and remove CO_2 and other pollutants from indoor spaces, ensuring a continuous supply of fresh outdoor air (WHO 2010). Regular maintenance of ventilation systems and avoidance of overcrowding in enclosed spaces are effective strategies to manage CO_2 levels indoors. Overall, while CO_2 itself is not a harmful pollutant at typical indoor concentrations, it serves as an important indicator of IAQ and ventilation effectiveness. Maintaining appropriate CO_2 levels can contribute to improved occupant comfort, productivity, and overall health within indoor environments.

5.2.2 Carbon Monoxide

CO is a colorless, odorless gas produced by the incomplete combustion of fossil fuels (e.g. gas stoves, heaters, and car engines) (WHO 2010). It is highly toxic and can cause headaches, dizziness, nausea, confusion, and even death in severe cases, especially in poorly ventilated spaces (Prockop and Chichkova 2007; Tran et al. 2020).

It is produced as a byproduct of incomplete combustion of carbon-containing fuels, such as gas, oil, coal, and wood. The primary sources of indoor CO include faulty or poorly maintained combustion appliances, such as gas stoves, furnaces, water heaters, fireplaces, and gas-powered generators (Prockop and Chichkova 2007). This characteristic is particularly dangerous because people may be exposed to high levels of CO without realizing it until symptoms appear. Indoor sources of CO include gas appliances, improperly ventilated stoves and ovens, blocked chimneys, or flues, and using fuel-burning equipment in enclosed or poorly ventilated spaces (WHO 2010). Homes with attached garages are also at risk of CO infiltration if car engines are left running inside. Inadequate ventilation is a key factor contributing to elevated levels of CO indoors. Poorly ventilated spaces allow CO to accumulate, increasing the risk of exposure. During colder months, when heating systems are frequently used, the risk of CO poisoning increases if these systems are not functioning properly or have blocked vents. CO poisoning is preventable through regular maintenance of combustion appliances, ensuring proper ventilation, and installing CO detectors in homes and other enclosed spaces (WHO 2010).

CO is highly toxic to humans and animals. When inhaled, it binds to hemoglobin in the bloodstream, reducing its ability to carry O_2. As a result, vital organs, and tissues, particularly the brain and heart, do not receive enough O_2. Even at low concentrations, CO exposure can cause headaches, dizziness, nausea, weakness, and confusion (Ali et al. 2021; Kumar et al. 2023). Prolonged exposure or exposure to higher concentrations can lead to loss of consciousness, permanent brain damage, and, in severe cases, death (Ali et al. 2021; Kumar et al. 2023).

Due to the severe health risks associated with CO exposure, it is essential to take precautions to prevent its buildup indoors (WHO 2010). Regular inspection and maintenance of gas appliances and ventilation systems, as well as the installation of CO detectors, can significantly reduce the risk of CO poisoning and improve IAQ. If CO poisoning is suspected, immediate evacuation from the affected area and seeking medical attention are crucial (Varon et al. 1999).

5.2.3 Nitrogen Dioxide (NO_2)

NO_2 is a reddish-brown gas that results from the burning of fossil fuels, particularly in vehicles and gas appliances. NO_2 is one of the significant gases contributing to indoor air pollution. It is primarily formed by the combustion of fossil fuels, particularly in vehicles and gas-burning appliances. In indoor environments, NO_2 can originate from sources such as gas stoves, space heaters, and tobacco smoke. NO_2 can react with other pollutants in the air to form fine PM (PM2.5) and other secondary pollutants. Indoor activities such as cooking with gas stoves can be a significant source of indoor NO_2. Poorly ventilated spaces or improperly adjusted gas-burning appliances can lead to elevated NO_2 levels indoors (WHO 2010).

NO_2 is a respiratory irritant and can lead to irritation of the mucous membranes in the eyes, nose, and throat. Short-term exposure to elevated levels of NO_2 can cause symptoms such as coughing, wheezing, and shortness of breath, especially in individuals with pre-existing respiratory conditions like asthma. NO_2 exposure can worsen asthma symptoms and may increase the frequency and severity of asthma attacks (Manisalidis et al. 2020). People with asthma, especially children and the elderly, are particularly susceptible to the effects of NO_2. Long-term exposure to elevated NO_2 levels may reduce the respiratory system's ability to defend against respiratory infections, making individuals more susceptible to illnesses like bronchitis and pneumonia (Manisalidis et al. 2020).

Ensure adequate ventilation in indoor spaces, especially in areas where gas-burning appliances are used. Ventilation helps dilute and remove NO_2 and other indoor pollutants. Regularly inspect and maintain gas-burning appliances to ensure they are functioning properly and not emitting excessive amounts of NO_2 (WHO 2010). Use exhaust fans in the kitchen and bathrooms to expel indoor air pollutants, including NO_2, to the outside. Minimize the use of gas-burning appliances indoors when possible, or ensure proper ventilation when using them (WHO 2010). Consider using air purifiers with appropriate filters that can effectively remove NO_2 and other pollutants from the indoor air. By taking these measures, it is possible to reduce NO_2 levels and improve IAQ, leading to better respiratory health and overall well-being for occupants (WHO 2010).

5.2.4 Sulfur Dioxide (SO_2)

SO_2 is a gas produced by the burning of fossil fuels containing sulfur impurities, such as coal and oil. Exposure to SO_2 can irritate the respiratory system, leading to respiratory issues and exacerbating asthma and other respiratory conditions (Manisalidis et al. 2020; Tran et al. 2020; Kumar et al. 2023). SO_2 is not typically a significant concern in indoor air pollution because it is primarily an outdoor air pollutant. SO_2 is a gas that is produced by the combustion of fossil fuels containing sulfur compounds, such as coal and oil, and by certain industrial processes. The primary indoor sources of SO_2 include gas stoves or heaters that burn fuels containing sulfur, tobacco smoke (if the tobacco contains sulfur compounds), and indoor combustion processes where sulfur-containing fuels are used. However, in most indoor environments, the levels of SO_2 are relatively low compared to outdoor environments where industrial activities and traffic emissions are more prevalent. The reason for this is that indoor spaces are generally better ventilated, which helps in diluting and removing pollutants, including SO_2.

When indoor levels of SO_2 become elevated, it can lead to health issues, especially for individuals with respiratory conditions such as asthma or chronic obstructive pulmonary disease (COPD). SO_2 irritates the respiratory system and can cause symptoms such as coughing, wheezing, shortness of breath, and throat irritation (Manisalidis et al. 2020; Tran et al. 2020; Kumar et al. 2023). To minimize the impact of SO_2 on IAQ, it is essential to ensure proper ventilation in indoor spaces, especially when using gas appliances or engaging in activities that may produce SO_2. Regular maintenance and proper functioning of gas appliances can also help reduce emissions of SO_2 indoors. While SO_2 is not typically a major indoor air pollutant, other gases such as CO_2, CO, NO_2, and VOCs can be more significant concerns for IAQ. It is essential to take a holistic approach to IAQ management, addressing all potential sources of pollutants to create a healthier indoor environment.

5.2.5 Volatile Organic Compounds (VOCs)

VOCs are a group of organic chemicals that can off-gas from a wide range of household products, including paints, varnishes, cleaning agents, and furniture. They can cause eye, nose, and throat irritation, as well as contribute to long-term health effects. Some VOCs, like formaldehyde, are known as carcinogens.

VOCs play a significant role in indoor air pollution and can have adverse effects on both human health and the environment. VOCs are a group of organic chemicals that easily evaporate into the air at room temperature. They are released from various sources commonly found in indoor environments, contributing to indoor air pollution (Rumchev et al. 2007).

VOCs are present in a wide range of household products, such as paints, varnishes, adhesives, solvents, cleaning agents, air fresheners, and pesticides. When these products are used indoors, they release VOCs into the air, contributing to indoor pollution. Some building materials and furnishings, such as carpets, flooring, plywood, and particleboard, can emit VOCs over time. This is known as off-gassing, and it occurs when VOCs trapped within the materials are released into the indoor air. Personal care products, including perfumes, deodorants, hair sprays, and cosmetics, can also release VOCs into indoor air when applied or used indoors. Tobacco smoke is a significant source of VOCs indoors. Smoking indoors can lead to the release of a wide range of harmful chemicals, including VOCs, which can linger in the air and pose health risks to both smokers and non-smokers.

Short-term exposure to high VOC levels can cause eye, nose, and throat irritation, headaches, dizziness, and nausea (Rumchev et al. 2007). Long-term exposure to certain VOCs, such as formaldehyde and benzene, has been linked to more severe health issues, including respiratory problems, central nervous system disorders, and even cancer (Rumchev et al. 2007). Elevated levels of VOCs can significantly impact IAQ, leading to discomfort and reduced well-being for occupants. Poor IAQ can affect productivity, concentration, and overall health, particularly in enclosed spaces with limited ventilation (Rumchev et al. 2007). Some VOCs can contribute to the formation of ground-level ozone (O_3) and smog when they react with nitrogen oxides in the presence of sunlight. This can have detrimental effects on the environment and contribute to air pollution on a broader scale.

Adequate ventilation helps dilute and remove VOCs from indoor air. Ensure good airflow and use exhaust fans when using VOC-emitting products. Choose low-VOC or VOC-free alternatives for household products and building materials

(WHO 2010). Avoid smoking indoors to reduce exposure to harmful VOCs from tobacco smoke. Consider using air purifiers equipped with appropriate filters to help remove VOCs from the indoor air. Store household products containing VOCs in sealed containers and in well-ventilated areas to minimize emissions. By being mindful of the sources of VOCs and taking steps to reduce their presence indoors, it is possible to improve IAQ and create a healthier living environment (WHO 2010).

5.2.5.1 Common VOC in Indoor Air

VOCs are a group of organic chemicals that can easily evaporate into the air at room temperature. They are commonly found in indoor environments and can originate from various sources, including building materials, furniture, household products, and human activities (Kumar et al. 2023). Some common VOCs found in indoor air environments include the following:

Formaldehyde: Found in building materials like plywood, particleboard, and glues used in furniture and cabinets. It is also present in some household products and tobacco smoke (Bernstein et al. 2008).
Benzene: Found in tobacco smoke, gasoline, and some household products like glues, paints, and detergents (Bari et al. 2015).
Toluene: Often found in paints, paint thinners, adhesives, and some cleaning agents.
Xylene: Found in paint, varnishes, and certain household products.
Ethanol: Commonly found in alcoholic beverages and some cleaning products.
Acetone: Often found in nail polish remover, paint thinners, and some cleaning agents.
Acetaldehyde: Present in tobacco smoke, some household products, and automobile exhaust.
Trichloroethylene: Found in some cleaning products and solvents used in metal degreasing.
Perchloroethylene (PCE): Commonly used in dry cleaning processes.
Styrene: Found in some building materials, such as insulation, and some plastic products.
1,4-Dichlorobenzene: Found in some mothballs and room deodorizers.

It is important to note that VOC concentrations in indoor air can vary depending on the specific sources and the ventilation of the indoor space. Prolonged exposure to high levels of VOCs can have adverse health effects, including eye, nose, and throat irritation, headaches, dizziness, and in some cases, long-term health concerns (WHO 2010; Kumar et al. 2023). To reduce VOC levels indoors, it is recommended to use low-VOC or VOC-free products, ensure adequate ventilation, and keep indoor spaces clean and well-maintained. Air purifiers with activated carbon filters can also help remove some VOCs from the air (WHO 2010).

5.2.6 Radon (Rn)

Rn is a radioactive gas that can enter buildings through cracks in the foundation and accumulate to harmful levels. Prolonged exposure to high Rn levels increases the risk of lung cancer (Appleton 2012).

Rn is a naturally occurring radioactive gas that can be a significant contributor to indoor air pollution. It is formed from the decay of uranium, which is found in various concentrations in soil, rocks, and water (Kumar et al. 2023). Rn can enter buildings, including homes, offices, and other structures, through cracks in the foundation, gaps in walls, and other openings. Once inside, it can accumulate to potentially harmful levels.

The role of Rn in indoor air pollution is primarily related to its radioactivity and the health risks associated with prolonged exposure (Al Jassim and Isaifan 2018). Rn emits ionizing radiation as it decays into other radioactive particles known as Rn decay products (Degu Belete and Alemu Anteneh 2021). These decay products can attach to dust and other airborne particles in indoor air. When inhaled, these radioactive particles can damage the cells lining the respiratory tract and increase the risk of lung cancer. Rn is the second leading cause of lung cancer after smoking (Riudavets et al. 2022). Prolonged exposure to high Rn levels is associated with an increased risk of lung cancer, especially in smokers or individuals with a history of smoking (Darby et al. 2005). Even at lower levels, Rn exposure can contribute to lung cancer cases.

Rn can accumulate in higher concentrations indoors, particularly in poorly ventilated spaces or areas with a higher concentration of Rn in the surrounding soil (Yang et al. 2019). This is a concern because people spend a significant amount of time indoors, and exposure to high Rn levels over an extended period can have adverse health effects (Jones 1999).

Rn is odorless and colorless, making it difficult to detect without specialized testing (Fisher et al. 1991). This characteristic makes it essential for homeowners to test for Rn levels in their living spaces to determine if action is needed to mitigate elevated concentrations. To address the risks associated with Rn in indoor air, it is crucial to conduct Rn testing in homes and other buildings. If elevated Rn levels are detected, appropriate mitigation measures should be taken. Rn mitigation systems typically involve ventilating the area under the building's foundation to prevent Rn from entering and dispersing it safely into the outdoor air. Overall, understanding and addressing the role of Rn in indoor air pollution is vital for protecting the health of occupants and reducing the risk of lung cancer associated with Rn exposure (WHO 2010).

5.2.7 Ozone (O_3)

While O_3 in the upper atmosphere is beneficial, ground-level O_3 can be harmful. It forms indoors as a result of certain chemical reactions between VOCs and other pollutants. O_3 can irritate the respiratory system and worsen respiratory conditions, such as asthma (Boeniger 1995).

In indoor environments, O_3 plays a complex and controversial role in air pollution. O_3 itself is a naturally occurring gas and is beneficial in the upper atmosphere (stratosphere) as it forms the O_3 layer, which protects the earth from harmful ultraviolet (UV) radiation. However, at ground level, O_3 can be harmful to human health and the environment.

Indoor ozone pollution typically occurs through two main mechanisms:

Ozone Generation by Some Air Purifiers: Some air purifiers and ionizers on the market intentionally generate O_3 as a means of air cleaning. The idea is that O_3 will react with and neutralize pollutants in the air. However, these O_3-generating air purifiers can significantly increase indoor O_3 levels, leading to potential health risks (Chen et al. 2009).

Indoor Ozone Formation: O_3 can also form indoors as a byproduct of certain chemical reactions involving VOCs and other pollutants (Chen et al. 2009). For example, some household products like air fresheners, cleaning agents, and paints can release VOCs, which can react with O_3 precursors in the air and produce O_3 indoors.

The role of O_3 in indoor air pollution is concerning due to its potential health impacts:

Respiratory Irritation: O_3 is a powerful respiratory irritant and can cause throat irritation, coughing, and breathing difficulties even at relatively low concentrations (Manisalidis et al. 2020; Tran et al. 2020; Kumar et al. 2023).

Aggravation of Respiratory Conditions: O_3 exposure can exacerbate pre-existing respiratory conditions, such as asthma and COPD. It may lead to increased frequency and severity of asthma attacks (Manisalidis et al. 2020; Tran et al. 2020; Kumar et al. 2023).

Lung Damage: Prolonged exposure to elevated O_3 levels may cause inflammation and damage to the lungs, potentially increasing the risk of respiratory illnesses (Manisalidis et al. 2020; Tran et al. 2020; Kumar et al. 2023).

Harmful to Vulnerable Populations: Children, the elderly, and individuals with respiratory problems are particularly vulnerable to the adverse effects of O_3 exposure (Manisalidis et al. 2020; Tran et al. 2020; Kumar et al. 2023).

Because of these potential health risks, it is generally recommended to avoid using O_3-generating air purifiers in indoor spaces and to choose air purifiers with safe and effective filtration technologies instead (WHO 2010). Moreover, reducing the use of products that release VOCs and ensuring good ventilation in indoor spaces can help prevent the formation of O_3 and maintain a healthier indoor environment (WHO 2010).

To mitigate indoor air pollution caused by gases, proper ventilation is crucial. Adequate ventilation can help dilute and remove pollutants from indoor air. It is also essential to minimize the use of products that emit VOCs, ensure the proper functioning and maintenance of gas appliances, and consider using air purifiers with appropriate filtration systems to reduce harmful gas levels indoors (WHO 2010).

5.3 Particulate Matter

PM plays a significant role in IAQ. PM refers to tiny solid or liquid particles suspended in the air, which can be generated from various sources both indoors and outdoors.

5.3.1 PM2.5

PM2.5, or particulate matter, plays a significant role in indoor air pollution. These particles are tiny enough to be inhaled deep into the respiratory system and can have adverse effects on human health (Yang et al. 2020).

PM2.5 can originate from various indoor sources, such as cooking (especially when using solid fuels), tobacco smoke, candles, incense, and certain household products like aerosols and cleaning agents. It can also enter indoor spaces from outdoor sources, like vehicle emissions, industrial activities, and natural sources such as wildfires. Exposure to PM2.5 can lead to various health problems, particularly affecting the respiratory and cardiovascular systems (Tran et al. 2020). These tiny particles can cause or exacerbate respiratory issues like asthma, bronchitis, and COPD (Kumar et al. 2023). Additionally, PM2.5 has been linked to cardiovascular diseases, heart attacks, and even premature death in some cases (Ali et al. 2021). The concentration of PM2.5 indoors depends on several factors, including the presence and use of indoor sources, ventilation rates, outdoor air quality, and the building's design. In poorly ventilated or tightly sealed spaces, PM2.5 levels can accumulate, leading to higher exposure levels for occupants. Certain individuals are more susceptible to the health effects of PM2.5, including young children, the elderly, and people with pre-existing respiratory or cardiovascular conditions (Kumar et al. 2023). Prolonged exposure to elevated PM2.5 levels can be particularly harmful to these vulnerable populations. To reduce indoor PM2.5 levels, several strategies can be employed, such as improving ventilation to bring in fresh air from outdoors, using air purifiers with HEPA filters, reducing the use of indoor sources that generate PM2.5, and avoiding smoking indoors (WHO 2010). Regular monitoring of IAQ, including PM2.5 levels, can help identify potential issues and guide effective strategies for improvement. Many IAQ monitors are available in the market that can measure PM2.5 levels and provide real-time data. In conclusion, PM2.5 is a significant contributor to indoor air pollution, and its presence can have serious implications for human health. Taking proactive measures to reduce PM2.5 levels indoors is essential to creating a healthy living environment for occupants (WHO 2010).

5.3.2 PM10

PM10 particles can be solid or liquid and are commonly found in indoor air pollution. The role of PM10 in indoor air pollution is significant and can have various health implications for occupants.

PM10 in indoor air can come from both outdoor and indoor sources. Outdoor sources include dust, pollen, vehicle emissions, and industrial pollutants that can enter buildings through ventilation systems or open windows. Indoor sources include tobacco smoke, cooking fumes, heating appliances, and household products like aerosol sprays and cleaning agents. PM10 particles can penetrate deep into the respiratory system and pose health risks when inhaled. Short-term exposure to elevated PM10 levels can irritate the eyes, nose, and throat, and exacerbate existing respiratory conditions such as asthma and bronchitis (Ali et al. 2021). Long-term exposure to high PM10 concentrations is associated with chronic respiratory diseases, cardiovascular issues, and even premature death. PM10 particles are relatively larger compared to finer particles like PM2.5 (Leung 2015). While PM2.5 can penetrate even deeper into the lungs, PM10 can still reach the respiratory system and contribute to health problems. Indoor PM10 concentrations can vary depending on factors such as indoor sources, ventilation rates, and outdoor air quality. Activities like cooking and smoking can significantly increase PM10 levels indoors. Poor ventilation can lead to the accumulation of indoor pollutants, including PM10, exacerbating IAQ issues (Leung 2015).

Proper ventilation helps dilute and remove indoor pollutants, including PM10. Using exhaust fans in kitchens and bathrooms and maintaining air circulation can be beneficial. Minimizing the use of products that emit PM10, such as aerosols, and opting for cleaner alternatives can help improve IAQ (Tran et al. 2020). Using air purifiers with high-efficiency particulate air (HEPA) filters can effectively remove PM10 particles from indoor air. Dusting and cleaning surfaces can reduce the accumulation of PM10 and other particles in indoor environments. Controlling indoor PM10 levels is essential for maintaining a healthy indoor environment and reducing the risk of respiratory and cardiovascular health problems associated with indoor air pollution (WHO 2010).

5.4 Human Health Effects and Toxicity Mechanisms

Indoor air pollution can arise from various sources, including gases, VOCs, and particulate matter (PM2.5 and PM10). These pollutants can have harmful effects on human health, and their toxicity mechanisms can differ based on their chemical properties and exposure levels (Tran et al. 2020; Kumar et al. 2023). Here is an overview of the toxicity mechanisms of these indoor pollutants:

Gases: Common indoor gases include CO, NO_2, formaldehyde (HCHO), and radon (Rn). Their toxicity mechanisms are as follows:

Carbon Monoxide (CO): CO is a colorless, odorless gas that binds to hemoglobin in red blood cells, forming carboxy-hemoglobin, which reduces the O_2-carrying capacity of blood. This leads to tissue hypoxia, particularly affecting vital organs like the brain and heart. Prolonged exposure to high levels of CO can result in headaches, dizziness, confusion, cardiovascular problems, and even death (Tran et al. 2020; Kumar et al. 2023).

Nitrogen Dioxide (NO_2): NO_2 is a reddish-brown gas, often associated with combustion processes. It irritates the respiratory tract, leading to symptoms like coughing, wheezing, and shortness of breath. Prolonged exposure can worsen respiratory conditions like asthma and increase the risk of respiratory infections (Tran et al. 2020; Kumar et al. 2023).

Formaldehyde (HCHO): Formaldehyde is a VOC emitted from various household products and building materials. It is a potent respiratory irritant and can cause eye, nose, and throat irritation. Prolonged exposure to high levels may lead to more severe respiratory issues and potential carcinogenic effects (Manisalidis et al. 2020; Tran et al. 2020; Kumar et al. 2023).

Radon (Rn): Rn is a radioactive gas that can seep into buildings from the ground. It can accumulate in indoor spaces and, when inhaled, can damage lung tissue, increasing the risk of lung cancer (Tran et al. 2020; Kumar et al. 2023).

VOCs: VOCs are emitted by various products like paints, cleaning agents, air fresheners, and furniture. Their toxicity mechanisms can vary depending on the specific VOC. Common VOCs include benzene, toluene, xylene, and formaldehyde. Short-term exposure to high VOC levels can lead to eye, nose, and throat irritation, headaches, dizziness, and nausea. Long-term exposure to certain VOCs, such as benzene, is associated with an increased risk of cancer (Tran et al. 2020; Kumar et al. 2023).

Particulate Matter (PM2.5 and PM10): PM2.5 and PM10 are fine and coarse particles, respectively, suspended in the air. They can come from various sources such as cooking, smoking, burning of candles, and outdoor pollution entering the indoor environment (Tran et al. 2020; Kumar et al. 2023).

PM2.5: These fine particles can penetrate deep into the respiratory system and even enter the bloodstream. They can trigger or worsen respiratory conditions, such as asthma and chronic bronchitis, and increase the risk of cardiovascular diseases (Tran et al. 2020; Kumar et al. 2023).

PM10: While larger than PM2.5, PM10 can still reach the respiratory system's lower airways, leading to similar health effects as PM2.5, though to a slightly lesser extent (Manisalidis et al. 2020).

Overall, the toxicity mechanisms of these indoor pollutants involve irritation, inflammation, oxidative stress, and in some cases, the potential to cause genetic mutations leading to cancer. To mitigate indoor air pollution, it is essential to improve ventilation, use air purifiers, avoid smoking indoors, and choose low-emission products (Manisalidis et al. 2020).

5.5 Mitigation Strategies Reducing Indoor Gases and Particulate Matter

Mitigation strategies for indoor air polluting gases and PM are essential for maintaining a healthy indoor environment. Indoor air pollution can have serious health effects, particularly for vulnerable groups such as children, elderly, and individuals with respiratory conditions. Here are some effective strategies to reduce indoor air pollutants:

Source Control: The most effective approach is to eliminate or minimize the sources of indoor air pollution. Identify and address the sources of pollutants, such as tobacco smoke, cooking emissions, VOCs from cleaning products, and building materials (WHO 2010).

Ventilation: Proper ventilation is crucial to dilute and remove indoor air pollutants. Ensure that your home is adequately ventilated, especially in areas prone to pollution, such as the kitchen and bathrooms. Use exhaust fans and open windows regularly to allow fresh air to circulate and pollutants to be expelled (WHO 2010).

Air Purifiers: Consider using air purifiers with HEPA (High-Efficiency Particulate Air) filters to remove PM, dust, pollen, and other pollutants from the indoor air. Air purifiers can be particularly useful in areas with high outdoor pollution levels or for individuals with respiratory issues (WHO 2010).

Indoor plants: Certain houseplants can help improve IAQ by absorbing pollutants and releasing O_2. Some examples include spider plants, peace lilies, aloe-vera, and snake plants. However, keep in mind that plants alone may not be sufficient to address significant air pollution (WHO 2010).

Avoid Smoking Indoors: Strictly prohibit smoking inside your home. Second-hand smoke contains harmful pollutants and carcinogens, which can seriously compromise IAQ (WHO 2010).

Regular Cleaning: Dust, vacuum, and mop your home regularly to reduce the accumulation of PM and allergens. Use vacuum cleaners with HEPA filters to trap fine particles effectively (WHO 2010).

Control Humidity: Keep indoor humidity levels between 30% and 50% to prevent mold growth and reduce the release of VOCs from building materials and household products (WHO 2010).

Use Natural Cleaning Products: Opt for natural or low-VOC cleaning products to minimize the release of harmful chemicals into the air (WHO 2010).

Properly Maintain Appliances: Ensure that gas stoves, heaters, and other combustion appliances are properly maintained and vented to prevent the release of CO and other harmful gases (WHO 2010).

Radon Mitigation: Test your home for Rn, a naturally occurring radioactive gas that can seep into buildings from the ground. If high levels are detected, consult with professionals to implement Rn mitigation measures (WHO 2010).

Avoid Using Air Fresheners: Air fresheners and similar products often contain VOCs and other chemicals. Instead, try natural alternatives like baking soda, essential oils, or simply keeping your living space clean (WHO 2010).

Educate and Raise Awareness: Inform household members about IAQ and the importance of maintaining a healthy living environment. Encourage them to follow these mitigation strategies to ensure everyone's well-being. Remember that the effectiveness of these strategies may vary depending on the specific pollutants and sources present in your indoor environment. When dealing with complex air quality issues, it is advisable to seek professional advice and conduct IAQ assessments (WHO 2010).

5.6 Conclusion

To conclude, understanding the role of gases, VOCs, PM2.5, and PM10 in biological contamination is crucial for improving IAQ. It is important to note that while gases, VOCs, PM2.5, and PM10 can contribute to biological contamination indirectly, the primary sources of biological contaminants in indoor environments are typically living organisms like humans, pets, pests, and microorganisms themselves. Their potential to serve as carriers and facilitators of biological contaminants poses significant health risks to occupants. Addressing this issue requires comprehensive approaches that focus on reducing pollutant emissions, improving ventilation, and implementing effective air purification techniques. Continued research and awareness are necessary to develop sustainable practices that promote healthier indoor environments. Effective ventilation, proper maintenance, and regular cleaning practices are essential for reducing biological contamination in indoor areas. Implementing appropriate mitigation strategies can significantly reduce the risk of exposure to harmful microorganisms and pollutants, leading to healthier indoor environments. Continued research and public awareness are vital for ensuring the well-being of occupants in indoor spaces. Policymakers, building managers, and individuals alike must work together to create healthier indoor environments for current and future generations. By addressing these factors, we can strive to create safer and healthier indoor environments for everyone.

References

Al Jassim, M. and Isaifan, R. (2018). A review on the sources and impacts of radon indoor air pollution. *Journal of Environmental and Toxicology Studies 2* (1): 1–8. https://doi.org/10.16966/2576-6430.112.

Ali, M.U., Yu, Y., Yousaf, B. et al. (2021). Health impacts of indoor air pollution from household solid fuel on children and women. *Journal of Hazardous Materials* 416: 126127.

Appleton, J.D. (2012). Radon in air and water. In: *Essentials of Medical Geology: Revised Edition*, 239–277. Dordrecht: Springer Netherlands.

Azuma, K., Kagi, N., Yanagi, U., and Osawa, H. (2018). Effects of low-level inhalation exposure to carbon dioxide in indoor environments: a short review on human health and psychomotor performance. *Environment International* 121: 51–56.

Bari, M.A., Kindzierski, W.B., Wheeler, A.J. et al. (2015). Source apportionment of indoor and outdoor volatile organic compounds at homes in Edmonton, Canada. *Building and Environment* 90: 114–124.

Batterman, S.A. and Burge, H. (1995). HVAC systems as emission sources affecting indoor air quality: a critical review. *HVAC&R Research* 1 (1): 61–78.

Bernstein, J.A., Alexis, N., Bacchus, H. et al. (2008). The health effects of nonindustrial indoor air pollution. *Journal of Allergy and Clinical Immunology* 121 (3): 585–591.

Boeniger, M.F. (1995). Use of ozone generating devices to improve indoor air quality. *American Industrial Hygiene Association Journal* 56 (6): 590–598.

Chen, H.L., Lee, H.M., Chen, S.H. et al. (2009). Removal of volatile organic compounds by single-stage and two-stage plasma catalysis systems: a review of the performance enhancement mechanisms, current status, and suitable applications. *Environmental Science & Technology* 43 (7): 2216–2227.

Darby, S., Hill, D., Auvinen, A. et al. (2005). Radon in homes and risk of lung cancer: collaborative analysis of individual data from 13 European case-control studies. *BMJ* 330 (7485): 223.

Degu Belete, G. and Alemu Anteneh, Y. (2021). General overview of radon studies in health hazard perspectives. *Journal of Oncology* 2021: 1–8. https://doi.org/10.1155/2021/6659795.

Fisher, A., McClelland, G.H., Schulze, W.D., and Doyle, J.K. (1991). Communicating the risk from radon. *Journal of the Air & Waste Management Association* 41 (11): 1440–1445.

Harčárová, K., Vilčeková, S., and Balintova, M. (2020). Building materials as potential emission sources of VOC in the indoor environment of buildings. In: *Key Engineering Materials*, vol. 838, 74–80. Trans Tech Publications Ltd.

Jones, A.P. (1999). Indoor air quality and health. *Atmospheric Environment* 33 (28): 4535–4564.

Kumar, P., Kausar, M.A., Singh, A.B., and Singh, R. (2021a). Biological contaminants in the indoor air environment and their impacts on human health. *Air Quality, Atmosphere and Health* 14 (11): 1723–1736.

Kumar, P., Singh, A.B., and Singh, R. (2021b). Seasonal variation and size distribution in the airborne indoor microbial concentration of residential houses in Delhi and its impact on health. *Aerobiologia* 37 (4): 719–732.

Kumar, P., Singh, A.B., and Singh, R. (2022). Comprehensive health risk assessment of microbial indoor air quality in microenvironments. *PLoS One* 17 (2): e0264226.

Kumar, P., Singh, A.B., Arora, T. et al. (2023). Critical review on emerging health effects associated with the indoor air quality and its sustainable management. *Science of the Total Environment* 872: 162163.

Lee, H.J., Lee, K.H., and Kim, D.K. (2020). Evaluation and comparison of the indoor air quality in different areas of the hospital. *Medicine* 99 (52): 1–5.

Leung, D.Y. (2015). Outdoor-indoor air pollution in urban environment: challenges and opportunity. *Frontiers in Environmental Science* 2: 69.

Liu, W., Zhong, W., and Wargocki, P. (2017). Performance, acute health symptoms and physiological responses during exposure to high air temperature and carbon dioxide concentration. *Building and Environment* 114: 96–105.

Manisalidis, I., Stavropoulou, E., Stavropoulos, A., and Bezirtzoglou, E. (2020). Environmental and health impacts of air pollution: a review. *Frontiers in Public Health* 8: 14.

Prockop, L.D. and Chichkova, R.I. (2007). Carbon monoxide intoxication: an updated review. *Journal of the Neurological Sciences* 262 (1–2): 122–130.

Riudavets, M., Garcia de Herreros, M., Besse, B., and Mezquita, L. (2022). Radon and lung cancer: current trends and future perspectives. *Cancers* 14 (13): 3142.

Rumchev, K., Brown, H., and Spickett, J. (2007). Volatile organic compounds: do they present a risk to our health? *Reviews on Environmental Health* 22 (1): 39–56.

Tran, V.V., Park, D., and Lee, Y.C. (2020). Indoor air pollution, related human diseases, and recent trends in the control and improvement of indoor air quality. *International Journal of Environmental Research and Public Health* 17 (8): 2927.

Varon, J., Marik, P.E., Fromm, R.E. Jr., and Gueler, A. (1999). Carbon monoxide poisoning: a review for clinicians. *The Journal of Emergency Medicine* 17 (1): 87–93.

World Health Organization (2010). *WHO Guidelines for Indoor Air Quality: Selected Pollutants*. World Health Organization, Regional Office for Europe.

World Health Organization. (2013). Health effects of particulate matter: policy implications for countries in eastern Europe, Caucasus and central Asia. World Health Organization. Regional Office for Europe. https://iris.who.int/handle/10665/344854.

Yang, S., Goyette Pernot, J., Hager Jörin, C. et al. (2019). Radon investigation in 650 energy efficient dwellings in western Switzerland: impact of energy renovation and building characteristics. *Atmosphere* 10 (12): 777.

Yang, L., Li, C., and Tang, X. (2020). The impact of PM2.5 on the host defense of respiratory system. *Frontiers in Cell and Development Biology* 8: 91.

6

Indoor Contaminants Based on Fungi

Anchal Chaudhary[1], *Anamika Bhardwaj*[1], *Nikhil Sharma*[1], *Manandeep Kaur*[1], *Shalini Kaushik Love*[2], *and Arti Mishra*[2,3]

[1] *Department of Microbiology, Guru Nanak Dev University, Amritsar, Punjab, India*
[2] *Department of Botany, Hansraj College, University of Delhi, Delhi, New Delhi, India*
[3] *Umeå Plant Science Center, Department of Plant Physiology, Umeå University, Umeå, Sweden*

6.1 Introduction

There have been certain indoor contaminants that are present in the indoor environment. They are known as bio-contaminants (or biological contaminants). They are generally defined as the pollutants of our biological origin. These contaminants in the air present comprise major microorganisms including prokaryotic bacteria, eukaryotic fungi, unicellular algae, parasites (mites), viruses as well as insect detritus and their derivatives such as volatile organic compounds and toxins (mainly mycotoxins and endotoxins).

They are originated from plant materials such as allergens, pollens as well as odors, from animals such as cats, dogs, etc., and building materials. Carpets, tissues, heating modes such as humidity levels and temperature conditions, and ventillation levels can exacerbate their effects (mites).

Fungi are described as mainly single-celled or multi-celled microbes, which have long, microscopic strands of hyphae to whom various types of conidia, which are spore-developing structures are attached. Group of fungi is so diverse that it facilitates aerolization process. Some of the various respiratory conditions or problems may result from mold spores and their derivatives.

Fungal concentrations depend upon the environment of the area. For illustration, let us take an example of damp houses, which has usually more water activity (a_W). In this condition, we generally evidence a high spore count of the

Airborne Biocontaminants and Their Impact on Human Health, First Edition.
Edited by Rajeev Singh and Anamika Singh.

fungus (Nevalainen and Seuri 2005). *Penicillium* and *Aspergillus versicolor* are fungal strains that are well grown under relatively moderate to less water activity (a_W), while fungal strains such as *Fusarium, Trichoderma,* and *Phialophora* are well grown under high-water activity. Spores of fungi have capacity to germinate under 0.8–0.98 a_W (water activity), and are ubiquitous. Fungi are grown efficiently on books, cooking oil, stored food, construction materials, paint, etc. *Penicillium, Cladosporium, Trichoderma, Aspergillus, Alternaria,* and *Stachybotrys* are the most common genera of fungi that are found in the indoor air as per the review study on the fungal bioaerosol. They may cause serious health complications.

Fungi's tendency for growing on certainly any building material, be it natural or manmade, necessitates research into their development on such substrates, particularly when they are hygroscopic or damp. The most influential component for the growth of fungi is water activity (the quantity of free water accessible to fungal metabolism). Many types of materials, according to studies, are prone to growth if there is an adequate quantity of readily accessible water: gypsum boards, mortars, woods, wallpaper, and so on. Building materials can have variable bioreceptivities, or susceptibilities, to support and favor mold growth depending on many intrinsic criteria such as chemical composition, pH, dust content, and so on. Furthermore, their growth is governed by various environmental factors such as temperature and relative humidity, and their spread in the indoor environment can involve other factors such as air exchange rate, air movement, building structure and location, design, and ventilation system. However, this list is not exhaustive (Hallak et al. 2023).

Indoor environment is never completely devoid of microbial species and spores; even the cleanest rooms present out there have about 25 spores per cubic meter of air (Chirca 2019). In indoor structures, there are key sources of reproduction of microorganisms and their spread includes some electrical appliances such as humidifiers, ACs, fans, and coolers. *Geotrichum, Penicilium* are just a few examples of bacteria, fungi, and yeasts, which generally inhabit ACs, fans, and some other ventilation equipment (Dubey and Maheshwari 2008). Biological contaminant(s) is/are prevalent almost everywhere in the indoor environmental air according to (Chirca 2019), even in the most controlled environmental conditions. Examples include the modern hospital's operating room. About 90% or more of people's time was spent indoors, such as in their classrooms, homes, colonial structures such as stores, vehicles, and offices, and in the university premises (Moldoveanu 2015).

The detection of fungal species that inhabit structures is thus critical as it offers information on the sorts of biological particles most likely to be discharged into the air, as well as the related health concerns.

In this chapter, we assessed the literature available on the contamination of fungus of interior materials and indoor air in structures, hospitals, etc. The objective was to emphasize any linkages, if any, between microbial development on interior construction building materials and subsequent air pollution, as well as the resultant danger to human health.

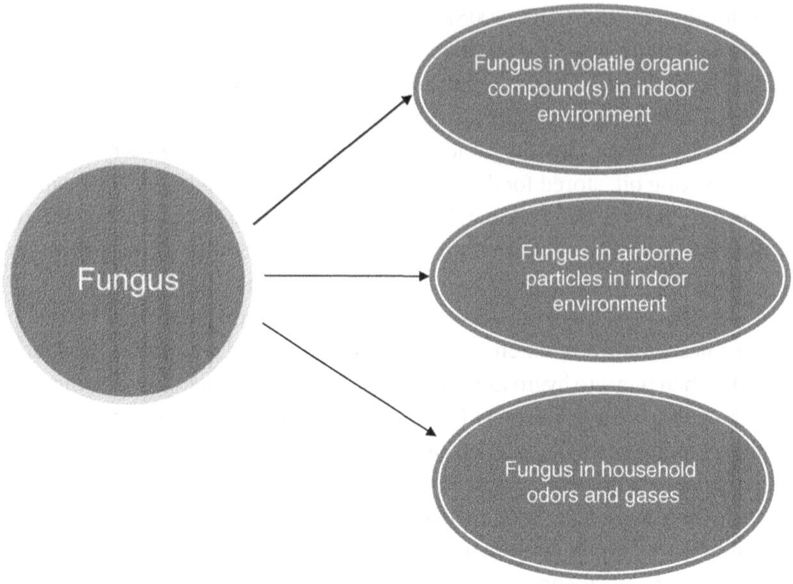

Figure 6.1 Possible fungal contaminants from indoor environment.

The first division of this chapter introduces the fungal particles, fungal diseases in humans and different factors leading to fungal colonization from indoor environment. The possible fungal contaminants from indoor environment are presented (Figure 6.1). The second division is an overview of data on fungal quantitation and the methods used for the quantitation process. In different parts such as structures, hospitals, and workspaces, most identified fungi that can be quantitated are presented. Finally, the third and the last division aims at the methodologies for the analysis of fungi, and their removal from the indoor environment.

6.2 Fungal Diseases

Fungal ailments turned out to be a chief medical issue in the other half of the twentieth century when developments in present medicine collected with HIV widespread brought about large figures of individuals with reduced immunity. Fungal diseases are tough to bring about because they are likely to be prolonged, hard to diagnose, and challenging to wipe out with antifungal medications. In the near prospect, the occurrence of fungal illnesses is anticipated to intensify as there will be extra hosts with reduced immunity, and medication resistance will certainly upsurge after assortment by antifungal medicine usage.

The existence of fungal ailments inside plant life has undeviating and unintended possessions on human well-being as pathogenic fungi are accountable for epidemics that diminish the harvest of bulk yields, for instance, rice, wheat, and corn that nourish billions of general public (Castanheira 2016; Almeida 2019).

Moreover, fungal contamination yields consequences in creation of deadly metabolites, polluting grain used for animal food, impacting the well-being and safety of livestock and cattle, as well as adulteration of administered nutrients for human ingestion.

6.2.1 Mycosis

It is a very common infection of the fungal species in the animal and human species. This disease can be classified into the following on the basis of response of the host to the pathogen as well as infected tissues:

- Cutaneous mycosis
- Subcutaneous mycosis
- Systemic mycosis
- Superficial mycosis

(Fung and Hughson 2003)

Those who are exposed to it may get systemic mycoses such as cryptococcosis, histoplasmosis, and coccidioidomycosis as a result of bird droppings and construction dust (Fung and Hughson 2003). In immunocompromised individuals, such as those with AIDS, cancer chemotherapy patients, and leukemia patients, mycosis is a more severe infection (Prussin and Marr 2015).

6.2.2 Mycotoxicosis

It is brought about by the exposure to 2° metabolites delivered by the parasites (Seltzer 1994). Based on the environmental conditions such as dampness, temperature, presence of O_2 and CO_2, and relative humidity, only a selective amount of molds may result in mycotoxin production (Moldoveanu 2015). The isolated mycotoxins go about as strong cancer/carcinogenic agents in the tenants or the occupants. Ochratoxin A, a mycotoxin that is produced by the fungal species, *Aspergillus flavus* and *A. parasiticus* is a human carcinogen that causes cirrhosis and liver cancer (Moldoveanu 2015).

6.2.3 Respiratory Symptoms

Sinusitis is a type of common cold that is reported to be in the form of either water damage or the visible molds in the homes. It is basically generated by the paranasal sinus inflammation. Based on a study, patients were reported with an association present between the nasal polyps and the skin showing reactivity toward

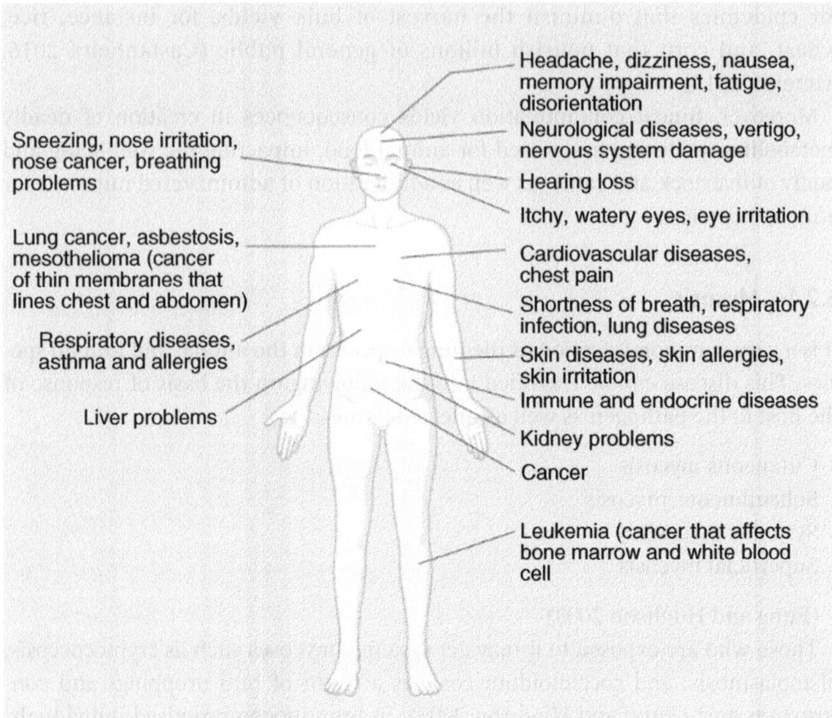

Figure 6.2 Indoor air pollutants and related human health hazards.

C. albicans who were set to be exposed to the indoor environment or pollution (Burge and Rogers 2000). In another study of a patient who was arrested with a past history of respiratory disease arrest, their confirmed skin prick test for the microbe *A alternata* found to be 91%. In this case, the proportion of 99 matched control with respect to asthma and no past history of any respiratory disease or arrest was only 31% (Downs et al. 2001). Therefore, there may be cases of severe asthma in the younger ones by the mold, *A. Alternaria* (Figure 6.2).

6.3 Factors Affecting the Colonization of Fungi

There are some of the factors that are important in the fungal growth of the building structures or materials. These include the following:

- Nutrient concentration
- Temperature
- Moisture content

1) Water activity; a_w is a term given to express the availability of water. The genus or the strain of fungus is the one that decides the requirement for any moisture. Generally, about 0.95–0.99 a_w favors the growth of fungus. Whereas, for xerophillic fungi, available water of 0.65–0.90 is required, and for yeasts, available water of 0.8–0.99 is required.
2) In the building materials or structures, the nutrients that are present in the home water and dust are the ones that favor the growth of fungus. Dust-accumulated galvanized steel, residue of the lubricant oils, or even the fiber glasses present in these building structures are the ones that permit fungal growth.
3) For the mesophilic fungi, the temperature condition of about 20–250 °C in the buildings permits the growth. There is also a fact that the temperature condition below this optimum condition of temperature slows down the fungal growth.
4) 5–6.5 is the best range of pH for the growth of fungus in the buildings (Vacher et al. 2010).
5) Some critical parameters of fungal growth are oxygen and sufficient light in indoor conditions.
6) Another parameter is the collection of nonorganic and organic household wastes. In many countries, this is a very common practice, which involves the storage of indoor organic household wastes such as remains of food, fruits, and vegetables in an area where there is no free space in the building's apartments till they have been disposed of. This will act as spores of the fungus source inside the building structures (house) (Figure 6.3).

6.4 Fungal Quantitation

For the determination of fungus in the provided samples in the case of nonculture-based practices, we can do the microscopic counting of cells or of the spores. There are several types of microscopy, which we can perform in order to identify the fungus. They are epifluorescence microscopy, scanning electron microscopy, and light microscopy. This purely depends upon the type of preparation of the sample just like for the morphological identification, we generally use light microscopy method (Moularat et al. 2008). Parts or metabolites of parasites can likewise be utilized to quantify growths populace in a climate. Extra cell polysaccharides can be distinguished by unambiguous examines for fractional ID of parasitic genera in indoor conditions (Jovanovic et al. 2004). A wide scope of contagious fungal antigens can be recognized based on the polyclonal antibody, yet it cannot distinguish the spores (Mitchell et al. 2007). Few of the methods such as karyotyping, PCRs, and

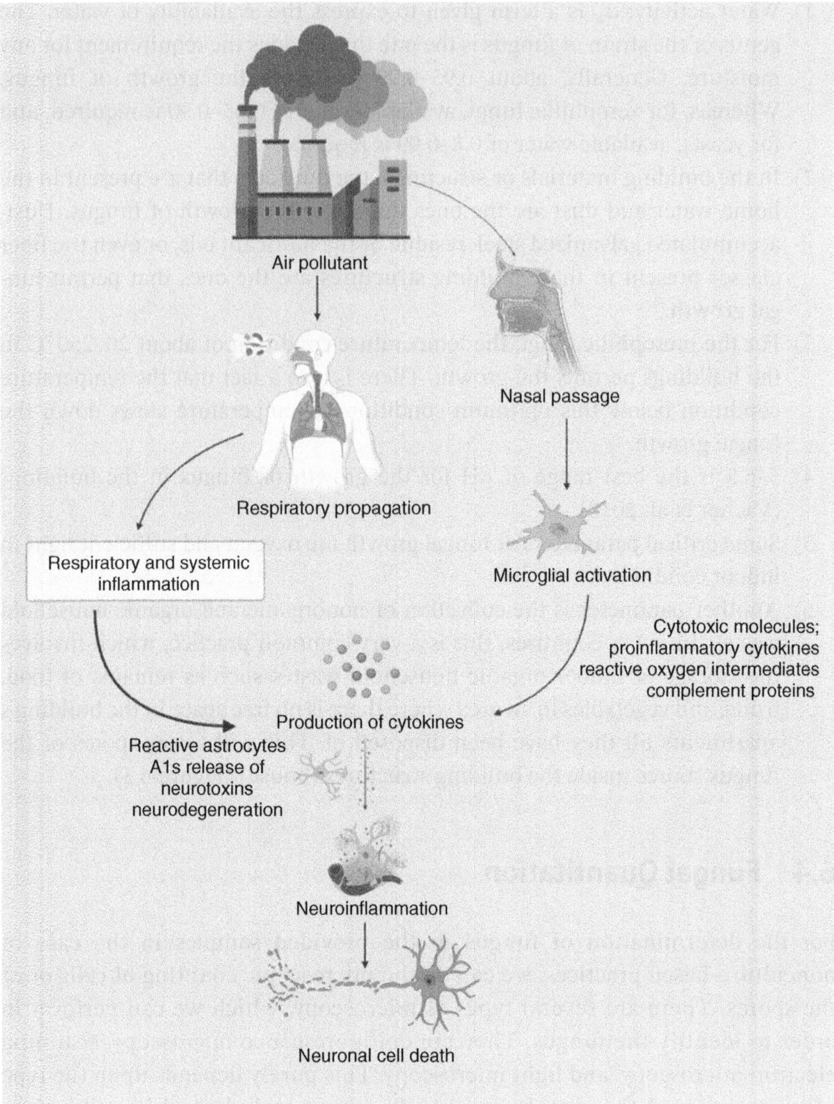

Figure 6.3 Impact of air pollutants on the brain.

species-/genus-specific probes are included in the molecular methods that are required for the fungal quantitation (Dean et al. 2005). For this purpose, i.e. for the DNA fingerprinting and analysis of restriction enzyme, mtDNA (mitochondrial-DNA) can be used for the identification of fungus.

6.5 Methods for Quantitation

The most reliable method of determining the indoor fungal load is "air sampling." Though there are readily several accessible techniques for sampling, only a handful have been frequently and often used in the literature. Some of the methods along with their assessment have been described in Table 6.1. The most common techniques are as follows are discussed in the subsequent sections.

6.5.1 Liquid Impingement

This technique involves the entering of the particles into the sampling devices, which are then channelled or directed via nozzles into a liquid medium underneath them. Large-sized particles impact the medium owing to their relative inertia. Particles with little buoyancy may adhere to the airflow streamlines and avoid acquisition.

6.5.2 Impaction

Recovery of the particles upon semisolid (Agar), or solid (adhesive-coated surface) collecting media is involved in the sampling of air by this technique known as "impaction." A jet is formed by an array of nozzles, resulting in particle deposition on a sample medium that is perpendicular to them. A considerable pressure drop between the nozzles' input and output causes the particles' velocity to rise. Because of the high velocity, particles with high inertia are impinging on the sample medium because they are unable to follow the airflow streamlines. Low-inertia particles, on the other hand, can avoid capture because they are subjected to centrifugal forces that are insufficient to compel their divergence from the streamlines.

Table 6.1 Sampling techniques and their assessment.

Method	Assessment
Culture-based	
Impactor	Organisms collected on the culture medium
Air filtration	Organisms collected on filter
Liquid impinger	Organisms collected in the collection fluid
Non-culture based	
Liquid impinge	Simple microscopy
Air filtration	Flow cytometry; electron microscopy
Air sampling	Microscopy

6.5.3 Electrostatic Precipitation

This technique is based on the accumulation of charged particles via electrostatic forces. Bio-aerosols get charged while at the precipitator's inlet. Particles are then subjected to an electromagnetic field, which enforces them for collision with the charged plates. When compared to the impaction technique, the impact's velocity stays relatively moderate, resulting in reduced microbe stress and, hence, loss of viability.

6.5.4 Gravitational Settling/Sedimentation

A sample media, often agar, is introduced in the area to be studied for an extended length of time. As a result of interior air currents, there is a collision of particles and impingement in the media. Large particles, due to their gravitational force and inertia, may be gathered more easily than tiny particles, which may be transported by the inside airflow and, therefore, avoid acquisition. This method can only offer limited amount of knowledge on the activity of fungus in the space undergoing inspection and may not be suited for assessing indoor fungal growth.

6.5.5 Filtration

Filtration is based on particle recovery on a porous media. By virtue of airflow, particles are pushed to travel through the media. Large particles are unable to depart the porous media due to their features in terms of shape, size, and so on, along with the size of the pores of filters and the air velocity, whereas tiny particles avoid acquisition. Mechanisms involving electrostatic attraction, inertial forces, and diffusion help gather the larger debris on the media and remove small particles through the filters (Efthymiopoulos et al. 2023).

6.6 Determination of Fungal Components in Indoor Conditions

Any estimation of any grouping for any agent or exposure evaluation comprises two similarly significant stages:

1) Collection of the sample
2) Examination of the agents being referred to

Fungi and their contagious materials can be tested (sampled) from the air, the surfaces, the building structure's samples, the dust in the house, and the last option generally picked in situations where the doubt of mold development on that particular material should be affirmed.

Air-borne organisms such as fungi are usually collected using instruments, which are the applications of molecule estimation methods, particularly created for assortment of biological particles. The standards generally utilized in these strategies are collection using centrifugation, electrostatic assortment, inertial impaction, assortment on a filter, and impingement in the liquid.

In the methodologies that utilize the culturing strategies for examination, assortment might happen directly in the impingers, i.e. liquid medium or on the impactors, i.e. agar surface. Tests gathered on channels are then suspended into a liquid and used for an examination of some kind or the other. The examples that are accumulated utilizing these methods are quantitative as the volume of the air, which has gone through the instrument is unequivocally known, and the result can be imparted as the obsession per cubic meter of air. There are a couple of expansive introductions and reviews open on the techniques for assortment.

House dust which comprises essentially of skin scales, textile fibers, and hair, likewise contains numerous microorganisms, both bacteria and fungi, and pollens, insects, and various other remains from the sources of plants and animals. Generally, the floor, mattress and carpet samples are collected using vacuuming a characterized region and for a foreordained time frame to normalize the sampling procedure (Giovannangelo et al. 2007). The gathered measure of dust is weighed, followed by the extraction of the fungal particle from the matrix of sample prior to the examination. Outcomes are typically expressed per gram of the residue, or in the form of surface heaps of growths, or at least/m^2. In allergy-related examinations, the mattress's residual (dust) samples are frequently utilized because of the proximity of the sample to the zone of breathing during rest, and because of their great reproduction ability when contrasted with the floor and the samples taken from air (Noss et al. 2010).

Passive collectors are developed for the collection of the settled dust from cupboards and shelves, i.e. from elevated surfaces. In this approach of sampling, the impact of the foot traffic and the more noteworthy commitment of bigger particles are stayed away from. A more exact assessment of the exposure to the respiratory parts may be accomplished rather than samples that are vacuumed from floor. The passive collectors likewise empower more sampling time as compared to work-escalated active sampling, also they are known to reflect the long-haul air-borne exposure environment (Noss et al. 2010).

6.7 Methodologies for Fungal Analysis

Microscopical techniques are being widely used for counting and recognizing parasitic particles, particularly for allergens' checking and in plant pathology. Some of the spore types may be recognized, and measured straightforwardly from tests. However, as a general rule, minuscule counting does not give ordered segregation. For indoor conditions, direct microscopy is the technique, which may be utilized in indoor air quality-related building examinations. The subsequent technique is refined, which takes advantage of the reasonability of the contagious spore(s) or the propagules that were present in the example. Hatching happens under normalized situations, which is followed by the counting and microscopical recognizable proof of apparent parasitic settlements. This strategy can give – adequate mycological skill – low-level taxonomical separation, yet just for a small part of the local area (Prezant et al. 2008).

Culturing is a special method because the additional medium that is provided, along with the incubation temperature and lighting conditions, controls the formation of colonies of microorganisms that can be seen on the medium. In fact, even access to the available space on the agar plate may limit growth, especially for slower-growing organisms. Culturing nonetheless provides a quantitative and subjective assessment of the organisms present in the sample and is capable of developing under the conditions present. Certain fungal types might be targeted for the estimation by using various medium and incubation conditions. The fungal colonies may be contained for further testing after the culturing. Mycobiota and fungal concentrations have been depicted in the media extensively using cultivation techniques.

For investigation strategies rather than the culturing strategy, analyte is typically extracted from the matrix of a sample, which is followed by exposure to the chemical compound, enzymatic, immunochemical, or DNA-based examinations to measure the biomass of the fungus or different characteristics of the sample mycobiota. Utilizing synthetic marker, how much fungal biomass can be assessed as (1-3)-beta-D-glucan or ergesterol (Saraf et al. 1997)? Glucan's biological activity can be detected using "limulus bioassay" (Foto et al. 2004).

The action of the protein, *N*-acetylhexosaminidase (NAHA) has been utilized in a commercial accessible application, that is MycometerR for evaluating how much developing mold in a surface, air, or material test. Immunochemical strategies are used in measuring contagious allergens. Fungal products, like microbial unstable natural mixtures (MVOC) and mycotoxins, are examined for the most part with the mass spectrometry-based methods, which are combined with predetachment in a gas or fluid chromatography (Bloom et al. 2007).

6.8 Practices that Contribute to the Biotic Indoor Pollution

In order to save energy during off-peak business hours, it is a common practice to switch cooling (A/C) units off. This could lead to water accumulation, an increase in relative humidity, and a temperature that favors the growth of fungi.

It is crucial to repeatedly clean the air conditioning channels or rooms in spaces where the air conditioning units are left off for extended periods of time. It is advised to keep the A/C systems running continuously. According to the manufacturer, the temperature can be regulated or programmed to save electricity. A/C channels should occasionally be replaced or cleaned since parasite invasion and residue buildup can cause the channel to become blocked. Since soil can act as a store for organisms, pruned plants kept in airconditioned spaces may pose a risk to the residents. This accounts for the disconnection of pathogenic growths from the soils of trimmed plants housed in air conditioned spaces. In areas with carpets, regular shampooing and vacuuming are necessary because floor coverings can harbor bacteria that can spread quickly (Khan and Karuppayil 2012).

Ductless minisplit A/Cs' indoor units may be readily removed and disinfected at a cheap cost and with relatively less labor by washing at high pressure with water and detergents. In a case study, where fungal burdens have been determined using culture or next-generation sequencing (NGS) acquired by swabbing the interior of A/Cs or collecting the blow-out air. In this case, the examiners used fungal-specific quantitative polymerase chain reaction (PCR) technique, followed by mycobiome (myco meaning fungus) analysis on samples acquired from all the internal portions of the A/Cs along with the cleaning effluents prior to and after washing at high pressure in order to develop an approach to evaluate the fungal pollution in A/Cs and to investigate the efficacy of washing at high pressure in fungal disinfection (Shiraishi et al. 2023).

6.9 Removing Fungal Material and Cleaning

Cleaning is another important method to get rid of and reduce the levels of microbiological material in the interior environment, together with ventilation. The fungal burden in household dust has an impact on cleaning effectiveness. In particular, regular vacuuming diminishes centralization of contagious agents in floor dust. Then again, vacuuming and clearing the floor increment the airborne centralization of parasites briefly because of their resuspension. A less explored question is whether the decreased parasite concentrations in household dust also translate to lower concentrations in indoor air. In addition to regular household cleaning, there are specific cleaning jobs like expertly cleaning ventilation pipes. After form remediation, meticulous cleaning of the surfaces, furniture, and ventilation conduits is essential to ensure that residents return safely. Circumstances requiring significantly more concentrated endeavors are the cleanup activities in the wake of flooding and tropical storms (Nevalainen et al. 2015).

6.10 Conclusion and Future Perspectives

According to the viewpoint of indoor air rehearses, the issue of form development is straightforward which warns that it should be stayed away from by people. Nonetheless, a more significant gamble evaluation of indoor growths and parasitic items expands the extent of the test by a few significant degrees. We ought to perceive here the extraordinary chances to utilize multidisciplinary approaches in settling these difficulties. Research that can be used in everyday practice will result in real progress. With the use of modern logical devices and the large range of global skills available, this test will be passed with success.

The application of appropriate biocides is required to suppress or prevent development of fungus on the construction materials. Unfortunately, many microbes have developed resistance to chemicals used, so mixtures of biocides, including combinations of different fungicides, are frequently used to fix up the deficiencies in fungicidal spectrum, therefore limiting selective pressure that is associated with the selection of resistance. In order to circumvent these constraints, research is focusing on the novel and even more efficient anti-microbial compounds' development for use in the preservation of building materials (Campana et al. 2020).

References

Al Hallak, M., Verdier, T., Bertron, A. et al. (2023). Fungal contamination of building materials and the aerosolization of particles and toxins in indoor air and their associated risks to health: a review. *Toxins 15* (3): 175.

Almeida, F., Rodrigues, M.L., and Coelho, C. (2019). The still underestimated problem of fungal diseases worldwide. *Frontiers in Microbiology 10*: 214.

Bloom, E., Bal, K., Nyman, E. et al. (2007). Mass spectrometry-based strategy for direct detection and quantification of some mycotoxins produced by Stachybotrys and Aspergillus spp. in indoor environments. *Applied and Environmental Microbiology 73* (13): 4211–4217.

Burge, H.A. and Rogers, C.A. (2000). Outdoor allergens. *Environmental Health Perspectives 108* (Suppl 4): 653–659.

Campana, R., Sabatini, L., and Frangipani, E. (2020). Moulds on cementitious building materials – problems, prevention and future perspectives. *Applied Microbiology and Biotechnology 104*: 509–514.

Castanheira, M., Messer, S.A., Rhomberg, P.R., and Pfaller, M.A. (2016). Antifungal susceptibility patterns of a global collection of fungal isolates: results of the SENTRY Antifungal Surveillance Program (2013). *Diagnostic Microbiology and Infectious Disease 85* (2): 200–204.

Chirca, I. (2019). The hospital environment and its microbial burden: challenges and solutions. *Future Microbiology 14* (12): 1007–1010.

Dean, T.R., Roop, B., Betancourt, D., and Menetrez, M.Y. (2005). A simple multiplex polymerase chain reaction assay for the identification of four environmentally relevant fungal contaminants. *Journal of Microbiological Methods 61* (1): 9–16.

Downs, S.H., Mitakakis, T.Z., Marks, G.B. et al. (2001). Clinical importance of Alternaria exposure in children. *American Journal of Respiratory and Critical Care Medicine 164* (3): 455–459.

Dubey, R.C. and Maheshwari, D.K. (2008). *A Textbook of Microbiology*. S. Chand Publishing.

Efthymiopoulos, S., Aktas, Y.D., and Altamirano, H. (2023). Common sampling techniques for the assessment of indoor fungal growth. In: *Common Sampling Techniques for the Assessment of Indoor Fungal Growth*. ScienceOpen.

Foto, M., Plett, J., Berghout, J., and Miller, J.D. (2004). Modification of the Limulus amebocyte lysate assay for the analysis of glucan in indoor environments. *Analytical and Bioanalytical Chemistry 379*: 156–162.

Fung, F. and Hughson, W.G. (2003). Health effects of indoor fungal bioaerosol exposure. *Applied Occupational and Environmental Hygiene 18* (7): 535–544.

Giovannangelo, M.E., Gehring, U., Nordling, E. et al. (2007). Levels and determinants of β (1→ 3)-glucans and fungal extracellular polysaccharides in house dust of (pre-) schoolchildren in three European countries. *Environment International 33* (1): 9–16.

Jovanovic, S., Felder-Kennel, A., Gabrio, T. et al. (2004). Indoor fungi levels in homes of children with and without allergy history. *International Journal of Hygiene and Environmental Health 207* (4): 369–378.

Khan, A.H. and Karuppayil, S.M. (2012). Fungal pollution of indoor environments and its management. *Saudi Journal of Biological Sciences 19* (4): 405–426.

Mitchell, C.S., Zhang, J., Sigsgaard, T. et al. (2007). Current state of the science: health effects and indoor environmental quality. *Environmental Health Perspectives 115* (6): 958–964.

Moldoveanu, A.M. (2015). Biological contamination of air in indoor spaces. In: *Current Air Quality Issues*. IntechOpen.

Moularat, S., Robine, E., Ramalho, O., and Oturan, M.A. (2008). Detection of fungal development in closed spaces through the determination of specific chemical targets. *Chemosphere 72* (2): 224–232.

Nevalainen, A. and Seuri, M. (2005). Of microbes and men. *Indoor Air 15*: 58–64.

Nevalainen, A., Täubel, M., and Hyvärinen, A. (2015). Indoor fungi: companions and contaminants. *Indoor Air 25* (2): 125–156.

Noss, I., Doekes, G., Sander, I. et al. (2010). Passive airborne dust sampling with the electrostatic dustfall collector: optimization of storage and extraction procedures for endotoxin and glucan measurement. *Annals of Occupational Hygiene 54* (6): 651–658.

Prezant, B., Weekes, D.M., and Miller, J.D. (ed.) (2008). *Recognition, Evaluation, and Control of Indoor Mold*. Aiha.

Prussin, A.J. and Marr, L.C. (2015). Sources of airborne microorganisms in the built environment. *Microbiome 3*: 1–10.

Saraf, A., Larsson, L., Burge, H., and Milton, D. (1997). Quantification of ergosterol and 3-hydroxy fatty acids in settled house dust by gas chromatography-mass spectrometry: comparison with fungal culture and determination of endotoxin by a Limulus amebocyte lysate assay. *Applied and Environmental Microbiology 63* (7): 2554–2559.

Seltzer, J.M. (1994). Biological contaminants. *Journal of Allergy and Clinical Immunology 94* (2): 318–326.

Shiraishi, Y., Harada, K., Maeda, C. et al. (2023). A method to evaluate and eliminate fungal contamination in household air conditioners. *Indoor Air 2023*.

Vacher, S., Hernandez, C., Bärtschi, C., and Poussereau, N. (2010). Impact of paint and wall-paper on mould growth on plasterboards and aluminum. *Building and Environment 45* (4): 916–921.

7

Health Implications Related With the Presence of Waste Sorting Plants and Biological Contaminants

Samina Mazahar

Department of Botany, Dyal Singh College, University of Delhi, Delhi, New Delhi, India

7.1 Introduction

The advancement and use of modern tools and techniques undoubtedly contribute to the progress of the nation. Besides, it also has numerous drawbacks leading to health concerns and environmental pollution. WSPs and biological contaminants possess several negative impacts on human health. The major components of bio-contaminants (biological contaminants) are microbes (bacteria, viruses, and numerous unicellular microorganisms), algae, fungi, insects' debris, mites, etc. (Moldoveanu 2015). Hence biological contaminants are pollutants of biological origin such as from plants (allergens, pollen, odors), animals, pets, building materials, carpets, fluctuation in temperature, humidity, poor ventilated areas (Kumar et al. 2021). They are present everywhere and even in highly controlled environmental conditions, indoor air (Chirca 2019). Report suggests that indoor air is never free from microorganisms and spores, the clean room contains about 25 spores/m^3 (Chirca 2019). Hence, both outdoor and indoor environments contribute to the spread of microorganisms and need to be taken care of to stay healthy and in less contaminated environments.

The WSPs also possess numerous adverse effects on the health of employees working there and are the major exposure to chemical particulates and bioaerosols (Kalwasinska and Burkowska 2013). Study suggests increasing cases of numerous health problems in the workers of WSPs such as irritation, inflammation, pulmonary diseases, and allergies (Malta-Vacas et al. 2012). Humans are always exposed to biological aerosols as they are always present in the air. Biological aerosols include microorganisms (bacteria, viruses, fungi) and their

Airborne Biocontaminants and Their Impact on Human Health, First Edition.
Edited by Rajeev Singh and Anamika Singh.

products (bacterial and fungal spores, pollen and plant parts) (Reponen et al. 2011). Untreated household waste can become toxic and hazardous and can have long-term and increasing impacts on environment and human health (Gutberlet and Uddin 2017). The accumulation of untreated waste and poor waste management facilities (landfills, incinerators, and dumps) adversely affect the health of local communities (Gutberlet 2011; Uddin et al. 2016).

Hence, people dealing with biological contaminants and working in WSPs are at high health risk. The management of basic infrastructure, awareness of handling hazardous waste and use of protective equipment should be a must for those handling garbage (Gutberlet and Uddin 2017). A study suggests that more than one-third of the global urban population exists in informal settlements (UN-Habitat 2010; Arimah 2012) and is poorly linked to basic facilities (Hardoy et al. 2001). Such areas are a threat to environment and human health due to presence of contaminated water, soil, and polluted air (methane and other gases) due to open dumping of solid waste (King et al. 2006).

7.2 Waste Sorting Plants

The WSPs are growing each year at a rate of 3–4% (Poole and Basu 2017). Study suggests that in India, about 143,449 metric tons (MT) of municipal solid waste are generated on a daily basis, and only approximately 111,000 MT is being collected and out of which only 35,602 MT is treated (Kumar et al. 2017). The injury rate (fatal or nonfatal) and the work-related illness of employers are increasing every year. The components of potential health risks are exposure to harmful chemicals and metals such as recycling of batteries, electrical goods, fluorescent lights, lead, and mercury; exposure to bioaerosol; manual handling of factory goods, etc. (Poole and Basu 2017). Major emission and exposure to bioaerosols at WSPs occur during the transport, unloading, and processing of waste and hence the concentration of microorganisms varies significantly in WSPs (Bragoszewska et al. 2019). Figure 7.1 shows the numerous sources of solid waste. A study suggests that wet fractions of municipal waste contain 109 viruses or bacteria per gram of waste (Breza-Boruta 2016). Bacteria are the major component of bioaerosols in WSPs and possess a major adverse impact on the health of the employees (Bragoszewska et al. 2019). The employees working in WSPs are exposed to health hazards due to direct contact with municipal waste and animal and plant waste coupled with high temperature and moisture during summer and spring seasons (Douwes et al. 2003). Also, waste sorting employees are exposed to such microbial strains in bioaerosols that are resistant to antibiotics and this is a matter of great concern. For example, *Neisseria shayeganii* is found to show resistance to 29.7% of the antibiotics tested (Bragoszewska et al. 2019).

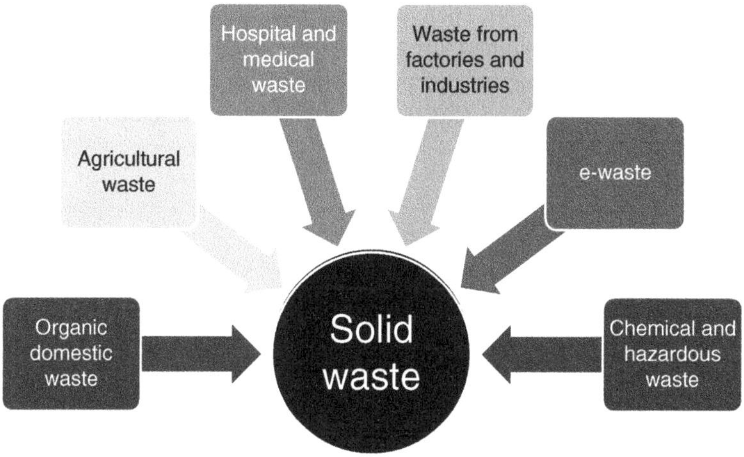

Figure 7.1 Sources of solid waste.

7.3 Biological Contaminants

The indoor environment is full of bio-contaminants and its monitoring and assessment are important to lessen the risk of potential irritants and toxins and the adverse effect on human health and environment. Numerous abiotic factors like moisture, temperature, relative humidity, insulation, etc., regulate the existence of biological contaminants (Kumar et al. 2021). Bio-aerosols are airborne organic volatile compounds released from living organisms, ranging in size from 0.1 to 100 μm in diameter (Seltzer 1994). Depending upon the particle size, the bio-aerosols are always present in the atmosphere as they remain suspended in the air if the particle size is small and settle down on surfaces if the particle size is large. The common sources of bio-aerosols are sewage, soil, and water and include components such as bacteria, viruses, fungi, pollens, etc. Public places such as hospitals, libraries, cinema halls, railway stations, and crowded and closed enclaves are the most pathogen-prone areas and also are the mode of transmission of pathogens.

7.4 Major Bio-Contaminants

7.4.1 Viruses

Indoor air is the medium for the growth, multiplication, and infection of viruses. The contamination caused due to the droplets of sneezing, coughing, talking, etc., in the indoor environment spreads millions of airborne viruses in the indoor environment (Diglisic et al. 1999). There are numerous deadly viruses that cause severe infections and are transmitted through humans. For example, *Variola* virus is an obligate parasite causing chickenpox, *Influenza* virus causes respiratory infections, and *Coronavirus* (SARS-CoV-2) causes coronavirus disease in humans (Kumar et al. 2021).

7.4.2 Bacteria

Bacteria are present both in indoor and outdoor environments and have diverse existence. They are mostly saprophytic and found in the microflora (nose, ear, mouth, etc.) and originating from outdoor air (Moldoveanu 2015). The bacteria present in the upper respiratory tract are often removed by near contact such as coughing, talking, and sneezing. Heterotrophic bacteria such as *Legionella* and *Mycobacteria* are found in water pipes and water reservoirs of cooling systems (Kumar et al. 2021). Indoor environment has numerous species such as *Micrococcus, Corynebacterium, Staphylococcus,* and *Streptococcus* (Kauser et al. 2016). *Legionella* spp. is a facultative parasite present in hot water and causes Pneumonia. *B. anthracis* causes anthrax and is mostly found in veterinarians and animal handlers. Similarly, *M. tuberculosis* is an obligate parasite causing tuberculosis and it is transmitted by humans (Kumar et al. 2021).

7.4.3 Fungi

Fungi have diverse existence and are known to be present in humid and moist habitats. The amount of fungal spores is high in such areas. They are known to germinate in water activity (aw) between 0.80 and 0.98 (WHO 2009). For example, *Trichoderma, Phialophora, Aspergillus fumigates, Yeast, Stachybotrys,* and *Fusarium* are present in places with high water activity, whereas *Penicillium* and *Aspergillus versicolor* grow well at moderate to low water activity area (Kumar et al. 2021). Research suggests that concentration of fungal spores can range between 100 and 100,000 spores m^3 in the outdoor area (Moldoveanu 2015). The fungal spores give rise to numerous respiratory problems (American Air and Water 2020). Fungi can grow on stored food, cooking oil, wood, paint, books, construction materials, leather goods, etc. (WHO 2009; Rautiainena et al. 2019). For instance, the *Penicillium* spp. found in buildings containing molds is a facultative parasite causing Penicilliosis. *Histoplasma* spp. causes Histoplasmosis and is present in bird droppings. Species of *Alternaria* are saprophytes and are known to cause rhinitis and asthma through damp surfaces and outdoor air (Kumar et al. 2021).

7.4.4 Mycotoxins and Endotoxins

Mycotoxins are toxic fungal products that alter RNA synthesis and cause DNA damage, hence leading to death in animals and humans. For example, acute pulmonary hemorrhage is caused by the mycotoxin produced by *Stachybotrys chartarum* (Sharma et al. 2011a, b) and aflatoxin produced by *Aspergillus flavus* and *A. parasiticus* causes cancer (Kumar et al. 2021). Endotoxins are secreted by bacteria during adverse conditions. They are present in the cell walls of gram-negative bacteria. High levels of endotoxins cause asthma and asthma-like problems (Michel et al. 1996). It is found in edible waste storage, low-ventilated areas, pets, etc. (Kumar et al. 2021).

7.4.5 Allergens

Allergens cause IgE-mediated allergic reactions in humans. Allergens cause numerous respiratory diseases. Numerous fungi such as *Cladosporium, Trichoderma, Stemphylium, Curvularia, Alternaria, Epicoccum, Ganoderma, Aspergillus,* and *Penicillium* are potent allergens (Kausar 2018). In indoor environment, major sources are poultry, bakery, sugar, and granary industry. Also, insects and their by-products, and mosquito debris are reported as potential allergens (Kausar et al. 2007).

7.5 Effect of Biological Contaminants and Waste Sorting Plants on Health

The biological contaminants and WSPs are the root cause of various infectious diseases, allergies, and health-related complications. Several infectious diseases such as chickenpox, measles, and influenza are spread through air and repeated exposure to allergens causes various allergic reactions (Seltzer 1994; Chirca 2019; Rautiainena et al. 2019). Multiple exposure to bio-contaminants by the health workers in the WSPs causes allergies, asthma, and various respiratory problems. Major health-related issues are coughing, dizziness, fever, shortness of breath, lethargy, watery eyes, digestive problems, etc. (Stetzenbach et al. 2004; UNEPA 2020). Figure 7.2 shows the major health hazard caused due to constant exposure to waste products.

7.5.1 Bacterial Diseases

Numerous diseases caused by bacteria are airborne and transmitted through respiratory droplets such as coughing and sneezing. The spores of *Bacillus anthracis* cause highly infectious disease anthrax. It is a deadly disease spread intentionally

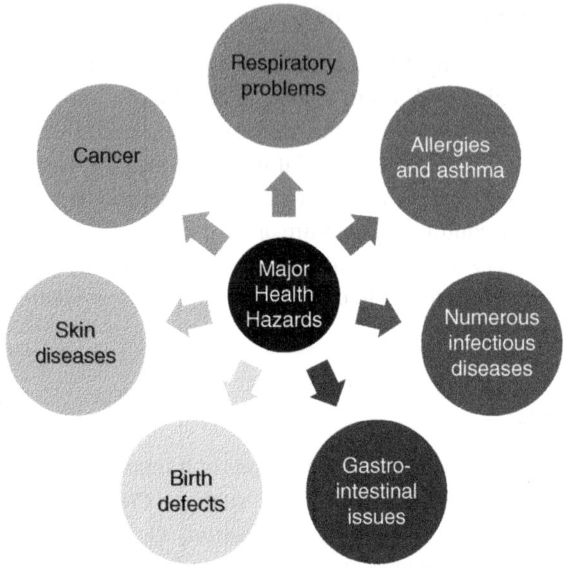

Figure 7.2 Major health hazards caused due to constant exposure to waste products.

through injections, skin, lungs, and intestine and its outbreak is associated with bioterrorism besides work-related transmission (Vieira et al. 2017). Another common bacterial disease is tuberculosis, which is caused by *Mycobacterium tuberculosis* that leads to high mortalities throughout the world. It is a communicable disease and is passed on to a healthy individual from infected persons via bioaerosols through sneezing, coughing, and talking (Srikant et al. 2008). Leprosy is another infectious bacterial disease caused by slow-growing intracellular bacteria *Mycobacterium laprae*. It is communicated through respiratory droplets from the infected person (Kumar et al. 2021). The major affected organs are upper respiratory mucosa, lymph nodes, skin, nails, and superficial nerves (Anathanarayan and Panikar 2009). People living in industrial and at workplace are often found infected with Legionellosis (community-acquired pneumonia) caused by *Legionella pneumophila*. It is found in air-conditioners, biofilms of cooling towers, and artificial water systems. It can be prevented by the use of disinfectants such as monochloramine (USEPA 2020).

7.5.2 Diseases Caused by Viruses

Viral infections are easily spread through air. Severe acute respiratory syndrome (SARS) is a viral respiratory contagious disease caused by SARS-associated coronavirus that leads to poor health, mortality, and severe pneumonia (Srikant et al. 2008).

It is spread by aerosol generation used in hospitals (Yu et al. 2004). Recently, the entire World suffered huge losses due to coronavirus disease (COVID-19) caused by coronavirus (SARS-CoV-2). According to the report of WHO (2020), four lakh deaths and seven million confirmed cases of infections caused by coronavirus had been reported. Various other viruses such as influenza virus (H1N1), respiratory syncytial virus (RSV), enteric viruses of intestine, hantavirus (from rodent's feces), rubella, measles, chickenpox, and mumps are easily spread through air (Yu et al. 2004).

7.5.3 Disease Spread Through Fungi

Mycotoxicosis is a fungal disease caused by exposure to toxins produced by certain fungi. They interfere with transcription and translation activities (Fung and Hughson 2001). Mycotoxins isolated from fungi act as potential carcinogens in people exposed to them (Kumar et al. 2021). Exposure to other fungi such as *Aspergillus flavus* (releasing aflatoxin) and *A. parasiticus* causes liver cancer (cirrhosis). Further, human carcinogens, namely ochratoxin A, are found in food containing mycotoxin (American Air and Water 2020). Another fungal infection is mycosis found in humans and animals. It is found in people working at construction sites as present in building dust and also found in bird droppings causing systemic mycosis (WHO 2009). Mycosis adversely affects persons with low immunity (Prussin and Marr 2015).

7.6 Conclusion

The segregation, handling, and proper disposal of waste along with awareness to the employees of sorting plants regarding exposure and adverse effects of biocontaminants, harmful chemicals, and radioactive compounds in the WSPs is an important task and needs to be addressed. Finding an effective potential solution regarding handling of waste is a challenging mission but proper management and use of eco-friendly technology should be adopted to lessen the exposure to occupational illness and follow sustainable ways of disposing of the waste products.

References

American Air and Water (2020). Mold, mold spores and indoor air quality. *American Air and Water* https://www.americanairandwater.com/mold.

Ananthanaryan, R. and Paniker, C.K.J. (2009). *Text of Microbiology*, 8e. Universities Press.

Arimah, B.C. (2012). *Slums as Expressions of Social Exclusion: Explaining the Prevalence of Slums in African Countries*. Nairobi: UN-Habitat.

Bragoszewska, E., Biedron, I., and Hryb, W. (2019). Air quality and potential health risk impacts of exposure to bacterial aerosol in a waste sorting plant located in the mountain region of Southern Poland, around which there are numerous rural areas. *Atmosphere* 10: 360. https://doi.org/10.3390/atmos10070360.

Breza-Boruta, B. (2016). The assessment of airborne bacterial and fungal contamination emitted by a municipal landfill site in Northern Poland. *Atmospheric Pollution Research* 7: 1043–1052.

Chirca, I. (2019). The hospital environment and its microbial burden: challenges and solutions. *Future Microbiology* 14 (12): 1007–1010. https://doi.org/10.2217/fmb-2019-0140.

Diglisic, G., Rossi, C.A., Doti, A., and Walshe, D.K. (1999). Seroprevalence study of *Hantavirus* infection in the community-based population. *Maryland Medical Journal* 48: 303–306.

Douwes, J., Thorne, P., Pearce, N., and Heederik, D. (2003). Bioaerosol health effects and exposure assessment: progress and prospects. *The Annals of Occupational Hygiene* 47: 187–200.

Fung, F. and Hughson, W.G. (2001). Health effects of indoor fungal bioaerosol exposure. *Applied Occupational and Environmental Hygiene* 18 (7): 535–544.

Gutberlet, J. (2011). Waste to energy, wasting resources and livelihoods. *Integrated Waste Management* 1: 219–236. Sunil Kumar (Ed.), ISBN: 978-953-469-6, In Tech.

Gutberlet, J. and Uddin, S.M.N. (2017). Household waste and health risks affecting waste pickers and the environment in low- and middle-income countries. *International Journal of Occupational and Environmental Health* 23 (4): 299–310. https://doi.org/10.1080/10773525.2018.1484996.

Hardoy, J.E., Mitlin, D., and Satterthwaite, D. (2001). *Environmental Problems in an Urbanizing World: Finding Solutions for Cities in Africa, Asia and Latin America*. London: Earthscan Publications.

Kalwasinska, A. and Burkowska, A. (2013). Municipal landfill sites as sources of microorganisms potentially pathogenic to humans. *Environmental Science. Processes & Impacts* 15: 1078–1086.

Kausar, M.A. (2018). A review on respiratory allergy caused by insects. *Bioinformation* 14 (9): 540–553.

Kausar, M.A., Vijayan, V.K., Bansal, S.K. et al. (2007). Mosquitoes as sources of inhalant allergens: clinico immunologic and biochemical studies. *The Journal of Allergy and Clinical Immunology* 120: 1219–1221.

Kausar, M.A., Arif, J.M., Alanazi, S.M.M. et al. (2016). Assessment of microbial load in indoor environment of university and hospitals of Hail. *Ksas Biochemical and Cellular Archives* 16 (1): 177–183.

King, A., Burgess, S., Ijomah, W. et al. (2006). Reducing waste: repair, recondition, remanufacture or recycle? *Sustainable Development* 14 (4): 257–267.

Kumar, S., Smith, S.R., Fowler, G. et al. (2017). Challenges and opportunities associated with waste management in India. *Royal Society Open Science* 4 (3): https://doi.org/10.1098/rsos.160764.

Kumar, P., Kausar, M.A., Singh, A.B., and Singh, R. (2021). Biological contaminants in the indoor air environment and their impacts on human health. *Air Quality Atmosphere and Health* 14: 1723–1736. https://doi.org/10.1007/s11869-021-00978-z.

Malta-Vacas, J., Viegas, S., Sabino, R., and Viegas, C. (2012). Fungal and microbial volatile organic compounds exposure assessment in a waste sorting plant. *Journal of Toxicology and Environmental Health, Part A* 75: 1410–1417.

Michel, O., Kips, J., Duchateau, J. et al. (1996). Severity of asthma is related to endotoxin in house dust. *American Journal of Respiratory and Critical Care Medicine* 154: 1641–1646.

Moldoveanu, A.M. (2015). Biological contamination of air in indoor spaces. In: *Current Air Quality Issues* (ed. F. Nejadkoorki), 489–514. IntechOpen https://doi.org/10.5772/59727.

Poole, C.J.M. and Basu, S. (2017). Systematic review: occupational illness in the waste and recycling sector. *Occupational Medicine* 67: 626–636. https://doi.org/10.1093/occmed/kqx153.

Prussin, A.J. and Marr, L.C. (2015). Sources of airborne microorganisms inthe built environment. *Microbiome* 3 (1): 1.

Rautiainena, P., Hyttinenb, M., Ruokolainenb, J. et al. (2019). Indoor air-related symptoms and volatile organic compounds in materials and air in the hospital environment. *International Journal of Environmental Health Research* 29 (5): 479–488. https://doi.org/10.1080/09603123.2018.1550194.

Reponen, T., Willeke, K., Grinshpun, S., and Nevalainen, A. (2011). Biological particle sampling. In: *Aerosol Measurement: Principles, Techniques, and Applications*, 3e (ed. P. Kulkarni, P.A. Baron, and K. Willeke), 549–570. New York, NY: Wiley.

Seltzer, J.M. (1994). Biological contaminants. *The Journal of Allergy and Clinical Immunology* 94 (2 Pt 2): 318–326.

Sharma, R., Deval, R., Priyadarshia, V. et al. (2011a). Singh AB indoor fungal concentration in the homes of allergic/asthmatic children in Delhi India. *Allergy and Rhinology* 2: 21–23.

Sharma, R., Gaur, S.N., Singh, V.P., and Singh, A.B. (2011b). Association between indoor fungi in Delhi homes and sensitization in children of respiratory allergy. *Medical Mycology* 50 (3): 281–290.

Srikanth, P., Sudharsanam, S., and Steinberg, R. (2008). Bio-aerosols in indoor environment: composition, health efects and analaysis. *Indian Journal of Medical Microbiology* 26 (4): 302–312.

Stetzenbach, L.D., Buttner, M.P., and Cruz, P. (2004). Detection and enumeration of airborne biocontaminants. *Current Opinion in Biotechnology* 15 (3): 170–174. https://doi.org/10.1016/j.copbio.2004.04.009.

Uddin, S.M.N., Li, Z., Adamowski, J.F. et al. (2016). Feasibility of 'greenhouse system' for household greywater treatment in nomadic-cultured communities in periurban ger areas of Ulaanbaatar, Mongolia: way to reduce greywater-borne hazards and vulnerabilities. *Journal of Cleaner Production* 114: 431–442.

UN-Habitat (2010). *State of the World's Cities 2010/2011: Bridging the Urban Divide. Overview and Key Findings. United Nations Human Settlement Program*. Nairobi: UN-Habitat.

United States Environmental Protection agency (EPA) Chloramine in Drinking Water (2020). Drinking water requirements for states and public watewr systems. https://www.epa.gov/dwreginfo/chloramines-drinking-water (accessed 9 June 2020).

Vieira, A.R., Salzer, J.S., Traxler, R.M. et al. (2017). Enhancing surveillance and diagnostics in anthrax-endemic countries. *Emerging Infectious Diseases* 23, Supplement to December: https://doi.org/10.3201/eid2313.170431.

WHO (2009). *WHO Guidelines for Indoor Air Quality: Dampness and Mould.* Copenhagen: World Health Organization.

Yu, I.T.S., Li, Y., Wong, T.W. et al. (2004). Evidence of airborne transmission of the severe acute respiratory syndrome virus. *The New England Journal of Medicine* 350: 1731–1739.

8

Neurological Health Hazards Associated with Biological Contaminants

Shweta[1#], Himanshi Yadav[1#], Jaldhi[1#], Anurag Thapliyal[1#], Amrita Bakshi[2], Anamika[2], and Shashank Kumar Maurya[1]

[1] *Biochemistry and Molecular Biology Laboratory, Department of Zoology, Faculty of Science, University of Delhi, Delhi, New Delhi, India*
[2] *Department of Zoology, Ramjas College, University of Delhi, Delhi, New Delhi, India*

8.1 Introduction

Any work environment issue that puts employees' health at risk is referred to as an "occupational hazard." It involves both long-term and short-term risks at the workplace. Workers are exposed to potentially fatal hazards at work because there are insufficient laws and regulations in organizations. Because of numerous other competing health concerns, occupational health and protection against work-related diseases are not prioritized in many developing nations. It has been shown that there have been occupational dangers brought on by issues with inadequate surveillance, under-reporting, and cultural and socioeconomic barriers for a long time in many industrialized countries, necessitating attention and awareness of occupational health. To comprehend the common risk factors connected with developing countries and provide remedies, it is necessary to investigate the numerous occupational risks (Figure 8.1).

Different categories of occupational hazards include the following:

1) **Safety Hazards:** Any circumstance, substance, or thing that has the potential to cause harm to a worker falls under this category, and includes working at heights, spills on floors, machinery with moving components, cramped quarters, steep stairs, and exposed electrical wires.

# Equal contribution.

Airborne Biocontaminants and Their Impact on Human Health, First Edition.
Edited by Rajeev Singh and Anamika Singh.

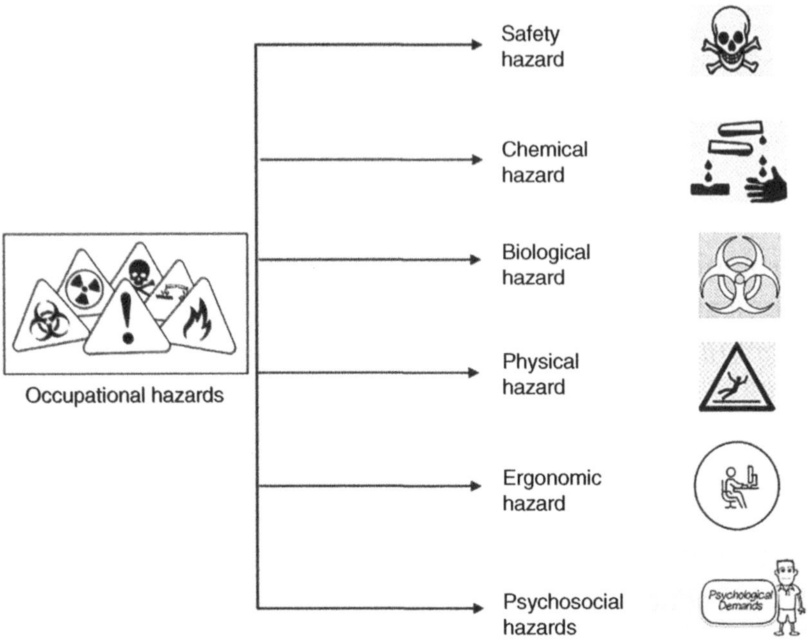

Figure 8.1 Types of occupational hazards.

2) **Chemical Hazards:** Different workplaces may contain a variety of toxic substances and pollutants, such as carbon monoxide, flammable liquids, acids, cleaning solutions, and environmental smoke.
3) **Biological Hazards:** Workers may come into contact with biological health risks such as blood, mold, fungi, viruses, insect bites, and animal droppings in some environments like farms, zoos, hospitals, or medical offices.
4) **Physical Hazards:** Environmental dangers including radiation, prolonged sun exposure, extremely hot or low temperatures, and loud noise can injure the body without touching them.
5) **Ergonomic Hazards:** These hazards include long-term sitting or standing in uncomfortable positions or continuously doing the same activities that can cause damage and sickness.
6) **Psychosocial Hazards:** The mental, emotional, and physical health of employees is endangered by workplace violence, discrimination, lack of respect, sexual harassment, and other situations. Lack of job satisfaction, instability, bad interpersonal relationships, work pressure, and uncertainty are examples of psychosocial risks. Behavioral and psychological changes, such as hostility, aggression, anxiety, sadness, drunkenness, drug addiction, and absenteeism arise due to these hazards.

Trauma, including post-traumatic stress disorder (PTSD), anxiety, depression, loss of dignity, lack of confidence in others, early aging, decreased self-esteem, loss of autonomy, injuries, suicide attempt, absenteeism, and musculoskeletal injuries are all potential outcomes of workplace hazards. Understanding the link between occupational health hazards and neurological problems is therefore critical for providing improved intervention. There are different prevention and treatment strategies available to deal with neurological health threats, including maintaining a regular sleep pattern, increasing physical activity, and practicing mindfulness to reduce work-related stress. Heavy metal toxicity can be treated by chelating these metals in order to lower their toxicity.

8.2 Occupational Hazards and Neurological Disorders

In general, it is observed that exposure to physical, chemical, biological, and psychological risk factors at work may have an adverse effect on workers' health, happiness, and safety by causing short- or long-term changes to complex neuronal circuitry. In view of therapeutic approaches that can restore normal brain function, it is therefore likely to be helpful to examine the relationship between occupation, risk factors, and neurological disorders.

8.3 Occupational Hazards Associated with Risk Factors Linked to Neurological Disorders

Working in an unsafe environment can risk employees' physical and mental health, which can have an impact on their families and close social networks.

1) **Pesticides:** Exposure to pesticides in pesticide/insecticide industry workers is associated with its metabolization leading to enhanced generation of free radical species. It leads to increased lipid peroxidation and decreased antioxidant capacity subsequently turning to diseased conditions (Lucero and Muñoz-Quezada 2021). These radicals act on the mitochondria to trigger the apoptotic cascade and are involved in the process of neurodegeneration. Moreover, alterations in mitochondrial permeability transition and disruption of calcium homeostasis due to oxidative damage lead to apoptosis and cell death. As pyrethroids and other classes of pesticides accumulate in biological membranes due to their lipophilic nature, it leads to oxidative damage (Kori et al. 2018). The compounds produced as a result of oxidative stress such as acrolein, malondialdehyde, and 4-hydroxynonenal (HNE) have been found to disturb cell homeostasis and alter the physiological function (Shoeb et al. 2014).

Due to the greater likelihood of delayed neurological effects from such pesticides, occupational exposures to a moderate amount of organophosphate (OP) insecticides that would not cause any overt toxicity have drawn a lot of attention. In their evaluation of the potential link between OP exposure and overt neurodegeneration, Sanchez-Santed and colleagues found a significant correlation between AD risk and OP exposure, moderate evidence of the link between OPs and PD, while a weak evidence of OPs and ALS (Sánchez-Santed et al. 2016).

S. no.	Pesticide	Mechanism of neurotoxicity
1.	Carbamates	Inhibition of acetylcholinesterase (AChE)
2.	Fungicides	Oxidative stress, decline in level of antioxidant enzymes
3.	Herbicides	Oxidative damage to neural cells
4.	Pyrethroids	Inhibition of voltage-gated sodium channels
5.	Organophosphates	Inhibition of AChE

2) **Heavy Metals:** Metals that weigh greater than $5\,g/cm^3$ are referred to as heavy metals, for example, arsenic (As), cadmium (Cd), lead (Pb), and mercury (Hg). There are around 40 elements, which fall within the group of heavy metals, having a significant impact as toxicity agents and development of neurological problems (Sharma and Agrawal 2005). There are two ways that might hamper metabolic processes. First, heavy metal accumulation in the brain, heart, liver, kidneys, and other organs interferes with their normal operation. Second, the removal of vital minerals from their source causes heavy metals to impair biological functioning (Singh et al. 2011). Studies in the fields of epidemiology and medicine have demonstrated a link between exposure to heavy metals and neurological conditions like AD, autism spectrum disorder (ASD), ALS, Gulf War syndrome, Guillain-Barré disease, Huntington's disease, multiple sclerosis, PD, and Wilson's disease (Chen et al. 2016). For instance, Pb accumulation causes numerous neurological and non-neurological issues like headaches, encephalopathy, memory loss, hallucinations, dullness, irritability, poor attention span, anemia, nausea, muscular tremor, and saturnism, which are caused by occupational exposure to Pb (Mohammadyan et al. 2019). Similarly, Cd can pass through the olfactory bulb and the cerebrospinal fluid (CSF) barrier to enter the brain (Bakulski et al. 2020).

According to a study, occupational exposure to Cd may be linked to the neurological symptoms of chronic fatigue syndrome and myalgic encephalomyelitis (Pacini et al. 2012). Herbicide and pesticide manufacturing, mining, smelting, glass and semiconductor manufacturing industry workers, and other occupations like carpentry that involve the removal of or exposure to structures or materials treated with arsenate wood preservatives, all are occupational risks to As exposure. Meta-analysis revealed that exposure to As impaired children's memory, language, and performance domains, albeit to a lesser extent, and that this exposure was associated with neural development and behavioral disorder. Almost all heavy metals, including Pb, Cd, Hg, and As, cause cellular damage by releasing free radicals. In addition, there is a significant impact of these heavy metals on the release of synaptic neurotransmitters, leading to neurotoxicity, impairment of body balance, tremors, hearing and visual issues, memory loss, low IQ, learning impairments, and other symptoms (Rodríguez-Barranco et al. 2013; Sharma et al. 2014).

3) **Stress and Other Psychosocial Risks:** Workplace stress and psychosocial risks such as lack of job satisfaction, instability, bad interpersonal relationships, workload, and uncertainty can have a significant impact on the autonomic nervous system, which can further compromise cognitive function and alertness. Long and erratic work shifts are among the factors that contribute to stress and are frequently linked to changes in physiologic sleep–wake rhythms. An atypical amount or quality of sleep, commonly brought on by shift work at the office, is linked to several ailments and people who do not get enough sleep are more likely to miss work, get into accidents, and are less productive (Alfonsi et al. 2021). In addition to affecting cognitive function and affective states (emotional mood, intellectual activity, degree of consciousness, and other psychological variables), occupational stress also affects mental health. With common symptoms including irritability, decreased job satisfaction, melancholy, nervousness, and anxiety, the mental impacts of occupational stress can range from moderate subjective symptoms to substantial mental abnormalities and even psychosis (Yi et al. 2022).
4) **Polycarbonate Plastics and Bisphenol A:** BPA, which is produced from plastic products in food items, is known to behave as an exogenous endocrine disruptor. It can traverse the blood–brain barrier (BBB), resulting in numerous cognitive dysfunctions, neurobehavioral disorders, and neurodegenerative diseases (Szymanska et al. 2018). BPA can affect the neuroendocrine system by interfering with the hypothalamic–pituitary–gonadal axis (HPG axis) and hypothalamic–pituitary–adrenal axis (HPA axis or HTPA axis). BPA competes with corticoid receptors at the HPA axis, controlling receptor expression and causing neurobehavioral problems. BPA can directly affect the pituitary at the HPT axis, causing thyroid-stimulating hormone (TSH) secretion to be disrupted and resulting in metabolic and neurodevelopmental abnormalities (Ma et al. 2019). It has been discovered that BPA exposure causes anxiety,

a higher likelihood of autistic traits, decreased memory and learning, and changes in social behavior. In the hippocampus, it decreases synaptic density, lengthens synaptic clefts, shortens active zones, and decreases post-synaptic density. BPA also impairs mitochondrial activity, which increases endoplasmic reticulum stress and raises the levels of cytochrome c. The Bcl-2 gene family and caspase-dependent signaling pathway cause cell apoptosis as a result, which causes neurodegenerative disorders, such as AD and PD (Wang et al. 2019).

5) **Polyvinyl Chloride:** Vinyl chloride toxicity is caused by a variety of mechanisms, including the depletion of oxidative fuel, disturbed ionic balance, disruption of neurotransmission, production of free radicals, release of additional highly toxic metabolites, and antigenic activity. The brainstem is largely unaffected by vinyl chloride toxins, but the cerebral cortex and cerebellum are severely affected. Inhalation exposures to vinyl chloride, rather than oral exposures, have been linked to neurological consequences. Dizziness, weariness, drowsiness, headaches, exhilaration, irritability, anxiousness, sleep disorders, nausea, visual and auditory disruptions, and loss of consciousness are only a few of the neurological effects of inhalation (Verma et al. 2023).

To sum up, these occupational hazards pose a great risk to health, and possible interventions are highly needed to combat the effects as well as to prevent any such diseases (Figure 8.2).

8.4 Neurological Diseases Associated with Occupational Hazards

Literature suggests that occupational hazards can cause a plethora of neurological diseases. These occupational hazards interfere with homeostatic functions and may have long-term impacts on people, such as abnormal behavior, cognitive decline, depression, and mobility disorders. Parkinson's, Alzheimer's, and ALS diseases are neurological illnesses caused by altered membrane and lipid rafts as a result of prolonged exposure to toxicants and workplace dangers (Figure 8.3).

The neurological diseases include the following:

1) **Neurochemical Alterations and Behavioral Abnormalities:** Neurochemical alterations can be caused due to several chemicals that workers get exposed to in industries. For instance, pesticides such as OPs have been found to affect behavioral features in workers and the effects include heightened anxiety, cognitive deficits, and sociosexual and appetitive disorders (Kamel and Hoppin 2004). Other effects include impaired motor coordination, reduced verbal memory, and other neurological and neuropsychiatric

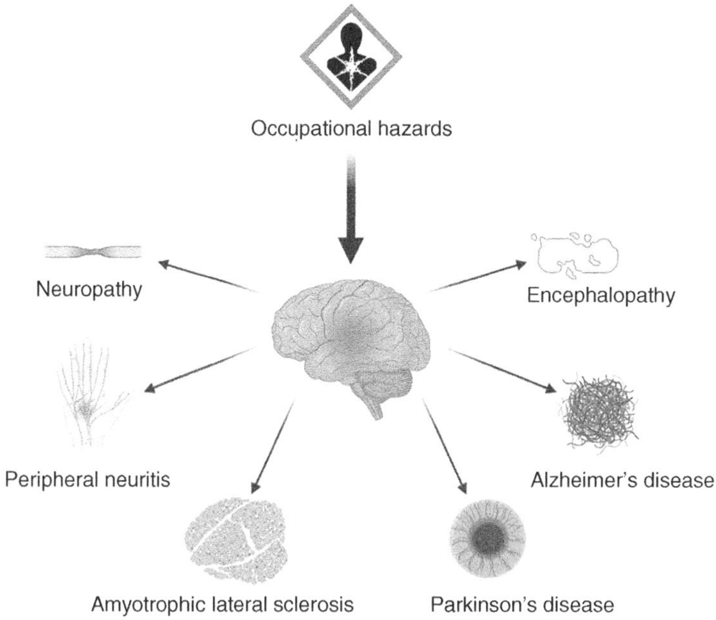

Figure 8.2 Diagrammatic representation showing the influence of occupational health hazards on neurological health.

effects. ASD in the children of farm workers due to take-home exposure from parents and the location of their home relative to production fields has also been reported by Roberts et al. (Roberts et al. 2007). Current experimental evidence from research studies indicates that oxidative stress is one of the most significant mechanisms of pesticide toxicity and that all classes of pesticides cause oxidative stress, RNS, and ROS in many cell types and animal models (Sule et al. 2022). Most pesticides including OPs, carbamate, and organochlorine insecticides as well as fungicides and fumigants are neurotoxic.

2) **Peripheral Neuritis:** Due to manual labor involving repetitive motions involving entrapment or compression, the peripheral nerves can get damaged physically. Early symptoms of peripheral neuritis include complaints of numbness, tingling, or weakness. Furthermore, in many cases, due to the working conditions involving extreme manual labor, peripheral nerve entrapment and segmental demyelination may take place. Moreover, an imbalance in neuronal metabolism can cause axonal degeneration. Peripheral nervous system disorders such as cranial nerve palsy, reflex sympathetic dystrophy, and anosmia have also been reported in manufacturing industries using mixed solvents (E.-A. Kim and Kang 2010).

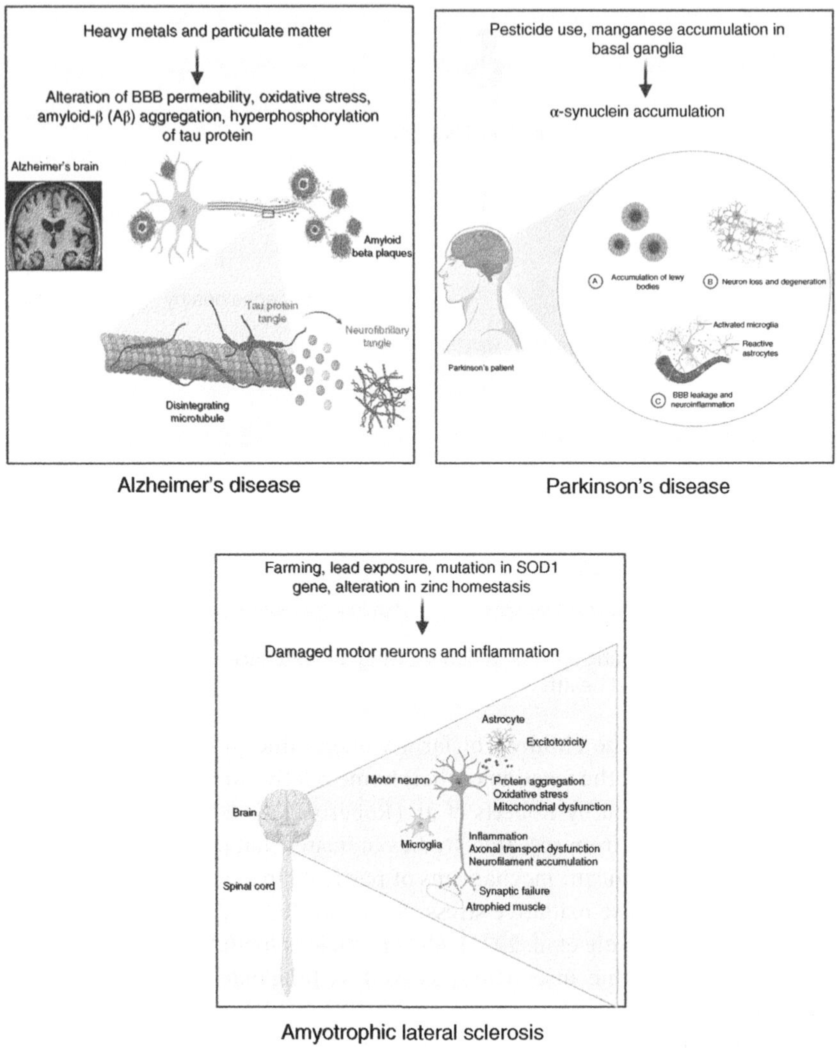

Figure 8.3 Diagrammatic representation showing the molecular changes in brain due to occupational health hazards in development and progression of neurological diseases.

3) **Neuropathy:** For many years, there have been concerns about metal toxicity on peripheral nerves, such as Pb and Hg. Focus on Pb neuropathy by physiologic-pathologic correlations of Fullerton, showed slowing in motor nerve conduction velocities in the sciatic nerves of guinea pigs fed with Pb acetate (Fullerton 1966). The larger fibers were more affected as compared to smaller fibers and both segmental demyelination and axonal degeneration

were observed. A common clinical manifestation is loss of sensation and tingling in the feet and hands and vibration sensations are usually impaired. In certain cases, such as neuropathy due to thallium exposure, the pain increases in severity as weakness and atrophy of muscles happen. The skin of the trunk region is painful to touch and muscles and nerves become tender.

In the case of neuropathy due to solvent toxins such as *n*-hexane, initial symptoms are characterized by paresthesia in the feet and fingers, with loss of sensation and touch. In some cases, the loss of sensation has been observed as high as the thigh and arm muscles. Peripheral nerves get damaged by lacerations or via stretch, repeated trauma, and vibration. The peripheral nerve damage gradually increases, severely affecting the worker's performance. This can result from repeated mechanical damage brought on by forces that are applied from the outside, such as when holding a tool in a specific position or when a body part is compressed or kept in a particular posture for an extended period, subacute and chronic syndromes of nerve dysfunction may also develop. The structure and function of the nerve fibers can change in response to intermittent and repeated impacts on the location where the nerves are present near the surface (Feldman 1987; Menorca et al. 2013).

4) **Encephalopathy:** Occupational exposure to heavy metals like Pb leads to symptoms such as encephalopathy, loss of memory, headache, poor attention span, and hallucination (Flora et al. 2012). Injuries and exposure to chemicals or heavy metals progress to the etiology of toxic encephalopathy brought by the work environment. The CNS is most easily accessed by nonpolar, lipid-soluble compounds (such as organic solvents), where neurons are particularly vulnerable because of their high lipid content and metabolic rates. Toxins with a lipophilic nature can easily harm both gray and white matter. There is a continuum of symptoms associated with toxin-mediated encephalopathy, from subtle clinical issues to overt subclinical deficiencies (Kim and Kim 2012).
5) **Sleep-Related Cognitive Effects:** Chronic partial sleep deprivation is a significant risk factor for several illnesses. Workplace cognitive failure, for instance, might happen when an individual exhibits poor attention span, loses focus on job-related information, or experiences lapses in recall for tasks and processes (Alhola and Polo-Kantola 2007). Reviews and meta-analyses, which have mostly focused on lab studies, show that sleep restriction (i.e. reduced sleep duration) and deprivation (i.e. complete lack of sleep) have negative effects on cognitive performance metrics like alertness, reaction time, reasoning, attention, and working memory (Chattu et al. 2018).

6) **Parkinson's Disease:** Loss of dopaminergic neurons in the *substantia nigra* is a common symptom of PD, a progressive neurodegenerative condition. Tremor, rigidity, sluggish movement, and unstable posture are some of the clinical signs of PD. Age is the biggest risk factor for the development of PD. However, its complex pathophysiology is still not fully understood (Tufekci et al. 2011).

   The presence of Lewy bodies, which mostly consist of the protein α-synuclein, is the pathogenic cause of PD (Sauvé et al. 2020). Many studies have found an association between PD with occupational pesticide exposure. Furthermore, metal buildup like manganese for workers in the welding rod-manufacturing industry results in bodily accumulation of metal known as manganism. It affects the basal ganglia region and induces Parkinson's-like symptoms. Mn-induced Parkinsonism patients show postural instability bradykinesia, and gait difficulties (Finkelstein and Jerrett 2007).
7) **Alzheimer's Disease:** AD is a neurological condition that progresses irreversibly starting with significant memory loss and ending with impairment. Neurofibrillary tangles and plaques are the neuropathological indicators of AD. The AD has been specifically linked to particulate matter (PM) 2.5 exposure (Dantzer et al. 2008). PM exposure may result in the generation of pro-inflammatory cytokines that lead to systemic and chronic respiratory inflammation, which may damage the BBB and result in long-term oxidative stress. Heavy metals such as Cd induce neurotoxicity by altering the permeability of BBB, resulting in amyloid-β (Aβ) aggregation (a protein linked with AD), and the production of tau neurofibrillary tangles (Moulton and Yang 2012). Several studies related to human aging are presenting evidence that suggests Cd exposure impairs cognitive function as observed with AD (Bakulski et al. 2020). Hg is also an etiological factor for AD because it causes hyperphosphorylation of tau protein, which is a component of neurofibrillary tangles. The most prominent features of AD are these structural and functional alterations. Additionally, the neurotransmitter abnormalities linked to AD and Hg-induced AD share many similarities, most notably the loss of acetylcholine, inhibition of serotonin binding to its receptors, and glutamate absorption (Siblerud et al. 2019).
8) **Amyotrophic Lateral Sclerosis:** ALS is a neurodegenerative disease that affects the motor neurons of the spinal cord and brain. Although the exact etiology of ALS remains unclear, environmental exposure has been considered one of the potential causes of ALS. Cases of ALS have been reported in manufacturing companies. Several studies have shown a correlation of ALS with farming as an occupation. The risk of developing ALS and motor neuron disease also increases with occupational exposure to Pb. Cd changes SOD1 through the stimulation of metallothionein (MT) expression, which may affect

the homeostasis of zinc (Zn) (Pfaender and Grabrucker 2014). Due to a potential interaction between Zn and elevated amounts of MT, the SOD1 enzyme's activity may decrease when Zn availability declines. On the other hand, an excess of Cd and a lack of Zn may impair SOD1 by causing it to misfold and partially aggregate (Vinceti et al. 2012).

S. no.	Occupation	Health effects	Reference
1.	Gardeners (3214) (male)	Parkinson's disease	(Kenborg et al. 2012)
2.	Farmers, horticulturists, and gardeners (2, 273, 872) (male and female)	Parkinson's disease	(Tüchsen and Jensen 2000)
3.	Pesticide applicators (216)	Organophosphate-induced delayed polyneuropathy	(Lotti and Moretto 2005)
4.	Pesticide applicators (30)	Neurobehavioral and neurocognitive effects	(Ismail et al. 2018)
5.	Children of agricultural workers in banana plantations	Neurobehavioral and neurocognitive effects	(van Wendel de Joode et al. 2016)
6.	Night shift nurses (2059)	Insomnia, chronic fatigue, anxiety, depression	(Øyane et al. 2013)
7.	Viscose rayon industry workers	Carbon disulfide neurotoxicity	(Kim et al. 1989)
8.	Workers in paint factories	Neurotoxicity	(Tsai et al. 1997)

Therefore, a thorough understanding of how occupational hazards can affect brain development and functions normally will advance research into reducing this risk as well as aid in the development of treatment.

8.5 Prevention and Treatment Strategies

There are several prevention and therapeutic strategies available to deal with neurological disorders caused by occupational hazards. These strategies are discussed in this section:

1) **Occupational Stress and Sleep Deprivation**
   i) A variety of psychological, behavioral, physical problems, and diseases are documented health risks associated with occupational stress (Quick and Henderson 2016). Workplace stress can be significantly reduced by using proper time and priority management. Stress-reduction techniques include

mindfulness (a state in which you actively examine current feelings and ideas without judging them), deep breathing exercises, and meditation. Taking a few minutes each day to concentrate on something easy, like breathing, walking, or a meal can help reduce stress. Burnout can be avoided by scheduling even a brief period of personal time throughout a busy day.

ii) Healthy eating also helps in reducing stress, and a healthy immune system and cell repair are supported by a balanced diet. It gives the extra energy needed to deal with stressful situations. According to studies, vegetables and diets high in polyunsaturated fats, particularly omega-3 fats, may assist in controlling cortisol levels. By promoting deep breathing, intelligent food selection, mealtime attention, and slow, thorough chewing, mindful eating techniques reduce stress. Physical training programs have become increasingly available to employees for management of stress. Walking and dancing are examples of aerobic exercises that speed up breathing and heart rate to increase the amount of oxygen that gets to the body's cells help in reducing levels of stress hormones, and this eases stress on the heart's muscles as well as other muscles. To prevent potential stress, it is crucial to establish clear boundaries between job and personal life.

iii) Stress can disrupt sleep and cause sporadic awakenings throughout the night. This may prevent a person from falling asleep more deeply, which is necessary for the body to repair itself, build new tissue, and maintain a strong immune system. Memory and emotional management are particularly benefited by rapid eye movement (REM) sleep. Slowing down around 30 minutes before bed can help in achieving 7–9 hours of sleep each night. Sleep deprivation produces memory loss and hippocampus ferroptosis; both can be reduced by melatonin through its binding capability to the MT2 receptor and activating ERK/Nrf2 signaling (Wang et al. 2021).

There is no particular drug that can relieve stress. However, some drugs can help control or lessen some of the stress-related symptoms and indicators. A wide range of drugs can be used to help in relieving and preventing stress. For instance, beta-blockers, antidepressants, and sedatives (also known as tranquillizers, hypnotics, and/or anxiolytics) have all been used to assist people in managing stress. Any medicine that is effective for reducing stress is also addictive. When consuming such substances, serious behavioral and health issues could arise and are even likely to do so if caution is not used. Sedative drugs depress (slow down or inhibit) CNS activity, resulting in a sense of calm, reduced anxiety and tension, sleepiness, and slowed breathing. Benzodiazepines are a class of sedatives that include alprazolam (Xanax), clonazepam (Klonopin), diazepam (Valium), lorazepam (Ativan), triazolam (Halcion), temazepam (Restoril), and chlordiazepoxide (Librium). BuSpar is an anti-anxiety drug. BuSpar is used to relieve the physical signs of anxiety, such as tense muscles, lightheadedness, and a racing heart. Usually, this medicine is recommended to treat anxiety temporarily. The most popular antidepressant drugs for this anti-anxiety and anti-stress purpose are Paxil, Prozac, and Lexapro. Commonly used herbs and nutritional supplements for reducing stress and anxiety include skullcap (Scutellaria lateriflora), chamomile (Anthemis nobilis), hops (Humulus Lupulus), motherwort (Leonurus cardiaca), and oats (oat straw) (Avena sativa).

2) **Neurotoxicity**
   i) Neurotoxicity induced by heavy metals can be prevented by quitting smoking, using filtered water, consuming more iron-rich food, maintaining adequate work hygiene, and avoiding future exposure. Heavy metal-induced neurotoxicity can be treated by chelation. Pb and Cd poisoning can be treated by chelation therapy via using CaNa2EDTA, British Anti-Lewisite (BAL), dimercaptopropane sulfonate (DMPS), dimercaptosuccinic acid (DMSA), and dimercaprol (DMCA). In this kind of treatment, a pill taken orally binds to the Pb and causes its excretion in urine (Lead Poisoning n.d.). The in vivo investigation revealed that Vitamin C has a protective effect against metal-induced toxicity. Selenium (Se) can dampen the toxicity induced by Hg and As and promote the development of selenoproteins, which include glutathione peroxidases and thioredoxin reductases to increase antioxidant activity. These proteins assist in lowering ROS generation.
   ii) Pesticides are employed in many industries; their widespread use can increase the risk of pesticide sickness. Atropine, pralidoxime (2-PAM), and benzodiazepines (e.g. diazepam) are the cornerstones of medical therapy in OP poisoning. In stomach toxicity of chlorinated hydrocarbon pesticides, phenobarbital (or) pentobarbital or azepam (5–10 mg) intravenous doses are given. Vitamins K1 and K2 are used to protect against Warfarin and Zn phosphide poisoning. Carbon disulfide and methyl bromide poisoning are treated with caffeine sodium benzoate. Cyanide poisoning is treated using methyl nitrite ampules. Calcium gluconate is used to alleviate toxicity caused by chlorinated insecticides, carbon tetrachloride, ethylene dichloride, and mercurial compounds.
   iii) Organic solvents are also agents of neurotoxicity. Selective serotonin reuptake inhibitors (SSRIs) and tricyclic antidepressants are two types of antidepressant drugs that are effective in treating patients with peripheral neuropathy symptoms as well as mood changes that result from organic solvent-induced neurotoxicity.

8.6 Conclusion and Future Directions

Modern society is increasingly exposed to many risk factors that could affect human neural regulation systems, necessitating a new sophisticated multidisciplinary approach to the research topic. Studies have shown how susceptible the brain is to industrial toxins, stress, and other hazards encountered in various occupations. There is an ardent need to accelerate research on occupational health with a well-rounded approach focusing on alleged neurotoxicants. Preventive approaches are required to deal with neurodevelopmental toxicity.

This chapter adds substantially to our understanding of the multitude of effects occupational hazards have and points toward the need for awareness programs tailored according to the audience. Industry administrators must take the proper procedures to manage work-related stress and take corrective action to lessen these risks. Moreover, intense focus must be placed on educating the general public regarding risks and leveraging local channels that guide decision-making. The likelihood of chemical exposure or interactions with physical variables at work should not be disregarded when making a differential diagnosis in individuals with neurological diseases. To evaluate whether neurological diseases are due to occupational neurotoxic exposure, physicians must be aware of the common signs and symptoms of toxic encephalopathy and work closely with occupational physicians.

Acknowledgments

SKM acknowledges the ICMR, New Delhi (EMRD/SG/11/2023-1452) and Institution of Eminence, University of Delhi (IOE/2021/12/FRP) for financial support. Figures were created using Biorender.com.

Conflict of Interest

The authors report that there are no competing interests to declare.

References

Alfonsi, V., Scarpelli, S., Gorgoni, M. et al. (2021). Sleep-related problems in night shift nurses: towards an individualized interventional practice. *Frontiers in Human Neuroscience 15*: 644570. https://www.frontiersin.org/articles/10.3389/fnhum.2021.644570.

Bakulski, K.M., Seo, Y.A., Hickman, R.C. et al. (2020). Heavy metals exposure and Alzheimer's disease and related dementias. *Journal of Alzheimer's Disease: JAD 76* (4): 1215–1242. https://doi.org/10.3233/JAD-200282.

Chattu, V.K., Manzar, M.D., Kumary, S. et al. (2018). The global problem of insufficient sleep and its serious public health implications. *Healthcare 7* (1): 1. https://doi.org/10.3390/healthcare7010001.

Chen, P., Miah, M.R., and Aschner, M. (2016). Metals and neurodegeneration. *F1000Research 5*: F1000 Faculty Rev-366. https://doi.org/10.12688/f1000research.7431.1.

Dantzer, R., O'Connor, J.C., Freund, G.G. et al. (2008). From inflammation to sickness and depression: when the immune system subjugates the brain. *Nature Reviews. Neuroscience 9* (1): 46–56. https://doi.org/10.1038/nrn2297.

Feldman, R.G. (1987). Occupational neurology. *The Yale Journal of Biology and Medicine 60* (2): 179–186.

Finkelstein, M.M. and Jerrett, M. (2007). A study of the relationships between Parkinson's disease and markers of traffic-derived and environmental manganese air pollution in two Canadian cities. *Environmental Research 104* (3): 420–432. https://doi.org/10.1016/j.envres.2007.03.002.

Flora, G., Gupta, D., and Tiwari, A. (2012). Toxicity of lead: a review with recent updates. *Interdisciplinary Toxicology 5* (2): 47–58. https://doi.org/10.2478/v10102-012-0009-2.

Fullerton, P.M. (1966). Chronic peripheral neuropathy produced by lead poisoning in guinea-pigs. *Journal of Neuropathology and Experimental Neurology 25* (2): 214–236. https://doi.org/10.1097/00005072-196604000-00003.

Ismail, A.A., Almalki, M., Agag, A. et al. (2018). Pesticide application and khat chewing as predictors of the neurological health outcomes among pesticide applicators in a vector control unit, Saudi Arabia. *The International Journal of Occupational and Environmental Medicine 9* (1): 32–44. https://doi.org/10.15171/ijoem.2018.1160.

Kamel, F. and Hoppin, J.A. (2004). Association of pesticide exposure with neurologic dysfunction and disease. *Environmental Health Perspectives 112* (9): 950–958. https://doi.org/10.1289/ehp.7135.

Kenborg, L., Lassen, C.F., Lander, F., and Olsen, J.H. (2012). Parkinson's disease among gardeners exposed to pesticides – a Danish cohort study. *Scandinavian Journal of Work, Environment & Health 38* (1): 65–69. https://doi.org/10.5271/sjweh.3176.

Kim, E.-A. and Kang, S.-K. (2010). Occupational neurological disorders in Korea. *Journal of Korean Medical Science 25* (Suppl): S26–S35. https://doi.org/10.3346/jkms.2010.25.S.S26.

Kim, Y. and Kim, J.W. (2012). Toxic encephalopathy. *Safety and Health at Work 3* (4): 243–256. https://doi.org/10.5491/SHAW.2012.3.4.243.

Kim, S.D., Song, D.B., and Yum, Y.T. (1989). Clinical analysis of 8 cases of chronic carbon disulfide poisoning in workers engaged in the viscose rayon industry.

Korean Journal of Occupational and Environmental Medicine 1 (2): 186–196. https://doi.org/10.35371/kjoem.1989.1.2.186.

Kori, R.K., Thakur, R.S., Kumar, R., and Yadav, R.S. (2018). Assessment of adverse health effects among chronic pesticide-exposed farm workers in Sagar district of Madhya Pradesh, India. *International Journal of Nutrition, Pharmacology, Neurological Diseases 8* (4): 153. https://doi.org/10.4103/ijnpnd.ijnpnd_48_18.

Lead Poisoning: Spot Symptoms and Limit Exposure-Lead Poisoning – Diagnosis & Treatment (n.d.). Mayo clinic. https://www.mayoclinic.org/diseases-conditions/lead-poisoning/diagnosis-treatment/drc-20354723 (accessed 13 July 2023).

Lotti, M. and Moretto, A. (2005). Organophosphate-induced delayed polyneuropathy. *Toxicological Reviews 24* (1): 37–49. https://doi.org/10.2165/00139709-200524010-00003.

Lucero, B. and Muñoz-Quezada, M.T. (2021). Neurobehavioral, neuromotor, and neurocognitive effects in agricultural workers and their children exposed to pyrethroid pesticides: a review. *Frontiers in Human Neuroscience 15*: 648171. https://www.frontiersin.org/articles/10.3389/fnhum.2021.648171.

Ma, Y., Liu, H., Wu, J. et al. (2019). The adverse health effects of bisphenol A and related toxicity mechanisms. *Environmental Research 176*: 108575.

Menorca, R.M.G., Fussell, T.S., and Elfar, J.C. (2013). Peripheral nerve trauma: mechanisms of injury and recovery. *Hand Clinics 29* (3): 317–330. https://doi.org/10.1016/j.hcl.2013.04.002.

Mohammadyan, M., Moosazadeh, M., Borji, A. et al. (2019). Investigation of occupational exposure to lead and its relation with blood lead levels in electrical solderers. *Environmental Monitoring and Assessment 191* (3): 126. https://doi.org/10.1007/s10661-019-7258-x.

Moulton, P.V. and Yang, W. (2012). Air pollution, oxidative stress, and Alzheimer's disease. *Journal of Environmental and Public Health 2012*: 472751. https://doi.org/10.1155/2012/472751.

Øyane, N.M.F., Pallesen, S., Moen, B.E. et al. (2013). Associations between night work and anxiety, depression, insomnia, sleepiness and fatigue in a sample of Norwegian nurses. *PLoS One 8* (8): e70228. https://doi.org/10.1371/journal.pone.0070228.

Pacini, S., Fiore, M.G., Magherini, S. et al. (2012). Could cadmium be responsible for some of the neurological signs and symptoms of Myalgic Encephalomyelitis/Chronic Fatigue Syndrome. *Medical Hypotheses 79* (3): 403–407. https://doi.org/10.1016/j.mehy.2012.06.007.

Pfaender, S. and Grabrucker, A.M. (2014). Characterization of biometal profiles in neurological disorders. *Metallomics: Integrated Biometal Science 6* (5): 960–977. https://doi.org/10.1039/c4mt00008k.

Quick, J.C. and Henderson, D.F. (2016). Occupational stress: preventing suffering, enhancing wellbeing †. *International Journal of Environmental Research and Public Health 13* (5): 459. https://doi.org/10.3390/ijerph13050459.

Roberts, E.M., English, P.B., Grether, J.K. et al. (2007). Maternal residence near agricultural pesticide applications and autism spectrum disorders among children in the California central valley. *Environmental Health Perspectives 115* (10): 1482–1489. https://doi.org/10.1289/ehp.10168.

Rodríguez-Barranco, M., Lacasaña, M., Aguilar-Garduño, C. et al. (2013). Association of arsenic, cadmium and manganese exposure with neurodevelopment and behavioural disorders in children: a systematic review and meta-analysis. *The Science of the Total Environment 454–455*: 562–577. https://doi.org/10.1016/j.scitotenv.2013.03.047.

Sánchez-Santed, F., Colomina, M.T., and Herrero Hernández, E. (2016). Organophosphate pesticide exposure and neurodegeneration. *Cortex; A Journal Devoted to the Study of the Nervous System and Behavior 74*: 417–426. https://doi.org/10.1016/j.cortex.2015.10.003.

Sauvé, J.-F., Locke, S.J., Josse, P.R. et al. (2020). Characterization of inhalable endotoxin, glucan, and dust exposures in Iowa farmers. *International Journal of Hygiene and Environmental Health 228*: 113525. https://doi.org/10.1016/j.ijheh.2020.113525.

Sharma, R.K. and Agrawal, M. (2005). Biological effects of heavy metals: an overview. *Journal of Environmental Biology 26* (2 Suppl): 301–313.

Sharma, B., Singh, S., and Siddiqi, N.J. (2014). Biomedical implications of heavy metals induced imbalances in redox systems. *BioMed Research International 2014*: 640754. https://doi.org/10.1155/2014/640754.

Shoeb, M., Ansari, N.H., Srivastava, S.K., and Ramana, K.V. (2014). 4-hydroxynonenal in the pathogenesis and progression of human diseases. *Current Medicinal Chemistry 21* (2): 230–237.

Siblerud, R., Mutter, J., Moore, E. et al. (2019). A hypothesis and evidence that mercury may be an etiological factor in Alzheimer's disease. *International Journal of Environmental Research and Public Health 16* (24): 5152. https://doi.org/10.3390/ijerph16245152.

Singh, R., Gautam, N., Mishra, A., and Gupta, R. (2011). Heavy metals and living systems: an overview. *Indian Journal Of Pharmacology 43* (3): 246–253. https://doi.org/10.4103/0253-7613.81505.

Sule, R.O., Condon, L., and Gomes, A.V. (2022). A common feature of pesticides: oxidative stress – the role of oxidative stress in pesticide-induced toxicity. *Oxidative Medicine and Cellular Longevity 2022*: 5563759. https://doi.org/10.1155/2022/5563759.

Szymanska, K., Makowska, K., and Gonkowski, S. (2018). The influence of high and low doses of bisphenol A (BPA) on the enteric nervous system of the porcine ileum. *International Journal of Molecular Sciences 19* (3): 917. https://doi.org/10.3390/ijms19030917.

Tsai, S.Y., Chen, J.D., Chao, W.Y., and Wang, J.D. (1997). Neurobehavioral effects of occupational exposure to low-level organic solvents among Taiwanese workers in paint factories. *Environmental Research 73* (1–2): 146–155. https://doi.org/10.1006/enrs.1997.3704.

Tüchsen, F. and Jensen, A.A. (2000). Agricultural work and the risk of Parkinson's disease in Denmark, 1981-1993. *Scandinavian Journal of Work, Environment & Health 26* (4): 359–362. https://doi.org/10.5271/sjweh.554.

Tufekci, K.U., Genc, S., and Genc, K. (2011). The endotoxin-induced neuroinflammation model of Parkinson's disease. *Parkinson's Disease 2011*: 487450. https://doi.org/10.4061/2011/487450.

Verma, R., Chakraborty, R., and Giri, P. (2023). Acute toxic encephalopathy in occupational exposure with polyvinyl chloride (PVC) fumes: a case series. *Neurology India 71* (3): 531. https://doi.org/10.4103/0028-3886.378675.

Vinceti, M., Bottecchi, I., Fan, A. et al. (2012). Are environmental exposures to selenium, heavy metals, and pesticides risk factors for amyotrophic lateral sclerosis? *Reviews on Environmental Health 27* (1): 19–41. https://doi.org/10.1515/reveh-2012-0002.

Wang, H., Zhao, P., Huang, Q. et al. (2019). Bisphenol-A induces neurodegeneration through disturbance of intracellular calcium homeostasis in human embryonic stem cells-derived cortical neurons. *Chemosphere 229*: 618–630. https://doi.org/10.1016/j.chemosphere.2019.04.099.

Wang, X., Wang, Z., Cao, J. et al. (2021). Melatonin alleviates acute sleep deprivation-induced memory loss in mice by suppressing hippocampal ferroptosis. *Frontiers in Pharmacology 12*: 708645. https://doi.org/10.3389/fphar.2021.708645.

van Wendel de Joode, B., Mora, A.M., Lindh, C.H. et al. (2016). Pesticide exposure and neurodevelopment in children aged 6–9 years from Talamanca, Costa Rica. *Cortex 85*: 137–150. https://doi.org/10.1016/j.cortex.2016.09.003.

Yi, X., Yang, J., Gao, X., and Li, F. (2022). The relationship between occupational stress, mental health and work ability of coal chemical workers in Xinjiang. *Frontiers in Psychiatry 13*: 903534. https://doi.org/10.3389/fpsyt.2022.903534.

9

Effect of Bioaerosols Exposure on Acute Exacerbation Among Individuals with COPD in Urban India

Chirashree Ghosh[1], Arun Kumar Yadav[1], Rajnish Awasthi[2], Arun Kumar Sharma[2], Shukla Das[2], B. D. Banerjee[2], and Palak Balyan[1]

[1] *Environmental Pollution Laboratory, Department of Environmental Studies, University of Delhi, Delhi, New Delhi, India*
[2] *Faculty in University College of Medical Sciences, University of Delhi, Delhi, New Delhi, India*

9.1 Introduction

Chronic obstructive pulmonary disease (COPD) is the third leading cause of death globally, killing 3.2 million people annually (Levine and Marciniuk 2022). According to WHO data for 2019, there was a considerable burden of COPD, with a total of 212.3 million reported cases (117.1 million affecting men and 95.2 million affecting women). Tragically, this led to approximately 3.3 million fatalities and accounted for the loss of 74.4 million disability-adjusted life years (DALYs). COPD ranked among the top five causes of DALYs lost, highlighting its significant impact on public health (WHO 2023). COPD is a progressive respiratory condition characterized by persistent airflow limitation and chronic inflammation of the airways. It is associated with significant healthcare burden, impaired quality of life, and increased healthcare costs. Understanding the risk factors and triggers for COPD exacerbations is crucial for effective management and prevention strategies. The common risk factors for COPD are smoking, biomass fuel, history of pulmonary tuberculosis, long-standing asthma, and particulate pollutants in the air (Hardie et al. 2002; Lindgren et al. 2009; Ni et al. 2015). In recent years, there has been growing recognition of the role of environmental factors in the development and progression of COPD. Environmental exposures such as air pollution, occupational hazards, and indoor air quality have been implicated in the

Airborne Biocontaminants and Their Impact on Human Health, First Edition.
Edited by Rajeev Singh and Anamika Singh.

pathogenesis of COPD and its exacerbations. Among these, bioaerosols, which are microscopic particles suspended in the air containing living organisms like bacteria and fungi, have gained attention as potential triggers for COPD exacerbations (Erkan et al. 2008; ElKorashy and El-Sherif 2014; Jubinville et al. 2018). Pathological bacteria and viruses have the capability of precipitating infections in the lungs. However, there is very limited literature on the role of bioaerosols as factors precipitating acute exacerbations among COPD patients. As the suspended particulate matter in air are a complex mixture of chemicals and microbes, it is difficult to delineate the role of microbes but superadded infections in COPD leading to acute exacerbations point toward strong association (Schikowski et al. 2005; Ni et al. 2015). However, to generate more evidence, prospective cohort studies need to be conducted. The present study aimed to investigate the association between aerosolized bacteria, aerosolized fungi, and COPD exacerbations in a cohort of patients diagnosed with COPD. The study also examined the influence of meteorological variables, including temperature and relative humidity, as well as particulate matter ($PM_{2.5}$) on COPD exacerbations. Understanding the interplay between these environmental factors and COPD exacerbations can provide valuable insights into preventive strategies and targeted interventions.

9.2 Methodology

A prospective cohort study was conducted for a period of two years extending to all four seasons in India (November 2015–October 2017). Fifty patients diagnosed with COPD by physicians of a tertiary care center in Delhi, India, were recruited according to Global Initiative for Chronic Obstructive Lung Disease (GOLD) criteria (Durheim et al. 2015). COPD was differentiated from asthma using bronchodilator-based reversibility testing of airway obstruction.

Inclusion Criteria: Individuals were eligible for inclusion if they were nonsmokers and had forced expiratory volume in 1 second (FEV_1) < 0.8 FEV_1 predicted and FEV_1 > 1.21 (check, this seems confusing, use any one criteria).

Exclusion Criteria: We excluded any individual with chest diseases other than COPD, COPD with thrombocytopenia with platelets less than 10 000/μl, psychological impairment, cancer patients and patients receiving immunosuppressive treatment, hepatic cirrhosis, chronic renal failure, autoimmune or connective tissue disorders, taking anti-TNF-a drugs, pregnant women, and heart failure.

Sample Size: Since this was a prospective cohort study, to be conducted over two years with once-a-month follow-up, 48 observations per patient, a sample size of 30 patients would have yielded 1440 observations, adequate for determining the association of most of the risk factors included in the study. Allowing for a 40% attrition rate in prospective studies, fifty cases of COPD were recruited from the medicine OPD ward of a tertiary care hospital in Delhi, India, for the study.

Method of Data Collection: Patients were recruited from the OPD and indoor beds of the medicine department of a tertiary care institution in Delhi and the nearest residential colony spread over 10 sq km and had a catchment population of 6000. After recruitment, the patient's residence was visited to assess the indoor environment with the help of a checklist. After obtaining consent, baseline information on variables like age, education and employment status, smoking status, house ventilation, location of kitchen, occupant density, etc., of all study participants was recorded using a structured questionnaire. The location of residence of patients was geocoded on a digital map of the study area using a GPS device. The entire area was divided into five grids on a digital map and a centroid of each grid was identified. A household closest to the centroid having a COPD patient was chosen in each grid for indoor and outdoor bioaerosol estimation. Starting from November 2014, the patients were grouped into five based on spatial proximity. Each patient was contacted through a personal visit once a fortnight to record episodes of acute exacerbation of COPD, if any, and simultaneously indoor and outdoor bioaerosol levels were measured fortnightly for each grid. Data collection was carried out in this manner up to October 2016. Data regarding weather parameters (temperature and relative humidity) and count of aerosolized microbes (bacteria and fungi) were collected for each cluster. Describe the method of bioaerosol sampling. Data of air pollution data parameters, namely $PM_{2.5}$, were obtained from state authority-owned monitoring station located in the local district of Delhi.

Statistical Analysis: Data of all patients was maintained on a spreadsheet using MS Excel. Geocoded data was added to the spreadsheet by transferring geocoded information from GPS device. The distribution of COPD exacerbation episodes across different levels of aerosolized microbes, meteorological variables, and particulate matter was visually represented using R studies. The entire data set was converted into time series data. The spatiotemporal analysis was conducted to find out an association between aerosolized bacteria, fungi, $PM_{2.5}$, temperature, and relative humidity using Intercooled Stata V9.0© for Windows.

9.3 Results

Characteristics of the 50 study participants are presented in Table 9.1. The participants had an average age of 60.28 ± 07.37 years, with almost equal distribution with respect to gender. Most of them were educated up to the primary level or more. Around one-third experienced passive smoking in routine and two-thirds had inadequate ventilation and overcrowding at home.

Table 9.1 Characteristics of study participants (n = 50).

Characteristics		**Mean ± SD or *N* (%)**
Age (years)		60.28 ± 07.37
Sex	Male	27 (54.0)
	Female	23 (46.0)
Educational attainment	Illiterate	05 (10.0)
	Primary	13 (26.0)
	Secondary	18 (36.0)
	Higher secondary	05 (10.0)
	College	09 (18.0)
Employment status	Unemployed	3 (6)
	Working	29 (58)
	Retired	18 (36)
Passive smoking		17 (34)
Inadequate ventilation at home		34 (68)
Overcrowding		30 (60)

9.3.1 Effect of Aerosolized Microbes on COPD Exacerbation

Aerosolized Bacteria: Spatial distribution of exacerbation episodes and the number of heat spots (Figure 9.1a–d) increased with the rise in aerosolized bacterial count. Large contiguous zones of heat spots were observed for a higher concentration of aerosolized bacteria. Three large contiguous and clustered heat spots with relatively larger geographical spread were noticed when aerosolized bacterial counts were >*190 CFU/plate*. The incidence density of COPD exacerbation episodes rose sharply with increase in aerosolized bacterial count as shown in Table 9.2. Incidence density was *six times higher* for aerosolized bacterial counts >*190 CFU/plate* (28.57 episodes per 100 person-fortnight) than for aerosolized bacterial count <*90 CFU/plate*. (4.90 episodes per 100 person-fortnight).

Aerosolized Fungi: Occurrence and incidence density of COPD exacerbation episodes in the different concentration ranges of aerosolized fungal count shown in Figure 9.2a–d and in Table 9.2. Highest number of heat spots were observed for aerosolized fungal count >131 CFU/plate. Spatial distribution had not altered significantly in different fungal count ranges. The spatial extent, number, and clustering of heat island, for COPD exacerbation episodes, did not vary with change in the aerosolized fungal count.

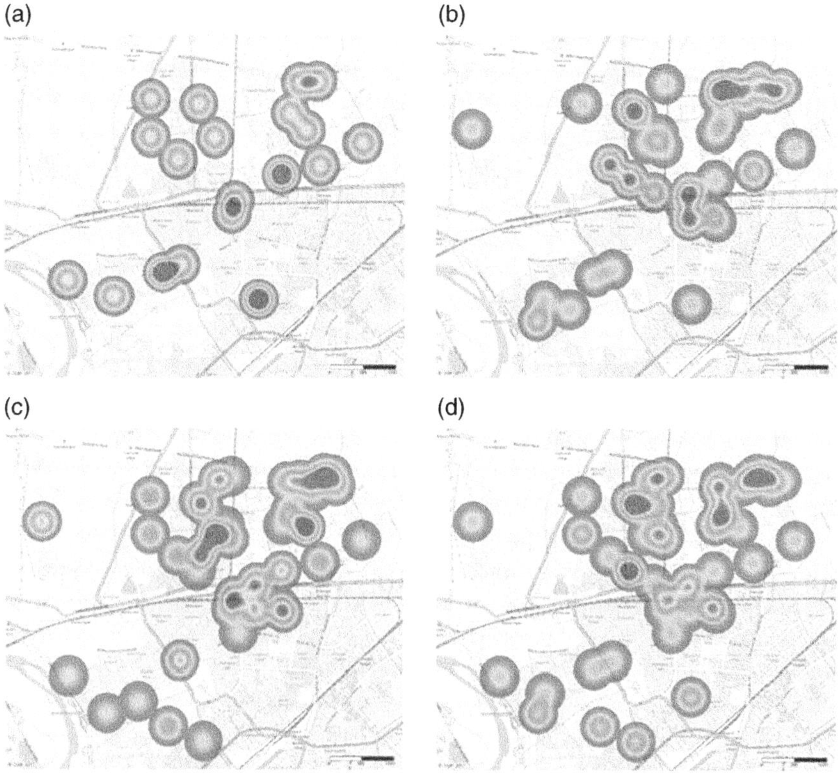

Figure 9.1 (a–d) distribution of COPD exacerbation episodes in different aerosolized bacteria concentration. (a) aerosolized bacteria<90 CFU/plate, (b) aerosolized bacteria = 90–140 CFU/plate, (c) aerosolized bacteria = 141–190 CFU/plate, (d) aerosolized bacteria>190 CFU/plate.

Table 9.2 COPD exacerbation episodes among study participants in different aerosolized bacteria and fungi concentration.

Aerosolized bacteria (CFU/plate)	Incidence density per 100 person-fortnight	Aerosolized fungi (CFU/plate)	Incidence density per 100 person-fortnight
<90	04.90	<57	05.17
90–140	05.73	57–94	13.27
141–190	15.38	94–131	09.48
>190	28.57	>131	14.25
Total	10.48	Total	10.48

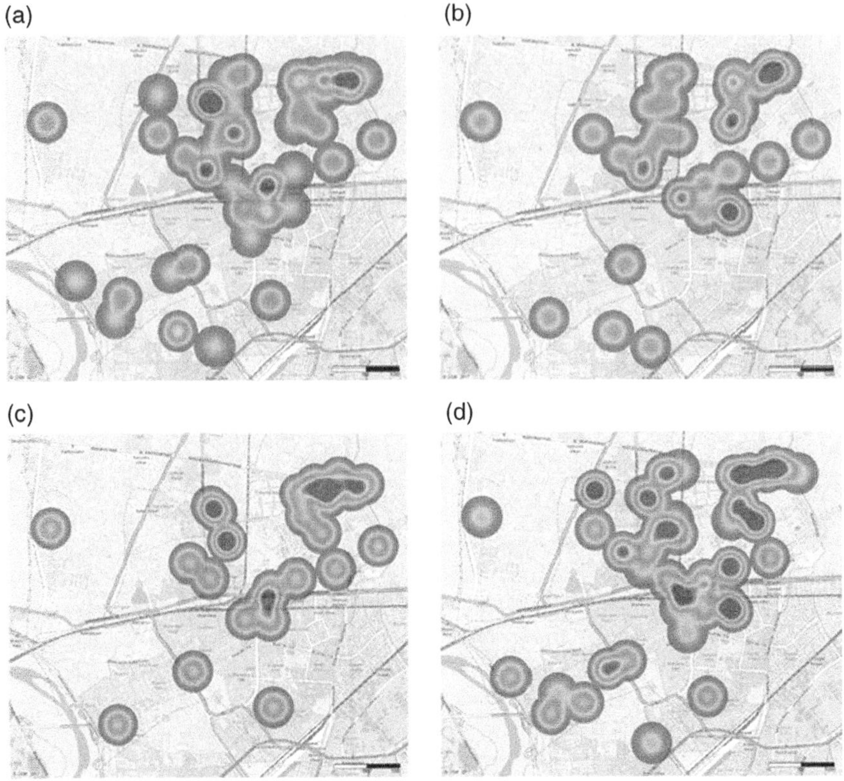

Figure 9.2 (a–d) distribution of COPD exacerbation episodes in different aerosolized fungal concentration. (a) aerosolized fungi<57 CFU/plate, (b) aerosolized fungi = 57–94 CFU/plate, (c) aerosolized fungi = 94–131 CFU/plate, (d) aerosolized fungi>131 CFU/plate.

9.4 Meteorological Variables

9.4.1 Temperature

Figure 9.3a–f and Table 9.3 show the occurrence of COPD exacerbation episodes in the cohort study in different temperature ranges. The number of heat spots increased with a fall in temperature. Only one clustered heat spot was observed for temperature range above 40 °C. The number of clustered heat spots increased with a fall in temperature (four each for a temperature range of 15–20 °C and 20–25 °C).

COPD exacerbation episodes also occurred in new areas in low-temperature ranges where no episodes were observed in higher temperatures. The incidence

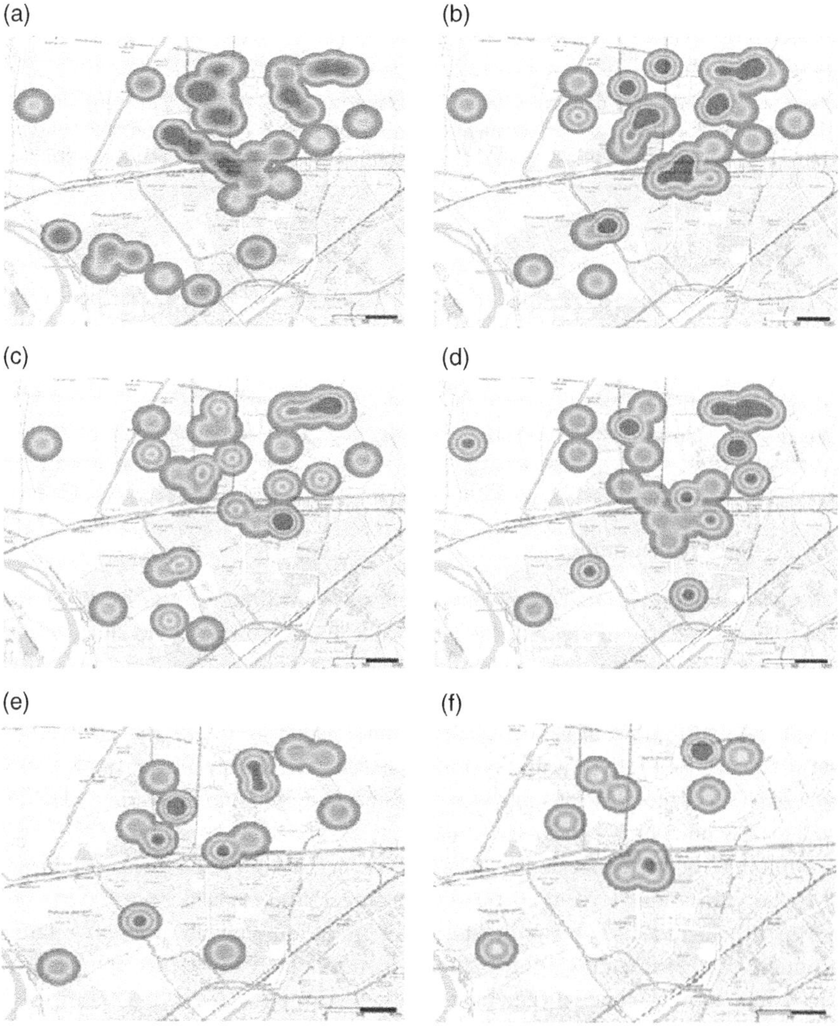

Figure 9.3 (a–f) distribution of COPD exacerbation episodes in different temperature ranges. (a) temperature 15–20 °C, (b) temperature 20–25 °C, (c) temperature 25–30 °C, (d) temperature 30–35 °C, (e) temperature 35–40 °C, (f) temperature >40 °C.

density of exacerbation episodes per fortnight was calculated and compared. Incidence density increased with a fall in temperature. Incidence density was about six times higher in the temperature range of 15–20 °C compared to incidence density in the temperature range of 35–40 °C and >40 °C.

Table 9.3 COPD exacerbation episodes among study participants in different temperatures, relative humidity, and $PM_{2.5}$ ranges.

Temperature range (°C)	Incidence density per 100 person-fortnight	Relative Humidity range (%)	Incidence density per 100 person-fortnight	$PM_{2.5}$ range	Incidence density per 100 person-fortnight
15–20	24.62	<50	07.94	<107	06.46
20–25	16.66	50–60	11.79	107–141	07.05
25–30	13.35	60–70	09.93	141–175	10.85
30–35	08.39	70–80	11.58	>175	19.72
35–40	03.85	>80	10.07	—	—
>40	03.56	—	—	—	—
Total	10.48	Total	10.48	Total	10.48

9.4.2 Relative Humidity

The occurrence of COPD exacerbation episodes in different relative humidity ranges are presented in Figure 9.4a–e and Table 9.3. The number and clustering of heat islands did not vary with a change in relative humidity. The highest numbers of heat spots were observed in the relative humidity range of 50–60%. The geographical distribution of COPD exacerbation episodes did not vary significantly with respect to relative humidity ranges except at relative humidity of <50%, where COPD exacerbation spots were few and scattered and clustering was limited to only one place.

Although the incidence density of COPD exacerbation episodes was maximum for relative humidity level of 50–60% (11.79 exacerbation episodes per 100 person-fortnight), incidence density did not differ much among the different relative humidity ranges except RH<50%. The incidence density for RH<50% was about three-fourths of the incidence density in other RH ranges.

9.4.3 Particulate Matter

The occurrence of COPD exacerbation episodes at different $PM_{2.5}$ concentrations is presented in Figure 9.5a–d and Table 9.3. An increase in the spatial distribution of COPD exacerbation episodes was observed with the rise in the $PM_{2.5}$ concentration. The number of clustered heat spots increased with the rise in the $PM_{2.5}$ concentration. Three isolated and one clustered heat spots were observed for $PM_{2.5} < 107\,\mu g/m^3$. Two isolated and four clustered heat spots were observed in the $PM_{2.5}$ concentration range of $141–175\,\mu g/m^3$, whereas only one isolated and

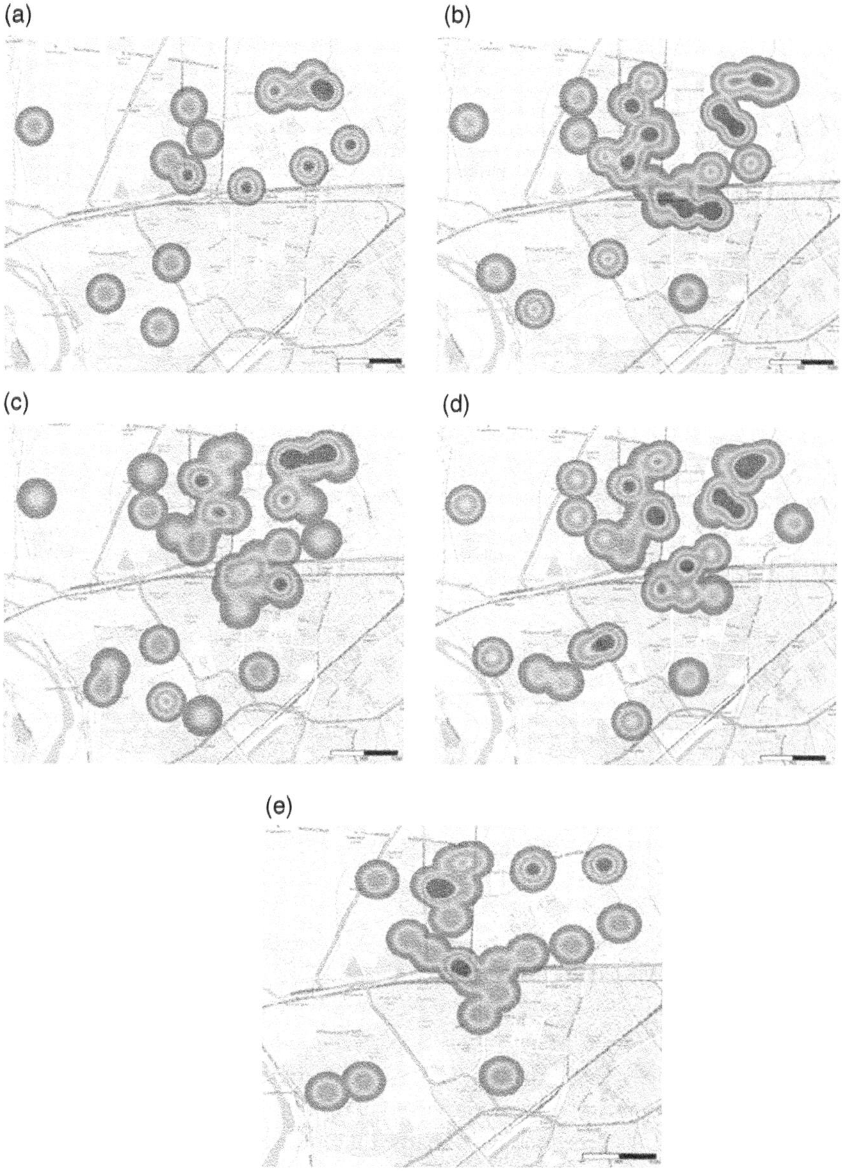

Figure 9.4 (a–e) distribution of COPD exacerbation episodes in different relative humidity ranges. (a) relative humidity<50, (b) relative humidity = 50–60, (c) relative humidity = 60–70, (d) relative humidity = 70–80.

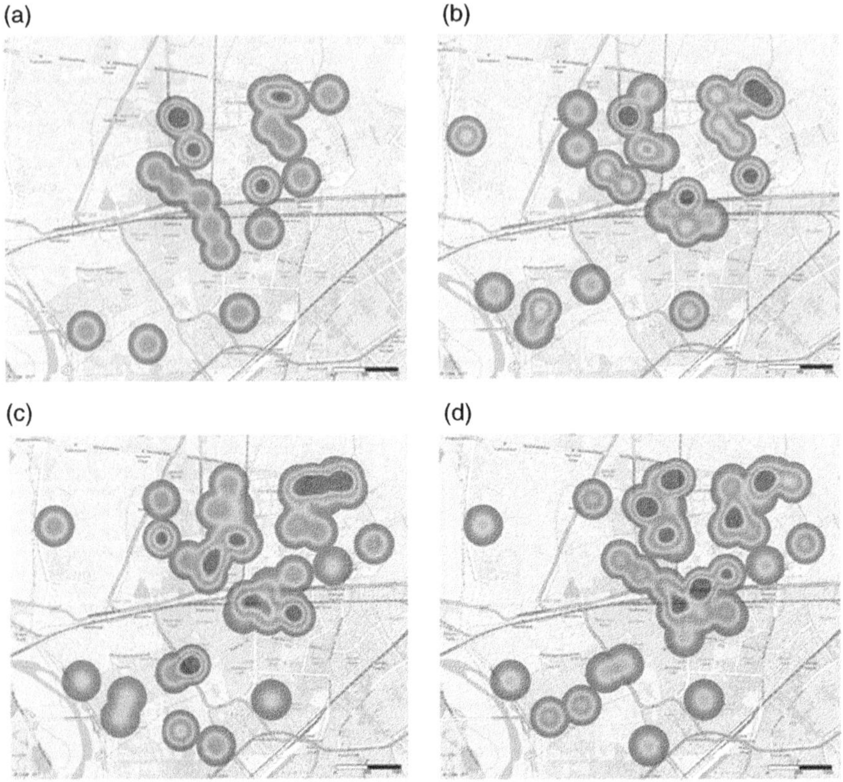

Figure 9.5 Distribution of COPD exacerbation episodes in different particulate matter concentrations. (a) $PM_{2.5} < 107$, (b) $PM_{2.5} = 107-141$, (c) $PM_{2.5} = 141-175$, (d) $PM_{2.5} > 175$.

six clustered heat spots were observed for $PM_{2.5}$ concentration >175 μg/m^3. The incidence density of COPD exacerbation episodes increased with the rise in the $PM_{2.5}$ concentration. Incidence density was approximately three times for $PM_{2.5}$ concentration >175 μg/m^3 (19.72 COPD exacerbation episodes per 100 person-fortnight) in comparison with the $PM_{2.5}$ concentration <107 μg/m^3 (6.46 COPD exacerbation episodes per 100 person-fortnight) (Table 9.3).

Regression analysis was done to study the quantitative effect of meteorological parameters, $PM_{2.5}$, aerosolized bacteria, and fungi on COPD exacerbations (Table 9.4). The standardized coefficient of temperature for COPD exacerbation episodes was negative, whereas standardized coefficients for $PM_{2.5}$ and aerosolized bacterial count were positive, i.e. the number of exacerbation episodes increases with a fall in temperature and rise in the aerosolized bacterial count. Relative humidity and aerosolized fungi had no significant effect on COPD

Table 9.4 Association between meteorological variables, $PM_{2.5}$, aerosolized bacteria fungi, and COPD exacerbations.

Parameters	Standardized coefficients	p-value	95% confidence interval Lower bound	Upper bound
Temperature	−0.49	0.01	−0.29	−0.08
Relative humidity	0.06	0.78	−0.08	0.11
$PM_{2.5}$	0.43	0.06	−0.00	0.07
Aerosolized bacteria	0.81	<0.01	0.03	0.09
Aerosolized fungi	0.07	0.64	−0.02	0.04

Significance level $p < 0.05$.

exacerbation. Temperature ($p = 0.01$) and aerosolized bacteria ($p < 0.01$) had a significant effect on COPD exacerbations. The effect of $PM_{2.5}$ on COPD exacerbation was nearly significant ($p = 0.06$) (Table 9.4).

9.5 Discussion

In this study, 50 COPD patients were followed for a period of two years. It was observed that increasing exposure to bacterial bioaerosol, higher levels of $PM_{2.5}$, and a fall in temperature were associated with a higher incidence of acute exacerbation episodes, while no conclusive association was observed with relative humidity and fungal bioaerosol level. Bacterial infection is reported as one of the major factors for exacerbation of COPD (Sethi 2000; Sethi and Murphy 2001; Sykes et al. 2007; Bathoorn et al. 2008; Martinez et al. 2013; Santos et al. 2016). Bafadhel et al. (2011) using unbiased analysis have identified bacteria as one of the four distinct biologic exacerbation clusters (bacteria-predominant, eosinophil-predominant, virus-predominant, pauci-inflammatory) (Bafadhel et al. 2011). Bacterial colonization of lower airways amplifies the chronic inflammatory response in COPD, leading to progressive airway obstruction and increased frequency of exacerbation (Sethi 2014). Thus, the previous studies mainly concentrated on the effect of bacterial presence in the respiratory microbiome and the severity of COPD. The effect of bacterial count in the surrounding air on COPD patients was not quantified earlier. Hence, the present study provides conclusive evidence that exposure to high bacterial load in the air increases the number of exacerbation episodes of COPD. However, this study limits itself to the concentration of microbes in the air. It paves the way for further research correlating the bacteria present in the surrounding air and those isolated from bronchoalveolar lavage (BAL) from COPD patients, which will provide more conclusive and comprehensive evidence, delineating the role of bacteria in the air to cause a COPD exacerbation episode. Our model estimates showing the increase in the incidence of exacerbation with a fall in temperature agree with those of Ferrari et al. (Ferrari et al. 2012) in Germany and Tseng et al. (Tseng et al. 2013) in Taiwan. The results of this work are also consistent with those of Almagro et al. (2015) who reported a fall in temperature to cause an increase in hospital visits due to COPD exacerbation. Although the association between $PM_{2.5}$ and the incidence of acute exacerbation was nearly significant ($p = 0.06$), this association could be important in this vulnerable population. This is important because $PM_{2.5}$ concentrations in Delhi, India, are very high throughout the year, being 3–10 times higher than WHO-recommended levels. Various studies conducted across the globe (Schikowski et al. 2005; Cortez-Lugo et al. 2008; Alahmari et al. 2015; Almagro et al. 2015; Bayram 2017) reported a positive association between COPD exacerbation episodes and a rise in particulate matter concentration. Observations from the present study can be correlated with the study of Cortez-Lugo et al. (2008) in which it was reported that a nonlinear relationship existed between COPD exacerbation and particulate matter concentration. The effect of particulate matter is very steep when the $PM_{2.5}$ concentration rises from low to moderate, but the effect is plateaued when the concentration rises from moderate to high. The present study was conducted in Delhi, where $PM_{2.5}$ levels were very high throughout the year, which can explain a nearly significant but nonlinear effect of $PM_{2.5}$ on COPD exacerbation. The present study did not observe any conclusive association between fungal bioaerosol exposure and incidence of acute COPD exacerbation. The findings in our study are similar to observations made by other researchers. Bafadhel et al. (2011), though observed a number of positive filamentous fungal cultures in COPD patients, but could not find any clinical significance. Fungi colonize and cause debilitating illness mainly to that subset of patients who have other comorbidities and decreased immunity (Blanco and Garcia 2008; Lionakis 2014). The virulence and pathogenicity of fungi are much inferior to that of bacteria. Hence, in COPD patients, fungal infections would occur only if immunity is very depressed or patient is using steroid-based inhalers as part of treatment. But by design, such patients were excluded in this study. So, conclusive evidence cannot be drawn about the role of fungi in exacerbation of COPD but it can be safely concluded that in otherwise uncompromised patients, fungus does not play a significant role. Such subset includes COPD patients with hematological malignancy, corticosteroid use, or diabetes mellitus. For example, the positive predictive value of a respiratory culture is eight times higher in neutropenic patients compared to immunocompetent COPD patients for the diagnosis of aspergillosis

(Perfect et al. 2001). The observed effect of relative humidity on the incidence of acute COPD exacerbation in the present study is consistent with the findings of Ferrari et al. (Ferrari et al. 2012) and Alahmari et al. (2015) who reported no effect of relative humidity on COPD exacerbations. The study by Almargo et al. (2015) reported an increase in COPD exacerbation episodes with a rise in relative humidity. However, when adjusted with other variables like temperature, air pollutants and RH failed to show any meaningful effect on exacerbation of COPD. A protective effect of relative humidity was observed by Tseng Che et al. (2013). They reported more number of COPD exacerbation episodes in low RH and a fall in COPD exacerbation with the rise in RH. It can be concluded that relative humidity per se may not have any role to play directly in acute exacerbation. Clinicians report that dry air increases the irritability of the respiratory tract causing exacerbation of symptoms. Hence, high RH may provide relief to the patients, but on the contrary, high RH is favorable for fungus and bacterial growth. Thus, it is difficult to establish the role of RH in such a complex multifactorial etiology of acute exacerbation of COPD. The robust study design ("cohort") in itself was the most important strength of the present study. Each patient was followed up for two years, thereby generating data for individual patients, which made the study more informative. Most other researchers have used "admissions to the emergency room" as an outcome measure of exacerbation of COPD. However, we must understand that most of the cases reported in the emergency room with acute exacerbation of COPD are unarguably either poorly controlled or uncontrolled. The findings of these subgroups have further been extrapolated to the entire cohort, thereby affecting the validity of such studies. Owing to the close monitoring and follow-up of each patient, this study was able to include those incidences of COPD exacerbations in the outcome, which did not require an emergency visit to the hospital. In addition, another benefit of this study was that the present study assesses the impact of exposures on the incidence of COPD exacerbation over all seasons. Lastly, we were able to follow up with the patients with respect to their exposure to various hitherto understudied potential confounders such as indoor tobacco smoke exposure, overcrowding, poor ventilation, absence of a separate cooking space, etc., allowing us to more robustly assess the impact of bioaerosol, $PM_{2.5}$ exposure and meteorological parameters. The present study has a number of limitations. This study used representative bioaerosol sampling (one sample for each cluster), meteorological parameters, and $PM_{2.5}$ exposure. The exposure observed was representative only and not the absolute level of exposure of the patients to the microbes (Jalaludin et al. 2004). Second, the levels of $PM_{2.5}$ in this study were quite high, limiting the power to detect statistically significant associations. Finally, effect of sociodemographic characteristics could not be evaluated in the present study because of study design as these characteristics do not vary with time.

9.6 Conclusion

Among this cohort of COPD cases in Delhi, we observed an increase in exacerbation episodes with the rise in aerosolized bacteria count and fall in temperature. The results of this study provide valuable insights into the association between aerosolized bacteria, aerosolized fungi, meteorological variables, $PM_{2.5}$, and COPD exacerbations. The findings suggest that increasing exposure to bacterial bioaerosols, higher levels of $PM_{2.5}$, and a fall in temperature are associated with a higher incidence of acute exacerbation episodes in COPD patients. However, no conclusive association was observed with relative humidity and fungal bioaerosol levels. These results align with previous research highlighting the significance of bacterial infection in exacerbating COPD symptoms. Bacterial colonization of the lower airways amplifies the chronic inflammatory response, leading to increased airway obstruction and exacerbation frequency. However, the impact of bacterial count in the surrounding air on COPD patients has received limited attention. This study contributes to filling this knowledge gap by exploring the role of aerosolized bacteria as a potential trigger for exacerbations. Moreover, the study investigates the influence of meteorological variables and $PM_{2.5}$ on COPD exacerbations. The temperature was found to have a significant negative effect on exacerbations, suggesting that colder temperatures may increase the risk of exacerbations. $PM_{2.5}$ showed a nearly significant positive association with exacerbations, indicating that higher levels of particulate matter in the air may contribute to exacerbation episodes in COPD patients. Further, long-duration research with a large number of COPD patients should be designed and conducted to understand the long-term effects of bioaerosol exposure on patients with respiratory morbidities to generate further evidence for advocating measures to reduce bioaerosol exposure.

Conflict of Interest

The authors declare that the research was conducted in the absence of any commercial or financial relationships that could be construed as a potential conflict of interest.

Disclosure of Interest

The authors of this research article do not have conflicts of interest.

Acknowledgments

This study was supported by the Ministry of Environment, Forest and Climate Change (MoEF&CC), India. The project team thanks University College of Medical Sciences (UCMS), Delhi for providing patients.

References

Alahmari, A.D., Mackay, A.J., Patel, A.R.C. et al. (2015). Influence of weather and atmospheric pollution on physical activity in patients with COPD. *Respiratory Research* 16 (1): 1–9.

Almagro, P., Hernandez, C., Martinez-Cambor, P. et al. (2015). Seasonality, ambient temperatures and hospitalizations for acute exacerbation of COPD: a population-based study in a metropolitan area. *International Journal of Chronic Obstructive Pulmonary Disease* 10: 899–908.

Bafadhel, M., McKenna, S., Terry, S. et al. (2011). Acute exacerbations of chronic obstructive pulmonary disease: identification of biologic clusters and their biomarkers. *American Journal of Respiratory and Critical Care Medicine* 184 (6): 662–671.

Bathoorn, E., Kerstjens, H., Postma, D. et al. (2008). Airways inflammation and treatment during acute exacerbations of COPD. *International Journal of Chronic Obstructive Pulmonary Disease* 3 (2): 217–229.

Bayram, H. (2017). Impact of air pollution on COPD; underlying mechanisms. *Tanaffos* 16 (Suppl 1): S10.

Blanco, J.L. and Garcia, M.E. (2008). Immune response to fungal infections. *Veterinary Immunology and Immunopathology* 125 (1–2): 47–70.

Cortez-Lugo, M., Moreno-Macias, H., Holguin-Molina, F. et al. (2008). Relationship between indoor, outdoor, and personal fine particle concentrations for individuals with COPD and predictors of indoor-outdoor ratio in Mexico city. *Journal of Exposure Science & Environmental Epidemiology* 18 (1): 109–115.

Durheim, M.T., Smith, P.J., Babyak, M.A. et al. (2015). Six-minute-walk distance and accelerometry predict outcomes in chronic obstructive pulmonary disease independent of global initiative for chronic obstructive lung disease 2011 group. *Annals of the American Thoracic Society* 12 (3): 349–356.

ElKorashy, R.I.M. and El-Sherif, R.H. (2014). Gram negative organisms as a cause of acute exacerbation of COPD. *Egyptian Journal of Chest Diseases and Tuberculosis* 63 (2): 345–349. https://doi.org/10.1016/j.ejcdt.2013.12.013.

Erkan, L., Uzun, O., Findik, S. et al. (2008). Role of bacteria in acute exacerbations of chronic obstructive pulmonary disease. *International Journal of COPD* 3 (3): 463–467.

Ferrari, U., Exner, T., Wanka, E.R. et al. (2012). Influence of air pressure, humidity, solar radiation, temperature, and wind speed on ambulatory visits due to chronic obstructive pulmonary disease in Bavaria, Germany. *International Journal of Biometeorology* 56: 137–143.

Hardie, J.A., Buist, A.S., Vollmer, W.M. et al. (2002). Risk of over-diagnosis of COPD in asymptomatic elderly never-smokers. *European Respiratory Journal* 20 (5): 1117–1122.

Jalaludin, B.B., O'Toole, B.I., and Leeder, S.R. (2004). Acute effects of urban ambient air pollution on respiratory symptoms, asthma medication use, and doctor visits for asthma in a cohort of Australian children. *Environmental Research* 95 (1): 32–42.

Jubinville, E., Veillette, M., Milot, J. et al. (2018). Exacerbation induces a microbiota shift in sputa of COPD patients. *PLoS One* 13 (3): 1–16.

Levine, S.M. and Marciniuk, D.D. (2022). Global impact of respiratory disease: what can we do, together, to make a difference? *Chest* 161 (5): 1153–1154. http://journal.chestnet.org/article/S0012369222000496/fulltext.

Lindgren, A., Stroh, E., Montnémery, P. et al. (2009). Traffic-related air pollution associated with prevalence of asthma and COPD/chronic bronchitis. a cross-sectional study in Southern Sweden. *International Journal of Health Geographics* 8 (1): 1–15.

Lionakis, M.S. (2014). New insights into innate immune control of systemic candidiasis. *Medical Mycology* 52 (6): 555–564.

Martinez, F.J., Erb-Downward, J.R., and Huffnagle, G.B. (2013). Significance of the microbiome in chronic obstructive pulmonary disease. *Annals of the American Thoracic Society* 10 (Supplement): S170–S179.

Ni, L., Chuang, C.C., and Zuo, L. (2015). Fine particulate matter in acute exacerbation of COPD. *Frontiers in Physiology* 6: 1–10.

Perfect, J., Cox, G.M., Lee, J.Y. et al. (2001). The impact of culture isolation of Aspergillus species: a hospital-based survey of aspergillosis. *Clinical Infectious Diseases* 33 (11): 1824–1833.

Santos, S., Marin, A., Serra-Batlles, J. et al. (2016). Treatment of patients with COPD and recurrent exacerbations: the role of infection and inflammation. *International Journal of Chronic Obstructive Pulmonary Disease* 11: 515–525.

Schikowski, T., Sugiri, D., Ranft, U. et al. (2005). Long-term air pollution exposure and living close to busy roads are associated with COPD in women. *Respiratory Research* 6 (1): 1–10.

Sethi, S. (2000). Infectious etiology of acute exacerbations of chronic bronchitis. *Chest* 117 (5): 380S–385S.

Sethi, S. (2014). Chronic obstructive pulmonary disease and infection. Disruption of the microbiome? *Annals of the American Thoracic Society* 11 (Supplement 1): S43–S47.

Sethi, S. and Murphy, T.F. (2001). Bacterial infection in chronic obstructive pulmonary disease in 2000: a state-of-the-art review. *Clinical Microbiology Reviews* 14 (2): 336–363.

Sykes, A., Mallia, P., and Johnston, S.L. (2007). Diagnosis of pathogens in exacerbations of chronic obstructive pulmonary disease. *Proceedings of the American Thoracic Society* 4 (8): 642–646.

Tseng, C.M., Chen, Y.T., Ou, S.M. et al. (2013). The effect of cold temperature on increased exacerbation of chronic obstructive pulmonary disease: a nationwide study. *PLoS One* 8 (3): e57066.

WHO (2023). Chronic obstructive pulmonary disease (COPD). https://www.who.int/news-room/fact-sheets/detail/chronic-obstructive-pulmonary-disease-(copd) (accessed 25 May 2023).

10

The Crucial Role of Pollen Grains, (1 → 3)-β-D-Glucan and Endotoxin in Triggering Respiratory Diseases

Khencha Aran[1], *Twinkle Chetia*[1], *Priti Giri*[2], *Bitupan Deka*[1], *and Ravi Kumar Goswami*[3]

[1] *Department of Botany, Cotton University, Guwahati, Assam, India*
[2] *Department of Botany, Maitreyi College, University of Delhi, Delhi, New Delhi, India*
[3] *Department of Zoology, Hindu College, University of Delhi, Delhi, New Delhi, India*

10.1 Respiratory Diseases

Respiratory diseases are a major health concern worldwide. It is caused by various factors (Shi et al. 2007). It is differentiated into acute respiratory disease (ARD) and chronic respiratory disease (CRD) (Miller 1973; WHO 2002). ARD refers to the unexpected onset of respiratory symptoms, namely congestion, cough, breathing difficulty, and sore throat. ARDs are caused by different infectious agents, which include fungi, bacteria, and viruses (Stetzenbach 2009; Kao et al. 2017). The most common ARDs are bronchitis, pneumonia, common cold, and influenza (Nobata et al. 2006; Sauro et al. 2006). In rare cases, ARDs can be critical and indeed life-threatening precisely, the vulnerable populations are at higher risk such as people with underlying medical conditions, elderly people, and infants (Smith et al. 2000; Bosma et al. 2010). On that account, seeking medical attention is important if any experience in ARDs, especially when the condition is observed to worsen or persist over time. Precautionary measures such as preventing close contact with the sick individual, practice of good hygiene, and vaccination can help lessen the risk of spreading ARDs and the risk of contracting (Troko et al. 2011; Warren-Gash et al. 2013; Benkouiten et al. 2014; Sim et al. 2014). CRD impacts the respiratory system and includes various forms of diseases like pneumonia, tuberculosis (TB), asthma, and COPD. Lung cancer is also included in this group. Asthma and COPD are among major prevalent respiratory conditions.

Airborne Biocontaminants and Their Impact on Human Health, First Edition.
Edited by Rajeev Singh and Anamika Singh.

When it comes to the treatment and prevention of respiratory diseases, vaccination is one of the significant flexible options. Vaccines boost immunity to respiratory disease (Gulati et al. 2021). Chitosan-coated vaccines are being utilized to treat respiratory diseases because they stick to mucus, increase cellular uptake, deliver vaccines to the right cells, prevent vaccines from breaking down, and are known for immune system stimulation (Kumar et al. 2016; Muralidharan et al. 2019).

10.2 Pollen Grains and Their Role in Triggering Respiratory Diseases

10.2.1 Pollen Grains

Pollen grain, a tiny microscopic structure, which plays a vital role in reproduction of angiospermic plants, is produced by the male reproductive organs in flowers containing the male gametes, which are required for the occurrence of fertilization (Bedinger 1992). Different plant pollens vary greatly in size, shape, and texture (Chen and Ju 2022). It often has a unique adaptation to their specific pollination states (Mahabalé 1968; Whitehead 1983; Hesse 2000). Dispersion of pollen grains from anther takes place by various modes by water, wind, and pollinators like insects, birds, and animals (Tauber 1967; Mahabalé 1968). Fertilization in plants occurs when pollen sticks to the stigma of a flower from the same species (Nepi and Pacini 1993; Abrol and Abrol 2012). Germination takes place and growth of a pollen tube to the ovary allows the male gamete to fertilize the female gamete to form a seed, the most important process for reproduction in plants (Cheung et al. 2010; Dresselhaus and Franklin-Tong 2013; Johnson et al. 2019). This is crucial for the maintenance of biodiversity and support of ecosystem. In addition to its function in reproduction, it has a notable impact on human health, specifically for those people who are allergic to pollen. When allergy-causing pollens are inhaled, it provokes respiratory symptoms and various allergic reactions, which make the pollen an environmental allergen (D'amato et al. 2005; Peden and Reed 2010). This reason makes the importance of assimilating the behavior and characteristics of pollen grains for the well-being of *Homo sapiens*.

Pollen grains play a fundamental role in triggering respiratory diseases, distinctively in people who are allergic to pollen (Traidl-Hoffmann et al. 2003; De Weger et al. 2013). When pollens are inhaled by a person with allergies, it causes a range of symptoms like watery eyes, coughing, itchiness, sneezing, congestions, and runny nose due to over-reacting of the person's immune system. Rhinitis and hay fever are common allergies caused by pollen (Bousquet et al. 2008). It affects people during definite time of the year when pollens are released by the plants. Weeds release pollens in autumn, grasses release pollens in summer, and trees release pollens in spring (Lewis et al. 1991; Port et al. 2006; Lo et al. 2019). Pollen also has the ability to provoke asthma attacks in certain people who are suffering from allergic asthma. This can happen due to inflammation of their airways when they are exposed to pollens. This can lead to different symptoms like wheezing, coughing, and difficulty in breathing (D'Amato et al. 2007, 2014, 2016; Harun et al. 2019; Cevhertas et al. 2020). Many factors like types of pollens, amount of pollens present in the air, specific sensitivity of an individual to the pollen, level of pollution, and many other environmental factors are responsible for the severity of the range of symptoms that can affect the respiratory system.

10.2.2 Pollen and Respiratory Diseases

The atmospheric surrounding is composed of a mass of particulate materials. It is known that pollens or their components could trigger allergic symptoms of respiratory diseases like hay fever, rhinitis, and asthma (Beggs 1998; Gaspar et al. 2020). Asthma and rhinitis are inflammatory disorders affecting airway mucosa, differing in clinical manifestation of disease and the area of inflammatory reaction (1, 2 of the reference). Many cases have been reported worldwide with health issues related to pollen (Salvaggio 1994; Salvaggio et al. 1971; D'Amato et al. 2005, 2014, 2016; Peden and Reed 2010; Cevhertas et al. 2020).

Allergy caused by pollen has a clinical effect worldwide; the effects cause inflammatory problems of the airways mucosa, which is steered by IgE-mediated response (Polosa et al. 2000; Skiepko et al. 2011). There has been a massive increase in the pollen-induced pervasiveness and extremity of allergic respiratory reactions (Burney et al. 1997; Van Cleemput et al. 2019). Allergic disorder triggered by pollen is a vital health issue, which affects a large percentage of the population (Amato et al. 2007). People allergic to pollens exhibit adverse reactions in contact with pollen allergens. Therapeutically, the most pertinent pollen-causing allergic responses are those of grasses (Brown et al. 1983; Reid et al. 1986; Boulet et al. 1993). Allergy caused by pollens of different plant species has been reported by many, few are of *Alnus incana* (Smith et al. 2007), *Ambrosia artemisiifoli* (Peternel et al. 2008; Grijincu et al. 2023), *Artemisia vulgaris* (Pastorello et al. 2002; Grewling et al. 2020), *Betula alba* (Carnes et al. 2009), *Betula verrucosa* (Ipsen and Løwenstein 1983), *Corylus avellana* (Lüttkopf et al. 2002; Lauer et al. 2004), *Cupressus sempervirens* (Ford et al. 1991; Shahali et al. 2017), *Olea europaea* (Lauzurica et al. 1988; Blanca et al. 1983), *Parietaria Judaica* (Colombo et al. 2003), *Platanus vulgaris* (Burbach et al. 2009; Fernández-González et al. 2010), *Chenopodium album* (Barderas et al. 2002; Severcan et al. 2021) and *Salsola kali* (Farnandez et al. 2015; Elvira-Rendueles et al. 2017). Pollen of *Plantago lanceolata* contains certain components that cause allergy, which may lead to clinical problems (Mehta and Wheeler 1991).

Munro et al. (1976) reported rye grass pollen as a major cause of significant deterioration in asthma-affected children during pollen season. Thunderstorm asthma (TA) is caused by instances of elevated pollen concentration that remain in the atmosphere for a prolonged period of time after the storm has hit the area (D'Amato et al. 2016; Hughes et al. 2020; Nickovic et al. 2023). Significant correlation is found among symptoms of rhinitis and asthma with pollen counts of plane-tree (Subiza et al. 1994).

10.2.3 How Pollen Triggers Respiratory Diseases

Environmental factors can affect our respiratory system in a number of ways (Gordon 2003). Pollen grains are one of those factors, which trigger the sensitization of allergic reaction (Schäppi et al. 1999). Presence of pores in pollen can accelerate dissemination of associated proteins through it to the upper airways and nose. Associate proteins like protease released by pollen when hydrated appear to play a significant role in inflammatory and immunological reactions in patients having hypersensitivity to pollen (Gaspar et al. 2020). Mostly pollens of weeds, trees, and grasses are well known to trigger allergic symptoms. Several proteases from pollens have been described in studies that have the potency in the involvement of pathogenicity of allergies causing respiratory complications. Specific allergen proteins in pollen of every plant report having variability in high degree among sensitivity in humans by distinct pollen type and so seasonal asthma and hay fever are basically linked with allergy to vast diversity of pollens consequently (Platts-Mills 1998). Mostly the pollen grains are difficult to enter over the upper respiratory tract due to their larger size than 10 μm to trigger symptoms of asthma (Salvaggio 1994). However, it was observed that when pollen grains are inhaled by a person, pollen grains are moistened/hydrated. They releasea range of proteins and allergens, for example, PGE2- and LTB-4-like substances, NADPH oxidases and proteases. This may act as a mediator for inflammation (Hassim et al. 1998; Widmer et al. 2000; Boldogh et al. 2005; Gunawan et al. 2008). Studies by Cortes et al. (2006), Matheson et al. (1995), Bagarozzi et al. (1996), and Matheson and Travis (1998) describe numerous proteases from pollen, which contributes to the pathogenesis of allergies, which causes respiratory problems. This factor creates a potential for the interaction of several cell receptors. In order to trigger the sensitization of allergic reactions through IgE production and to be processed by dendritic cells, the airborne allergens should have the ability to cross the epithelial barrier. It was suggested by Wan et al. (1999) that the degradation of endogenous protein may be due to the existence of exogenous proteolytic activity, which helps allergens reach the subepithelial immune cells. Along with allergens, pollen also carries enzymes. When pollen grains are inhaled, the pollen grains are hydrated and enzymes are released along the airways. Although pollen protease plays a primary function in pollen germination, it has a notable effect on allergic disorders (Hassim et al. 1998; Widmer et al. 2000; Boldogh et al. 2005; Cortes et al. 2006; Gunawan et al. 2008).

Studies on pollens of *Parietaria judaica*, mesquite, and ragweed found that pollens of these plants release proteolytic enzymes, which are involved in lung regulatory neuropeptide inactivation such as VIP (vasoactive intestinal peptide) and substance P (Matheson et al. 1995; Bagarozzi et al. 1996, 1998; Matheson and Travis 1998; Cortes et al. 2006). Hammad and Lambrecht (2008), in their study on dendritic and epithelial cells for innate immunity and adaptive immunity in asthma, explained that when allergens are inhaled, allergens must pass over the epithelial barrier to be processed by the dendritic cells present in the subepithelial region during the course of sensitization process. This is when the allergen peptides are presented by activated dendritic cells to naıve T lymphocytes, which elicit the specific IgE production. Dendritic cells sample the allergens by gaining access to dendritic cells by cleavage of proteins between epithelial cells from junctional complexes or by interepithelial pseudopods. Cortes et al. (2006) found the detachment of A549 human alveolar epithelial cells can be caused by degradation of intercellular adhesion proteins from adherent junction and tight junction caused by 98-kDa aminopeptidase from *Parietaria judaica*. Epithelial cell lines exhibit a loss of immunofluorescence labeling on tight junction proteins indicating that a disruption in the integrity of the epithelial barrier is found to cause by diffusates from pollens of *Betula pendula*, *Ambrosia trifida*, *Lilium longifolium*, and *Poa pratensis*, which are highly allergic. The effect was blocked by cysteine and serine protease inhibitors (Runswick et al. 2007). Allergic capabilities of pollen study found *Pinus sylvestris* having low allergenicity, *Cupressus sempervirens* having moderate allergenicity, *Olea europaea*, and *Dactylis glomerata* having high allergenicity (Jato et al. 2000; Amato et al. 2007; Heinzerling et al. 2009). All the plant pollens under study had an interesting relationship of having a partially overlapping season of pollination, for example, grasses and olive tree pollination period is April–July, which is perceived as foremost cause of seasonal respiratory allergies (Amato et al. 2007; Jato et al. 2007). Pollination of Cypress pollen, which is often known to induce rhinoconjunctivitis, occurs mainly in December–March (Amato et al. 2007; Jato et al. 2000). Study by Vinhas et al. (2011) revealed that pollens of *Olea europaea* and *Dactylis glomerata* are highly allergic and *Cupressus sempervirens* and *Pinus sylvestris* are low-moderately allergic with enough airborne counts during pollination period. Amato et al. (2007), Heinzerling (2009), Jato (2000), and Gastaminza (2009) found high molecular weight aminopeptidases and serine proteases in extract of all four species. Inhibitors of serine proteases (aprotinin, a1-antitrypsin, and SLPI), which are known to safeguard the lower respiratory tract against any endogenous proteases, did not block the pollen proteases of *Cupressus sempervirens, Dactylis glomerata, Olea europaea*, and *Pinus sylvestris*.

The said inhibitors are expected to undergo inactivation due to hydrolyzation by pollen protease (Bagarozzi et al. 1996; Cortes et al. 2006). This indicates that pollen released into the respiratory tract can activate pollen protease, thus implicating changes in chronic inflammation and allergic disorders. Protease of pollens can play a notable role in allergic asthma due to their conceivable involvement in neurogenic inflammation (Barnes and Dixon 1984; Barnes 1991). Decreases in pollen proteolytic activities have been proven to be caused by serine protease inhibitors, e.g. aminopeptidases in pollens of *Parietaria judica*, arginine-specific peptidase of ragweed, mesquite aminopeptidase, endopeptidase, trypsin-like serine protease, and diffusates of *C. obtusa, P. pratensis*, etc. (Matheson et al. 1995; Bagarozzi et al. 1996, 1998; Matheson and Travis 1998; Cortes et al. 2006; Runswick et al. 2007; Gunawan et al. 2008). An imbalance of the anti-inflammatory and pro-inflammatory effects within the lung caused by hydrolysis of the bioactive peptidase released by nerve endings due to pollen diffusates has been observed and this may worsen the hyper-responsiveness of bronchia. Pollen proteases from some species are found responsible for disrupting the transmembrane adhesion proteins like claudin-1 and E-cadherin and occluding by increasing the paracellular permeability in the epithelium. Pollen proteases are found effective in decreasing the epithelial integrity, damaging the transmembrane cell junction protein integrity (Gumbiner et al. 1988; Baraniu and Kim 2007; Vinhas et al. 2011) (Figure 10.1).

10.2.4 Respiratory Diseases Caused by Pollens

Introduction on basic science of allergy by Charles Blackley showed that pollen grains cause hay fever by performing studies on pollen grain counts, provocation of nasal and bronchial passage, and skin test. He however found it difficult to define the relation of asthma with pollens.

Reid (1986) suggested that exposure to pollen is linked to seasonal asthma in his study on seasonal asthma patients and seasonal rhinitis patients when he found a positive result in the skin test and immunoglobulin E(IgE) antibody for ryegrass. Many epidemiologic and clinical studies were conducted for the determination of allergic reactions to pollens (Agata et al. 1993; Kam and Hsieh 1994). The immune system's reaction to foreign proteins (i.e. allergens) involves both T cells and IgE antibodies. With repeated exposure to these substances, individuals can experience both rapid and prolonged allergic reactions. Pollen exposure of an asthmatic patient who is allergic to pollen can enhance bronchial obstruction, hyper-responsiveness, persistent inflammation of the airways, and several other triggers. Although, there is inscrutability in the mechanism of an inflammatory response and allergic effect when exposed to pollens (Hassim et al. 1998; Widmer et al. 2000; Boldogh et al. 2005; Gunawan et al. 2008).

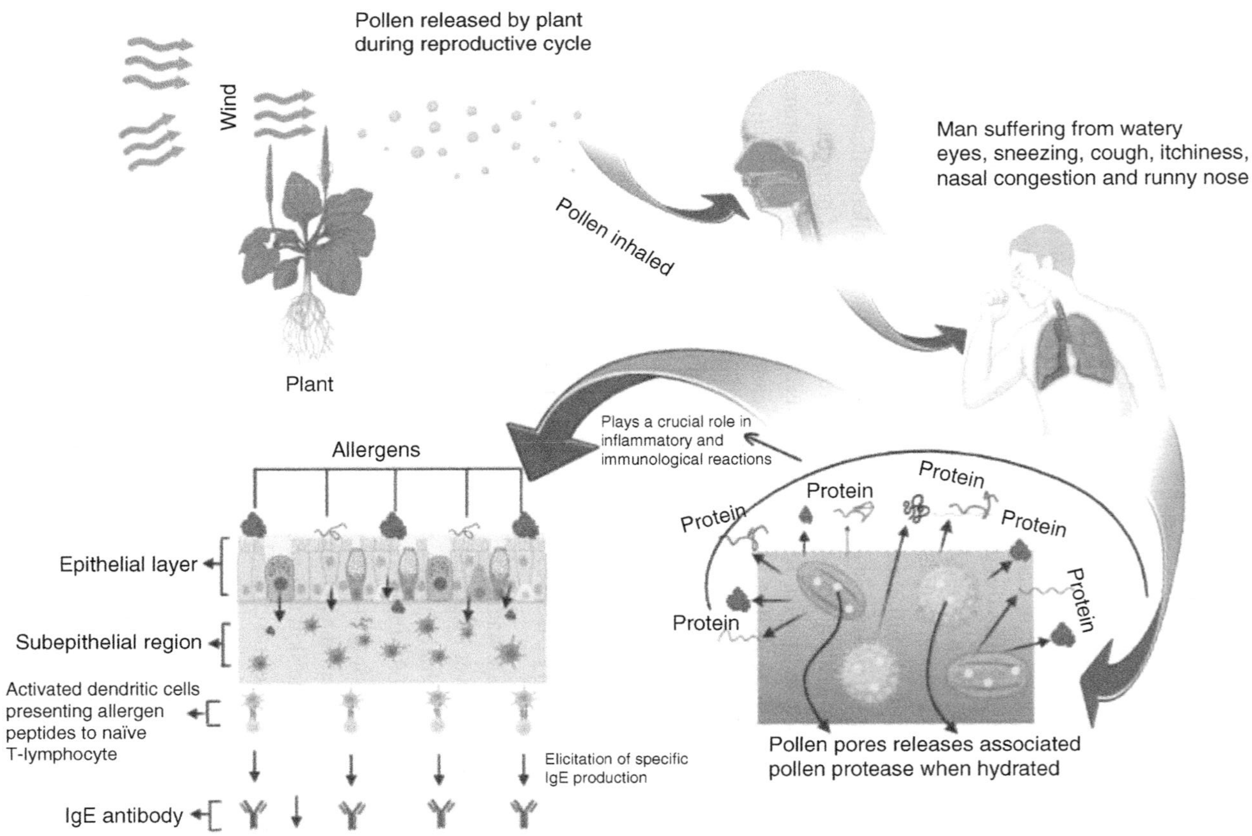

Figure 10.1 Diagram representing role of pollen grain in triggering respiratory diseases.

10.2.5 Mitigating the Impact of Pollen on Respiratory Health

Immunoblotting is developed for the counting of pollen allergen particles by Takashi et al. (1993a). This method was used for the examination of antigens in grass pollen (Takahashi et al. 1993b). He also examined two plant species, i.e. *Cryptomeria japonica* and *Oliumperenne* (Takahashi et al. 1993c). ELISA is used by Emberlin and McCartney (1996) for airborne allergen assessment. Burkard cyclone sampler is the new method used in immunological analysis of sampling airborne particles. Other requirements like minimization of distance within the population of study and location of monitoring were followed by Whittemore (1981). Consequently, if the result was positive indicating the presence of pollen allergens, it should be supplemented with reports of antigen/allergen (Takahashi et al. 1993c; Rantio-Lehtimakiet al. 1994a).

Different workers propose different ranges for monitoring the effect of pollen allergens, above 30 m from the roof of the hospital (Khot et al. 1988), above 15 m (Rossi et al. 1993), above 20 m (Subiza et al. 1994; Solomon 1984; Monn et al. 1997) suggested for the careful consideration of monitoring of height since some studies found difference in atmospheric composition at ground level and specific heights. Beggs (1998) suggested reassessing in broadening to include the duration of the season when the air contains pollen antigens. Atopic characteristics can be determined by intradermal testing and skin prick testing using pollen extracts in clinical settings (Beggs 1998), tests like MAST-CLA (multiple allergosorbent chemiluminescent assays) and RAST (radioallergosorbent test) are also known to indicate pollen allergy and allergens of mold spore (Agata et al. 1993; Kam and Hsieh 1994). IgG-ELISA (Immunoglobulin G – enzyme-linked immunosorbent assays were recently used for evaluating airborne allergens (Rantio-Lehtimaki et al. 1994; Emberlin and McCartney 1996). Polyclonal antibodies or monoclonal Bet v have been recommended for aerobiological monitoring due to high monology in major allergens of Butaceae family (Yli-Panula and Rantio-Lehtimäki 1994; Pehkonen and Rantio-Lehtimäki 1995). Inflammation in airway can be monitored using bronchial biopsy, lavage, and bronchoscopy with broncho alveolar, inflammatory mediators in exhaled breath, condensate [EBC]), a fraction of exhaled nitric oxygen (FENO), induced sputum, and exhaled nitric oxide measurement (Polosa et al. 2000).

10.2.6 Future Directions in Research on Pollen and Respiratory Diseases

Pollen proteases responsible for disrupting inter-cellular complexes and bioactive peptides are required to be characterized and identified. It is an essential step to look for a specific therapeutic target for the treatment and prevention of allergic disorders associated with pollinosis.

10.3 (1 → 3)-β-D-Glucan and Its Role in Triggering Respiratory Diseases

10.3.1 (1 → 3)-β-D-Glucan

(1 → 3)-β-D-glucan is a homopolymer of glucose, which is linked by β-glycosidic linkage between C1 and C2 of glucose monomers (Finkelman 2010). (1 → 3)-β-D-glucan is known to be present in cell walls of every fungus, animal, plant, and a few algae and bacteria (Williams 1997). Glucan is said to be a prominent constituent in the cell wall of fungi (pathogenic and saprophytic) with few exceptions like Blastomyces dermatitidis, Cryptococcus species, Mucor, and Rhizopus (Miyazaki et al. 1995; Inoue et al. 1996; Bowman and Free 2006; Girouard et al. 2007; Mennink-Kersten and Verweij 2006). Presence of (1 → 3)-β-D-glucan in fungi cell walls is in branched configuration (Finkelman 2010). The presence of (1 → 3)-β-D-glucan in pneumocystic cyst was described by Matsumoto et al. (1989). Within the fungal cell wall, (1 → 3)-β-D-glucan exists in insoluble structure. It transforms into random coil forms, triple helix, and single helix and exhibits soluble form (Bowman and Free 2006; Mennink-Kersten and Verweij 2006; Latgé 2007) and the soluble form can be competent of immune system modulation by leukocyte and phagocytosis inhibition. (1 → 3)-β-D-glucan derived from fungi is known to regulate various aspects of immunity (Browder et al. 1990; Ruiz-Herrera 1991; William et al. 1991, 1995; William and Browder 1994). (1 → 3)-β-D-glucan are found linked to lipids, proteins, and carbohydrates (Ruiz-Herrera 1991) in cell wall of fungi and is known to help in maintenance of integrity and rigidity of the cell wall in fungi as a primary function and hence maintaining the integrity of fungal cell wall structurally (Ruiz-Herrera 1991; Lesage and Bussey 2006), it is also known to be the specific biomarker for fungi (Williams 1997; De Sousa Lacerda et al. 2017). Immunological reactions of (1 → 3)-β-D-glucans were studied by Browder et al. (1990) and Williams et al. (1991, 1995). Many studies have shown (1 → 3)-β-D-glucans to have possible biomedical applications (Browder et al. 1990), these function like macrophage-stimulating agents (Riggi and Di Luzio 1961), and help in increasing the function of neutrophils, immunocytes, and macrophages (William et al. 1986, 1988).

10.3.2 Toxic Effects of (1 → 3)-β-D-Glucans

(1 → 3)-β-D-glucans isolated from yeast are usually found as microparticulates in insoluble form (Williams et al. 1991). However, there is a report of conversion of insoluble form to water-soluble form of (1 → 3)-β-D-glucans. The preservation of immunologic activity is observed and so the undesirable side effects are eliminated (Williams et al. 1988). Observation of granuloma formation in

macrophage-rich organs (spleen, lungs, and liver) due to hyperplasia and hypertrophy are associated with glucan microparticulates (William et al. 1985). Effects of (1 → 3)-β-D-glucans in symptoms of lung problems may include in combination with endotoxin (Fogelmark et al. 1994). (1 → 3)-β-D-glucans demonstrated in many studies of having the potential to exert as adjavant when amalgamated with a variety of parasitic vaccines (Holbrook et al. 1981; Williams et al. 1989; Williams 1997). Studies demonstrated glucan showing various effects. It is known to increase immunity (cell-mediated and humorial immunity) to antigens (Holbrook et al. 1981; Williams et al. 1989) and it also shows the combined effects on immunity (Browder et al. 1987; Williams et al. 1987).

10.3.3 Immune System-Modulating Properties of (1 → 3)-β-D-Glucan

(1 → 3)-β-D-glucan is known as dominant mediator for innate immune system responses as per the studies (Goodridge et al. 2009; Finkelman 2010). (1 → 3)-β-D-glucan binding has cellular recognition having a vast variety of sequela such as enhancement of phagocytosis, stimulation of T-cell, production of cytokine, and oxidative burst triggering. These are all the complex effects known. Complement Receptor-3 and Dectin-1 are the two main receptors for (1 → 3)-β-D-glucan (Xia et al. 1999; Willment et al. 2001). Complementary receptor-3 and Dectin-1 are the two important receptors (Xia et al. 1999; Willment et al. 2001). A variety of immune system cells consist of Dectin-1 (Kerrigan and Brown 2009) and when (1 → 3)-β-D-glucan is bound to Dectin-1, it leads to triggering of transduction of signal through a pathway, i.e. spleen tyrosine kinase pathway, which leads to gene transcription for chemokines, cytokines, and those proteins responsible for mediating inflammatory response and translocation of nuclear factor kappa B to the nucleus (Dennehy et al. 2008). Effects of (1 → 3)-β-D-glucan, when Dectin-1 mediates, seems to operate via cooperative interaction with toll-like receptors as well as independently (Werner et al. 2009). In presence of toll-like receptors, ligand (1 → 3)-β-D-glucan acts synergistically to increase the production of pro-inflammatory cytokine (Kikkert et al. 2007; Dennehy et al. 2008). (1 → 3)-β-D-glucan promotes antibody-dependent cellular cytotoxicity, and it is used as an adjuvant for cancer therapy (Hong et al. 2004; Modak et al. 2005; Hazama et al. 2009). Cook et al. (1980) and Bowers et al. (1986) showed induced condition of lethal shock in rodent models when exposed to (1 → 3)-β-D-glucan followed by exposure to endotoxin at sublethal levels. Studies by Takahashi et al. (2001) and Moriya et al. (2002) showed lethal effects when exposed to fragments of candida cell wall followed by sublethal endotoxin. Pneumocystis cell wall characterized by (1 → 3)-β-D-glucan is known to elicitate the macrophage-inflammatory protein-2 (MIP-2) and pro-inflammatory

mediators TNF-α (Vassallo et al. 2000). Dennehy et al. (2008, 2009) described the combined synergy in pro-inflammatory cytokine (TNF-α, IL-6, IL-23) elicitation through colligation of TLR and Dectin-1. Finkelman (2010) delineated the assistance of (1 → 3)-β-D-glucan along with diagnostic tests for Pneumocystis colonization and pathophysiology in distinguishing colonization and diagnostic tests from pathological infections. Properties of immunomodulators are suggested to be possessed by β-D-glucan due to observation in enhancement of the macrophage phagocytic activity. However, β-D-glucan could not exhibit the activity of cytotoxicity against glioblastoma cells (Rizzi et al. 2023). (1 → 3)-β-D-glucan also can be defined as fungal surrogate marker, providing the procedure for diagnosing invasive fungal infections and potential surveillance (Browder et al. 1990). Current data rather indicates using (1 → 3)-β-D-glucan as a surrogate for withdrawal of empirically started antifungal treatment rather than for early initiation (Rouzé et al. 2017; De Pascale et al. 2020; Dupuis et al. 2021) (Figure 10.2).

10.4 Endotoxin and Role in Triggering Respiratory Diseases

10.4.1 Endotoxins

Endotoxins, also called pyrogens, are complex lipopolysaccharides (LPS), found in outer cell walls of all Gram-negative bacteria. It is responsible for the cell wall to be in organized and stable form. Endotoxins are frequently found in environment, food, and some clinical products made from the bacteria (Kim et al. 2012; Das et al. 2014). Richard Pfeiffer (1858–1945) first used the term Endotoxin. He observed that Vibrio cholerae's poisonous feature remained for a long duration after the bacteria had undergone lysis. He hypothesized that Gram-negative and Gram-positive bacteria uniformly contain endotoxins, and later found evidence of these toxins in *Salmonella typhi* and *Haemophilus influenzae*. Eugenio Centanni (1863–1948) later discovered the ability of endotoxin to induce fever and the poisonous characteristics were inseparable, and he named this toxin "pyrotoxina" (Beutler and Rietschel 2003). Lipid A consisting of endotoxin is covalently bonded to the outer membrane molecule by a saccharide substitution, an O-specific side chain linked in smooth LPS (S-form LPS) and oligosaccharide with upto 15 monosaccharides (Holst et al. 1996; Chhetry et al. 2022). Crop dust, poultry, swine, and grain dust contain endotoxin (Thorne et al. 2010; Shivpuje et al. 2020). Endotoxin is pervasive and represents major bioaerosol components (Liebers et al. 2008). The most prominent pyrogen in parenteral medications and medical devices is endotoxin, which is ubiquitous in natural environment. The digestive system also contains endotoxins. Septic reactions characterized by symptoms like fever,

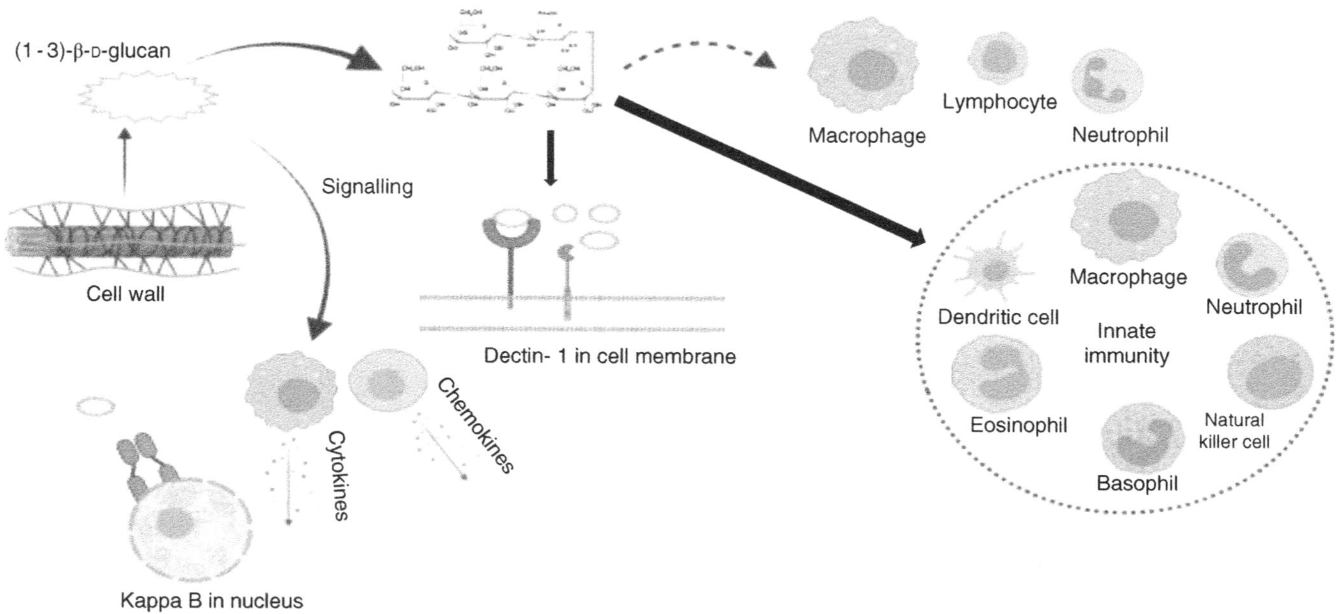

Figure 10.2 Diagram representing role of (1 → 3)-β-D-glucan in triggering respiratory diseases.

hypotension, nausea, shaking, and shock may result from their presence in the bloodstream (Lemke 1994). Complications such as adult respiratory distress syndrome (ARDS), disseminated intravascular coagulation (DIC), and endotoxin shock can also occur at high concentrations of endotoxin (Gorbet and Sefton 2004).

10.4.2 Endotoxin Exposures

Endotoxin exposure can happen in the workplace or the natural environment. Health effects demographics of those most at risk and pathways of exposure are all variable between the two. Most often, negative health outcomes were the result of occupational exposure, whereas the incidence of major health effects due to environmental exposure to endotoxin varies (Shamsollahi et al. 2019). Inhalation remains the most common and significant route to endotoxin exposure, particularly in industrial settings. Nonetheless, there are notable distinctions between these two categories, i.e. in occupational exposure and environmental exposure. Much more endotoxin is present in the occupational matrix than there is in the environment (Pankhurst et al. 2011). Furthermore, the manner of occupational exposure is periodic or discontinuous while environmental exposure is continuous. Exposed groups varied too. Unlike environmental exposures, in which only vulnerable demographics like infants and the elderly are put at risk, occupational exposures put healthy adults at risk. The health effects seen after each form of exposure, however, are what truly set them apart. Fever, wheezing, cough, asthma, and dyspnea symptoms are documented in most occupational endotoxin exposure investigations (Basinas et al. 2015).

10.4.3 Impact of Endotoxins in Respiratory Diseases

Farokhi et al. (2018) mentioned that the air around livestock farms has been found to have higher levels of endotoxin, which worries people who live nearby and there is clear evidence that working in an environment with high levels of airborne endotoxin causes inflammation of the lungs, loss of lung function, and also symptoms of the lungs. Organic dust toxic syndrome (ODTS) is a flu-like systemic inflammatory reaction, which causes chills, fever, chest tightness, myalgias, and dyspnea. All the symptoms occur four to eight hours after the exposure. For these reasons, endotoxins are crucial to ODTS formation (Von Essen 1997; Singh and Schwartz 2005; Radon 2006). Studies that looked at the effects of inhaling LPS also looked at how the function of the lungs and airways changed. These changes are not too big in healthy people, but they are bigger in asthmatics suggesting the inflammation caused by endotoxin makes the asthmatic inflammatory response even worse (Michel et al. 1989; Rylander et al. 1989; Rylander 2002) (Figure 10.3).

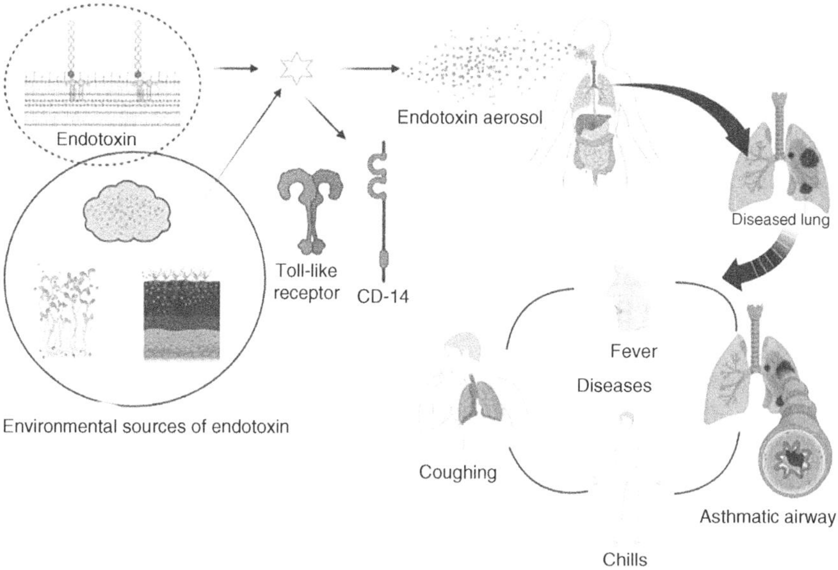

Figure 10.3 Diagram representing role of endotoxins in triggering respiratory diseases.

10.4.4 Mitigating the Impact of Endotoxin on Respiratory Health

10.4.4.1 Quantitative Endotoxin Testing Assay

***Limulus amebocytelysate* (LAL)-test:** Levin and Bang (1964) found endotoxin-mediated intravascular coagulation in the *Limulus polyphemus* in 1964. Later research showed implications of enzymatic mechanism and *Limulus*'s only circulating cells, amebocytes were the active compounds in endotoxin-mediated intravascular coagulation (Vanhaecke et al. 1987). The Limulus amebocyte lysate (LAL) assay is the gold standard for determining the endotoxin and in 1980, the US Food and Drug Administration adopted LAL-assay as the gold standard for detection of endotoxin. The turbidimetric method, which measures turbidity, the chromogen-kinetic approach, which measures change in color over time owing to enzymatic turnover of substrate, and endpoint assay, which measures gel clotting, are the three types of the LAL-test used for quantitative detection of endotoxin. The LAL-assay has a wide measuring range (from around 0.01–100 endotoxin units (EUs) per milliliter to 1 pg/ml to 10 ng/ml) and high sensitivity, especially the chromogen kinetic variants (Chun et al. 2006; Liebers et al. 2007, 2008).

Gas chromatography and mass spectrometry (GCMS): Measure of 3-hydroxy fatty acids from lipid A by GCMS may offer quantitative information regarding the endotoxin level of dust samples (Liebers et al. 2008). In contrast to LAL test, measurements do not show how active a sample is biologically.

Since different types of bacteria show different 3-OH fatty acids, this method cannot be used to figure out the exact amount of LPS (Binding et al. 2004). There is a strong association between the quantity of short-chain 3-hydroxy fatty acids and endotoxin as measured by the LALtest. Nevertheless, this was not the case with long-chain fatty acids. Also, the low sensitivity of this chemical technique restricts its usefulness (Rylander 2006).

10.5 Implications for Public Health Policy and Individual Behavior

Raising awareness to develop strategies to reduce exposure to the particles raises awareness about the impact of the particles on respiratory health. This involves initiative by issuing warning when pollen count is at peak, monitoring air quality, and aiding the use of protective equipment and air filters. People suffering from respiratory diseases should minimize their exposure to these particles. However, various treatments are available for people with asthma, including over the counter and prescribed medications such as antihistamines, decongestants, and corticosteroids. Immunotherapy, the process of regular exposure to a small number of allergens over time to desensitize the immune system. People can seek help from health care providers for medical attention promptly if their symptom worsens for the treatment options for managing respiratory symptoms. Public health officials could think about funding research on understanding the mechanism by which these particles trigger respiratory troubles and assessing the impact of measures taken to minimize exposure to these particles. Addressing the role of pollen grains, (1 → 3)-β-D-glucan and endotoxin play a crucial role in triggering respiratory diseases mandates the multifarious approach involving individual behavior change and public health policy. This uniting effort can facilitate improved health outcomes for people with respiratory diseases and can also facilitate the impact of these particles on respiratory diseases.

10.6 Conclusion

Pollen grains, (1 → 3)-β-D-glucan and endotoxin, can play a crucial role in triggering respiratory diseases, particularly for people with allergies or asthma. These are the common triggers for allergic rhinitis and asthma. Intensification of this condition can be caused when exposed to high levels of this particle. People suffering from respiratory diseases should take preventive measures by limiting their exposure to the particles, assuring their indoor environments are free of pollutants and allergens and avoiding outdoor activities during peak pollen season. Understanding the impact of pollen, (1→3)-β-D-glucan and endotoxin on respiratory health is important for both individuals and public health professionals, as it can help inform strategies for reducing exposure and managing symptoms. It also helps in preventing exacerbations of these conditions and improving public health. More research is required to fully understand the mechanisms of these particles triggering respiratory diseases.

References

Abrol, D.P. (2012). Pollination-basic concepts. In: *Pollination Biology: Biodiversity Conservation and Agricultural Production*, vol. 792, 37–54. New York: Springer.

Agata, H., Yomo, A., Hanashiro, Y. et al. (1993). Comparison of the MAST chemiluminescent assay system with RAST and skin tests in allergic children. *Annals of Allergy* 70: 153–157.

Amato, D., Amato, G., Cecchi, L. et al. (2007). Allergenic pollen and pollen allergy in Europe. *Allergy* 62 (9): 976–990.

Bagarozzi, D.A., Potempa, J., and Travis, J. (1998). Purification and characterization of an arginine-specific peptidase from ragweed (Ambrosia) pollen. *American Journal of Respiratory Cell and Molecular Biology* 18: 363–369.

Bagarozzi, D.A., Pike, R., Potempa, J., and Travis, J. (1996). Purification and characterization of a novel endopeptidase in ragweed (Ambrosia artemisiifolia) pollen. *The Journal of Biological Chemistry* 271: 26227–26232.

Baraniuk, J.N. and Kim, D. (2007). Nasonasal reflexes, the nasal cycle, and sneeze. *Current Allergy and Asthma Reports* 7: 105–111.

Barderas, R., Villalba, M., Lombardero, M., and Rodríguez, R. (2002). Identification and characterization of Che a 1 allergen from *Chenopodium* album pollen. *International Archives of Allergy and Immunology* 127 (1): 47–54.

Barnes, P.J. and Dixon, C.M. (1984). The effect of inhaled vasoactive intestinal peptide on bronchial reactivity to histamine in humans. *The American Review of Respiratory Disease* 130: 162–166.

Barnes, P.J. (1991). Neurogenic inflammation in airways. *International Archives of Allergy and Applied Immunology* 94: 303–309.

Basinas, I., Sigsgaard, T., Kromhout, H. et al. (2015). A comprehensive review of levels and determinants of personal exposure to dust and endotoxin in livestock farming. *Journal of Exposure Science and Environmental Epidemiology* 25 (2): 123–137.

Bedinger, P. (1992). The remarkable biology of pollen. *The Plant Cell* 4 (8): 879.

Beggs, P.J. (1998). Pollen and pollen antigen as triggers of asthma—what to measure? *Atmospheric Environment* 32 (10): 1777–1783.

Benkouiten, S., Brouqui, P., and Gautret, P. (2014). Non-pharmaceutical interventions for the prevention of respiratory tract infections during Hajj pilgrimage. *Travel Medicine and Infectious Disease* 12 (5): 429–442.

Beutler, B. and Rietschel, E.T. (2003). Innate immune sensing and its roots: the story of endotoxin. *Nature Reviews Immunology* 3 (2): 169–176.

Binding, N., Jaschinski, S., Werlich, S. et al. (2004). Quantification of bacterial lipopolysaccharides (endotoxin) by GC–MS determination of 3-hydroxy fatty acids. *Journal of Environmental Monitoring* 6 (1): 65–70.

Blanca, M., Boulton, P., Brostoff, J., and González-Reguera, I. (1983). Studies of the allergens of Olea europea pollen. *Clinical and Experimental Allergy* 13 (5): 473–478.

Boldogh, I., Bacsi, A., Choudhury, B.K. et al. (2005). ROS generated by pollen NADPH oxidase provide a signal that augments antigen-induced allergic airway inflammation. *The Journal of Clinical Investigation* 115 (8): 2169–2179.

Bosma, K.J., Taneja, R., and Lewis, J.F. (2010). Pharmacotherapy for prevention and treatment of acute respiratory distress syndrome: current and experimental approaches. *Drugs* 70: 1255–1282.

Bousquet, J., Khaltaev, N., Cruz, A.A. et al. (2008). Allergic rhinitis and its impact on asthma (ARIA) 2008. *Allergy* 63: 8–160.

Boulet, L.P., Turcotte, H., Boutet, M. et al. (1993). Influence of natural antigenic exposure on expiratory flows, methacholine responsiveness, and airway inflammation in mild allergic asthma. *Journal of Allergy and Clinical Immunology* 91 (4): 883–893.

Bowers, G.J., Patchen, M.L., MacVittie, T.J. et al. (1986). A comparative evaluation of particulate and soluble glucan in an endotoxin model. *International Journal of Soc Immunopharmacology* 8: 313–321.

Bowman, S.M. and Free, S.J. (2006). The structure and synthesis of the fungal cell wall. *BioEssays* 28: 799–808.

Browder, W.I., Williams, D.A., Pretus, H.E. et al. (1990). Beneficial effect of enhanced macrophage function in the trauma patient. *Annals of Surgery* 211 (5): 605–613.

Browder, W., Williams, D., Sherwood, E. et al. (1987). Synergistic effect of nonspecific immunostimulation and antibiotics in experimental peritonitis. *Surgery* 102: 206–214.

Brown, H.M., Jackson, F., Birch, D., and Alderson, M.R. (1983). Asthma and the weather. *The Lancet* 322 (8350): 630–631.

Burney, P., Malmberg, E., Chinn, S. et al. (1997). The distribution of total and specific serum IgE in the european community respiratory health survey. *The Journal of Allergy and Clinical Immunology* 99: 314–322.

Burbach, G.J., Heinzerling, L.M., Edenharter, G. et al. (2009). GA2LEN skin test study II: clinical relevance of inhalant allergen sensitizations in Europe. *Allergy* 64 (10): 1507–1515.

Carnes, J., Himly, M., Gallego, M. et al. (2009). Detection of allergen composition and in vivo immunogenicity of depigmented allergoids of *Betula alba*. *Clinical and Experimental Allergy* 39 (3): 426–434.

Cevhertas, L., Ogulur, I., Maurer, D. J. et al. (2020). Advances and recent developments in asthma in 2020. *Allergy* 75 (12): 3124–3146.

Chen, X. and Ju, F. (2022). Automatic classification of pollen grain microscope images using a multi-scale classifier with SRGAN deblurring. *Applied Sciences* 12 (14): 7126.

Cheung, A.Y., Boavida, L.C., Aggarwal, M. et al. (2010). The pollen tube journey in the pistil and imaging the in vivo process by two-photon microscopy. *Journal of Experimental Botany* 61 (7): 1907–1915.

Chhetry, B.S.K., Dewangan, K.N., Mahato, D.K., and Kumar, P. (2022). Endotoxins affecting human health during agricultural practices: an overview. *Applied Chemistry* 3 (1): 11–31.

Chun, D.T., Bartlett, K., Gordon, T. et al. (2006). History and results of the two inter-laboratory round robin endotoxin assay studies on cotton dust. *American Journal of Industrial Medicine* 49 (4): 301–306.

Colombo, P., Bonura, A., Costa, M.A. et al. (2003). The allergens of Parietaria. *International Archives of Allergy and Immunology* 130 (3): 173–179.

Cook, J.A., Dougherty, W.J., and Holt, T.M. (1980). Enhanced sensitivity to endotoxin induced by the RE stimulant, glucan. *Circulatory Shock* 7: 225–238.

Cortes, L., Carvalho, A.L., Todo-Bom, A. et al. (2006). Purification of a novel aminopeptidase from the pollen of *Parietariajudaica* that alters epithelial integrity and degrades neuropeptides. *The Journal of Allergy and Clinical Immunology* 118: 878–884.

Damato, G., Liccardi, G., D'amato, M., and Holgate, S. (2005). Environmental risk factors and allergic bronchial asthma. *Clinical & Experimental Allergy* 35 (9): 1113–1124.

D'Amato, G., Cecchi, L., D'Amato, M., and Annesi-Maesano, I. (2014). Climate change and respiratory diseases. *European Respiratory Review* 23 (132): 161–169.

D'Amato, G., Vitale, C., D'Amato, M. et al. (2016). Thunderstorm-related asthma: what happens and why. *Clinical and Experimental Allergy* 46 (3): 390–396. https://doi.org/10.1111/cea.12709.

Das, A.P., Kumar, P.S., and Swain, S. (2014). Recent advances in biosensor based endotoxin detection. *Biosensors and Bioelectronics* 51: 62–75.

Dennehy, K.M., Ferwerda, G., Faro-Trindade, I. et al. (2008). Syk kinase is required for collaborative cytokine production induced through dectin-1 and toll-like receptors. *European Journal of Immunology* 38: 500–506.

Dennehy, K.M., Willment, J.A., Williams, D.L., and Brown, G.D. (2009). Reciprocal regulation of IL-23 and IL-12 following co-activation of Dectin-1 and TLR signaling pathways. *European Journal of Immunology* 39: 1379–1386.

De Pascale, G., Posteraro, B., D'Arrigo, S. et al. (2020). (1, 3)-β-d-Glucan-based empirical antifungal interruption in suspected invasive candidiasis: a randomized trial. *Critical Care* 24: 1–10.

De Sousa Lacerda, C.M., Ferreira, I.M., Dos Santos, S.R. et al. (2017). (1 → 3)-β-d-glucan aptamers labeled with technetium-99 m: biodistribution and imaging in experimental models of bacterial and fungal infection. *Nuclear Medicine and Biology* 46: 19–24.

Dresselhaus, T. and Franklin-Tong, N. (2013). Male–female crosstalk during pollen germination, tube growth and guidance, and double fertilization. *Molecular Plant* 6 (4): 1018–1036.

Dupuis, C., Le Bihan, C., Maubon, D. et al. (2021). Performance of repeated measures of (1 → 3)-β-d-glucan, mannan antigen, and antimannan antibodies for the diagnosis of invasive candidiasis in ICU patients: a preplanned ancillary analysis of the EMPIRICUS randomized clinical trial. *In Open Forum Infectious Diseases US: Oxford University Press* 8 (3): 1–10.

De Weger, L.A., Bergmann, K.C., Rantio-Lehtimäki, A. et al. (2013). Impact of pollen. In: *Allergenic Pollen: A Review of the Production, Release, Distribution and Health Impacts* (ed. M. Sofiev and K.C. Bergman), 161–215. Springer.

Emberlin, J. and McCartney, A. (1996). Monitoring atmospheric allergenicity: an alternative to pollen and spore counting. In: *Clinical Applications of Aerobiology* (ed. S.M. Tariq and J. Emberlin), 57–67. B-1420 Braine-lÕAlleud: The Isle of Wight Asthma and Allergy Research Centre, D. Van Moerbeke for The UCB Institute of Allergy, Chemin du Foriest.

Elvira-Rendueles, B., Zapata, J.J., Miralles, J.C. et al. (2017). Aerobiological importance and allergic sensitization to Amaranthaceae under arid climate conditions. *Science of the Total Environment* 583: 478–486.

Farokhi, A., Heederik, D., and Smit, L.A. (2018). Respiratory health effects of exposure to low levels of airborne endotoxin–a systematic review. *Environmental Health* 17 (1): 1–20.

Fernández, J., Emilio, F., Montserrat, V. et al. (2015). Evolution of the incidence of pollen grains and sensitivity to pollen in the city of elche (Spain). *Asian Pacific Journal of Allergy and Immunology* 33: 196–202.

Fernández-González, D., González-Parrado, Z., Vega-Maray, A.M. et al. (2010). Platanus pollen allergen, Pla a 1: quantification in the atmosphere and influence on a sensitizing population. *Clinical and Experimental Allergy* 40 (11): 1701–1708.

Finkelman, M.A. (2010). Pneumocystis jirovecii infection: cell wall (1 → 3)-β-d-glucan biology and diagnostic utility. *Critical Reviews in Microbiology* 36 (4): 271–281.

Fogelmark, B.I.R.G.I.T.T.A., Sjöstrand, M., and Rylander, R.A.G.N.A.R. (1994). Pulmonary inflammation induced by repeated inhalations of beta (1, 3)-d-glucan and endotoxin. *International Journal of Experimental Pathology* 75 (2): 85.

Ford, S.A., Baldo, B.A., Panzani, R., and Bass, D. (1991). Cypress (*Cupressus sempervirens*) pollen allergens: identification by protein blotting and improved detection of specific IgE antibodies. *International Archives of Allergy and Immunology* 95 (2–3): 178–183.

Gaspar, R., de Matos, M.R., Cortes, L. et al. (2020). Pollen proteases play multiple roles in allergic disorders. *International Journal of Molecular Sciences* 21 (10): 3578.
Gastaminza, G., Lombardero, M., Bernaola, G. et al. (2009). Allergenicity and cross-reactivity of pine pollen. *Clinical and Experimental Allergy* 39 (9): 1438–1446.
Girouard, G., Lachance, C., and Pelletier, R. (2007). Observations on (1 → 3)-beta-d-glucan detection as a diagnostic tool in endemic mycosis caused by histoplasma or blastomyces. *Journal of Medical Microbiology* 56 (7): 1001–1002.
Gorbet, M.B. and Sefton, M.V. (2004). Biomaterial-associated thrombosis: roles of coagulation factors, complement, platelets and leukocytes. *Biomaterials* 25 (26): 5681–5703.
Gordon, C.J. (2003). Role of environmental stress in the physiological response to chemical toxicants. *Environmental Research* 92 (1): 1–7.
Goodridge, H.S., Wolf, A.J., and Underhill, D.M. (2009). B-glucan recognition by the innate immune system. *Immunological Reviews* 230: 38–50.
Grijincu, M., Huțu, I., Weber, M. et al. (2023). Physicochemical and immunological characterization of Amb a 12, a novel ragweed (*Ambrosia artemisiifolia*) pollen allergen. *Molecular Immunology* 157: 18–29.
Grewling, Ł., Bogawski, P., Kostecki, Ł. et al. (2020). Atmospheric exposure to the major Artemisia pollen allergen (Art v 1): seasonality, impact of weather, and clinical implications. *Science of the Total Environment* 713: 136611.
Gulati, N., Dua, K., and Dureja, H. (2021). Role of chitosan based nanomedicines in the treatment of chronic respiratory diseases. *International Journal of Biological Macromolecules* 185: 20–30.
Gunawan, H., Takai, T., Kamijo, S. et al. (2008). Characterization of proteases, proteins, and eicosanoid-like substances in soluble extracts from allergenic pollen grains. *International Archives of Allergy and Immunology* 147 (4): 276–288.
Gumbiner, B., Stevenson, B., and Grimaldi, A. (1988). The role of the cell adhesion molecule uvomoru-lin in the formation and maintenance of the epithelial junctional complex. *The Journal of Cell Biology* 107: 1575–1587.
Harun, N.S., Lachapelle, P., and Douglass, J. (2019). Thunderstorm-triggered asthma: what we know so far. *Journal of Asthma and Allergy* 2019 (12): 101–108.
Hassim, Z., Maronese, S.E., and Kumar, R.K. (1998). Injury to murine airway epithelial cells by pollen enzymes. *Thorax* 53: 368–371.
Hammad, H. and Lambrecht, B.N. (2008). Dendritic cells and epithelial cells: linking innate and adaptive immunity in asthma. *Nature Reviews Immunology* 8: 193–204.
Hazama, S., Watanabe, S., Ohashi, M. et al. (2009). Efficacy of orally administered superfine dispersed lentinan (β-1, 3-glucan) for the treatment of advanced colorectal cancer. *Anticancer Research* 29 (7): 2611–2617.
Heinzerling, L.M., Burbach, G.J., Edenharter, G. et al. (2009). GA2LEN skin test study I: GA^2LEN harmonization of skin prick testing: novel sensitization patterns for inhalant allergens in Europe. *Allergy* 64 (10): 1498–1506;Hesse, M. (2000).

Pollen wall stratification and pollination. *Plant Systematics and Evolution* 222: 1–17.

Hong, F., Yan, J., Baran, J.T. et al. (2004). Mechanism by which orally administered β-1,3-glucans enhance the tumoricidal activity of antitumor monoclonal antibodies in murine tumor models. *Journal of Immunology* 173: 797–806.

Holbrook, T.W., Cook, J.A., and Parker, B.W. (1981). Glucan-enhanced immunogenicity of killed erythrocyte stages of *Plasmodium berghei*. *Infection and Immunity* 32 (2): 542–546.

Holst, O., Ulmer, A.J., Brade, H. et al. (1996). Biochemistry and cell biology of bacterial endotoxins. *FEMS Immunology and Medical Microbiology* 16 (2): 83–104.

Hughes, D.D., Mampage, C.B.A., Jones, L.M. et al. (2020). Characterization of atmospheric pollen fragments during springtime thunderstorms. *Environmental Science & Technology Letters* 7 (6): 409–414. https://doi.org/10.1021/acs.estlett.0c00213.

Inoue, S. B., Qadota, H., Arisawa, M. et al. (1996). Signaling toward yeast 1, 3-β-glucan synthesis. *Cell structure and function* 21 (5): 395–402.

Ipsen, H. and Løwenstein, H. (1983). Isolation and immunochemical characterization of the major allergen of birch pollen (*Betula verrucosa*). *Journal of Allergy and Clinical Immunology* 72 (2): 150–159.

Jato, V., Rodríguez-Rajo, F. J., and Aira, M. J. (2007). Use of phenological and pollen-production data for interpreting atmospheric birch pollen curves. *Annals of Agricultural and Environmental Medicine* 14 (2).

Jato, M.V., Rodriguez, F.J., and Seijo, M.C. (2000). Pinus pollen in the atmosphere of Vigo and its relationship to meteorological factors. *International Journal of Biometeorology* 43: 147–153.

Johnson, M.A., Harper, J.F., and Palanivelu, R. (2019). A fruitful journey: pollen tube navigation from germination to fertilization. *Annual Review of Plant Biology* 70: 809–837.

Kam, K.L. and Hsieh, K.H. (1994). Comparison of three in vitro assays for serum IgE with skin testing in asthmatic children. *Annals of Allergy* 73: 329–336.

Kao, K.C., Chiu, L.C., Hung, C.Y. et al. (2017). Coinfection and mortality in pneumonia-related acute respiratory distress syndrome patients with bronchoalveolar lavage: a prospective observational study. *Shock (Augusta, Ga.)* 47 (5): 615.

Kerrigan, A.M. and Brown, G.D. (2009). C-Type lectins and phagocytosis. *Immunobiology* 214: 562–575.

Khot, A., Burn, R., Evans, N. et al. (1988). Biometeorological triggers in childhood asthma. *ClinicalAllergy* 18: 351–358.

Kikkert, R., Bulder, I., De Groot, E.R. et al. (2007). Potentiation of toll-like receptor-induced cytokine production by (1 → 3)-β-d-glucans, implications for the monocyte activation test. *Journal of Endotoxin Research* 13: 140–149.

Kim, S.E., Su, W., Cho, M. et al. (2012). Harnessing aptamers for electrochemical detection of endotoxin. *Analytical Biochemistry* 424 (1): 12–20.

Kumar, A., Vimal, A., and Kumar, A. (2016). Why chitosan? From properties to perspective of mucosal drug delivery. *International Journal of Biological Macromolecules* 91: 615–622.

Latgé, J.P. (2007). The cell wall: a carbohydrate armour for the fungal cell. *Molecular Microbiology* 66: 279–290.

Lauer, I., Foetisch, K., Kolarich, D. et al. (2004). Hazelnut (*Corylus avellana*) vicilin Cor a 11: molecular characterization of a glycoprotein and its allergenic activity. *Biochemical Journal* 383 (2): 327–334.

Lauzurica, P., Maruri, N., Galocha, B. et al. (1988). Olive (olea europea) pollen allergens—II Isolation and characterization of two major antigens. *Molecular Immunology* 25 (4): 337–344.

Lemke, H.D. (1994). Methods for the detection of endotoxins present during extracorporeal circulation. *Nephrology, Dialysis, Transplantation: Official Publication of the European Dialysis and Transplant Association-European Renal Association* 9: 90–95.

Lesage, G. and Bussey, H. (2006). Cell wall assembly in *Saccharomyces cerevisiae. Microbiology and Molecular Biology Reviews* 70 (2): 317–343.

Levin, J. and Bang, F.B. (1964). The role of endotoxin in the extracellular coagulation of Limulus blood. *Bulletin of the Johns Hopkins Hospital* 115: 265–274.

Lewis, W.H., Dixit, A.B., and Ward, W.A. (1991). Distribution and incidence of North American pollen aeroallergens. *American Journal of Otolaryngology* 12 (4): 205–226.

Liebers, V., Raulf-Heimsoth, M., Linsel, G. et al. (2007). Evaluation of quantification methods of occupational endotoxin exposure. *Journal of Toxicology and Environmental Health, Part A* 70 (21): 1798–1805.

Liebers, V., Raulf-Heimsoth, M., and Brüning, T. (2008). Health effects due to endotoxin inhalation. *Archives of Toxicology* 82: 203–210.

Lo, F., Bitz, C.M., Battisti, D.S., and Hess, J.J. (2019). Pollen calendars and maps of allergenic pollen in North America. *Aerobiologia* 35: 613–633.

Lüttkopf, D., Müller, U., Skov, P.S. et al. (2002). Comparison of four variants of a major allergen in hazelnut (*Corylus avellana*) Cor a 1.04 with the major hazel pollen allergen Cor a 1.01. *Molecular Immunology* 38 (7): 515–525.

Mahabalé, T.S. (1968). Spores and pollen grains of water plants and their dispersal. *Review of Palaeobotany and Palynology* 7 (4): 285–296.

Matheson, N.R. and Travis, J. (1998). Purification and characterization of a novel peptidase (IImes) from mesquite (*Prosopisvelutina*) pollen. *The Journal of Biological Chemistry* 273: 16771–16777.

Matheson, N., Schmidt, J., and Travis, J. (1995). Isolation and properties of an angiotensin II-cleaving peptidase from mesquite pollen. *American Journal of Respiratory Cell and Molecular Biology* 12: 441–448.

Matsumoto, Y., Matsuda, S., and Tegoshi, T. (1989). Yeast glucan in the cyst wall of *Pneumocystis carinii*. *The Journal of Protozoology* 36: 21S–22S.

Mehta, V. and Wheeler, A.W. (1991). IgE-mediated sensitization to English plantain pollen in seasonal respiratory allergy: identification and partial characterisation of its allergenic components. *International Archives of Allergy and Immunology* 96 (3): 211–217.

Mennink-Kersten, M.A. and Verweij, P.E. (2006). Non-culture-based diagnostics for opportunistic fungi. *Infectious Disease Clinics of North America* 20: 711–727.

Michel, O.L.I.V.I.E.R., Duchateau, J., and Sergysels, R.O.G.E.R. (1989). Effect of inhaled endotoxin on bronchial reactivity in asthmatic and normal subjects. *Journal of Applied Physiology* 66 (3): 1059–1064.

Miller, W.F. (1973). Aerosol therapy in acute and chronic respiratory disease. *Archives of Internal Medicine* 131 (1): 148–155.

Miyazaki, T., Kohno, S., and Mitsutake, K. (1995). Plasma (1 → 3)-beta-d-glucan and fungal antigenemia in patients with candidemia, aspergillosis, and cryptococcosis. *Journal of Clinical Microbiology* 33: 3115–3118.

Modak, S., Koehne, G., Vickers, A. et al. (2005). Rituximab therapy of lymphoma is enhanced by orally administered (1→3),(1→4)-d-β-glucan. *Leukemia Research* 29 (6): 679–683.

Monn, C., Carabias, V., Junker, M. et al. (1997). Small-scale spatial variability of particulate matter< 10 μm (PM10) and nitrogen dioxide. *Atmospheric Environment* 31 (15): 2243–2247.

Moriya, K., Miura, N.N., Adachi, Y., and Ohno, N. (2002). Systemic inflammatory response associated with augmentation and activation of leukocytes in candida/indomethacin administered mice. *Biological & Pharmaceutical Bulletin* 25: 816–822.

Munro, D.A., LeRoy, P.A., Hill, D.J. et al. (1976). Aero-allergens and childhood asthma in Melbourne. *Clean Air* 10: 42–44.

Muralidharan, A., Russell, M.S., Larocque, L. et al. (2019). Chitosan alters inactivated respiratory syncytial virus vaccine elicited immune responses without affecting lung histopathology in mice. *Vaccine* 37 (30): 4031–4039.

Nepi, M. and Pacini, E. (1993). Pollination, pollen viability and pistil receptivity in Cucurbita pepo. *Annals of Botany* 72 (6): 527–536.

Nickovic, S., Petković, S., Ilić, L. et al. (2023). Prediction of airborne pollen and sub-pollen particles for thunderstorm asthma outbreaks assessment. *Science of the Total Environment* 864: 160879.

Nobata, K., Fujimura, M., Ishiura, Y. et al. (2006). Ambroxol for the prevention of acute upper respiratory disease. *Clinical and Experimental Medicine* 6: 79–83.

Pankhurst, L.J., Deacon, L.J., Liu, J. et al. (2011). Spatial variations in airborne microorganism and endotoxin concentrations at green waste composting facilities. *International Journal of Hygiene and Environmental Health* 214 (5): 376–383.

Pastorello, E.A., Pravettoni, V., Farioli, L. et al. (2002). Hypersensitivity to mugwort (*Artemisia vulgaris*) in patients with peach allergy is due to a common lipid transfer protein allergen and is often without clinical expression. *Journal of Allergy and Clinical Immunology* 110 (2): 310–317.

Pehkonen, E. and Rantio-Lehtimäki, A. (1995). Sensitivity of IgG-ELISA for detecting airborne pollen antigens of Betulaceae. *Grana* 34 (5): 350–356.

Peternel, R., Music Milanovic, S., and Srnec, L. (2008). Airborne ragweed (*Ambrosia artemisiifolia* L.) pollen content in the city of Zagreb and implications on pollen allergy. *Annals of Agricultural and Environmental Medicine* 15 (1): 125–130.

Peden, D. and Reed, C.E. (2010). Environmental and occupational allergies. *Journal of Allergy and Clinical Immunology* 125 (2): S150–S160.

Platts-Mills, T.A. (1998). The role of allergens in allergic airway disease. *Journal of Allergy and Clinical Immunology* 101 (2): S364–S366.

Polosa, R., Ciamarra, I., Mangano, G. et al. (2000). Bronchial hyperresponsiveness and airway inflammation markers in nonasthmatics with allergic rhinitis. *European Respiratory Journal* 15 (1): 30–35.

Port, A., Hein, J., Wolff, A., and Bielory, L. (2006). Aeroallergen prevalence in the northern New Jersey-New York City metropolitan area: a 15-year summary. *Annals of Allergy, Asthma & Immunology* 96 (5): 687–691;Rantio-Lehtimaki, A., Viander, M., and Koivikko, A. (1994). Airborne birch pollen antigens in di¤erent particle sizes. *Clinical and Experimental Allergy* 24: 23–28.

Radon, K. (2006). The two sides of the "endotoxin coin". *Occupational and Environmental Medicine* 63 (1): 73–78.

Reid, M.J., Moss, R.B., Hsu, Y.P. et al. (1986). Seasonal asthma in northern California: allergic causes and efficacy of immunotherapy. *Journal of Allergy and Clinical Immunology* 78: 590–600.

Riggi, S.J. and Di Luzio, N.R. (1961). Identification of a reticuloendothelial stimulating agent in zymosan. *American Journal of Physiology-Legacy Content* 200 (2): 297–300.

Rizzi, J., Moro, T.R., Winnischofer, S.M.B. et al. (2023). Chemical structure and biological activity of the (1 → 3)-linked β-d-glucan isolated from marine diatom *Conticribraweissflogii*. *International Journal of Biological Macromolecules* 224: 584–593.

Rouze, A., Loridant, S., Poissy, J. et al. (2017). Biomarker-based strategy for early discontinuation of empirical antifungal treatment in critically ill patients: a randomized controlled trial. *Intensive Care Medicine* 43: 1668–1677.

Rossi, O.V.J., Kinnula, V.L., Tienari, J., and Huhti, E. (1993). Association of severe asthma attacks with weather, pollen, and air pollutants. *Horax* 48: 244–248.

Ruiz-Herrera, J. (1991). Biosynthesis of beta-glucans in fungi. *Antonie Van Leeuwenhoek* 60: 72–81.

Runswick, S., Mitchell, T., Davies, P. et al. (2007). Pollen proteolytic enzymes degrade tight junctions. *Respirology* 2007 (12): 834–842.

Rylander, R. (2002). Endotoxin in the environment—exposure and effects. *Journal of Endotoxin Research* 8 (4): 241–252.

Rylander, R. (2006). Endotoxin and occupational airway disease. *Current Opinion in Allergy and Clinical Immunology* 6 (1): 62–66.

Rylander, R., Bake, B.J.O.R.N., Fischer, J.J., and Helander, I.M. (1989). Pulmonary function and symptoms after inhalation of endotoxin 1, 2. *American Review of Respiratory Disease* 140: 981–986.

Salvaggio, J.E. (1994). Inhaled particles and respiratory disease. *Journal of Allergy and Clinical Immunology* 94: 304–309.

Salvaggio, J., Seabury, J., and Schoenhardt, E.A. (1971). New *Orleans asthma*. V. Relationship between Charity Hospital asthma admission rates, semiquantitative pollen and fungal spores counts, and total particulate aerometric sampling data. *Journal of Allergy and Clinical Immunology* 48: 96–114.

Sauro, A., Barone, F., Blasio, G. et al. (2006). Do influenza and acute respiratory infective diseases weigh heavily on general practitioners' daily practice? *The European Journal of General Practice* 12 (1): 34–36.

Schäppi, G.F., Taylor, P.E., Pain, M.C.F. et al. (1999). Concentrations of major grass group 5 allergens in pollen grains and atmospheric particles: implications for hay fever and allergic asthma sufferers sensitized to grass pollen allergens. *Clinical and Experimental Allergy* 29 (5): 633–641.

Severcan, E.U., Baskaya, N., Ertugrul, A., and Bostanci, İ. (2021). Chenopodium album is a remarkable allergen pointing to allergic rhinoconjunctivitis in children. *Revue Française d'Allergologie* 61 (6): 406–409.

Shahali, Y., Sutra, J.P., Hilger, C. et al. (2017). Identification of a polygalacturonase (Cup s 2) as the major CCD-bearing allergen in *Cupressus sempervirens* pollen. *Allergy* 72 (11): 1806–1810.

Shamsollahi, H.R., Ghoochani, M., Jaafari, J. et al. (2019). Environmental exposure to endotoxin and its health outcomes: a systematic review. *Ecotoxicology and Environmental Safety* 174: 236–244.

Shivpuje, S.H., Mehta, A.K., Patil, D.V., and Dharaiya, P.A. (2020). Evaluation of organic and inorganic dust concentration in different mechanized agricultural operations for wheat crop. *International Journal of Current Microbiology and Applied Sciences* 9: 2806–2813.

Shi, W., Bellusci, S., and Warburton, D. (2007). Lung development and adult lung diseases. *Chest* 132 (2): 651–656.

Sim, S.W., Moey, K.S.P., and Tan, N.C. (2014). The use of facemasks to prevent respiratory infection: a literature review in the context of the Health Belief Model. *Singapore Medical Journal* 55 (3): 160.

Singh, J. and Schwartz, D.A. (2005). Endotoxin and the lung: insight into the host-environment interaction. *Journal of Allergy and Clinical Immunology* 115 (2): 330–333.

Solomon, W.R. (1984). Aerobiology of pollinosis. *Journal of Allergy and Clinical Immunology* 74: 449–461.

Skiepko, R., Zietkowski, Z., Tomasiak-Lozowska, M.M. et al. (2011). Bronchial hyperresponsiveness and airway inflammation in patients with seasonal allergic rhinitis. *Journal of Investigational Allergology & Clinical Immunology* 21: 532–539.

Stetzenbach, L.D. (2009). Airborne infectious microorganisms. *Encyclopedia of Microbiology* 175–182.

Smith, K.R., Samet, J.M., Romieu, I., and Bruce, N. (2000). Indoor air pollution in developing countries and acute lower respiratory infections in children. *Thorax* 55 (6): 518–532.

Smith, M., Emberlin, J., Stach, A. et al. (2007). Regional importance of Alnus pollen as an aeroallergen: a comparative study of Alnus pollen counts from Worcester [UK] and Poznan [Poland]. *Annals of Agricultural and Environmental Medicine* 14 (1): 123–128.

Subiza, J., Cabrera, M., Valdivieso, R. et al. (1994). Seasonal asthma caused by airborne *Platanus* pollen. *Clinical and Experimental Allergy* 24: 1123–1129.

Takahashi, H., Ohno, N., Adachi, Y., and Yadomae, T. (2001). Association of immunological disorders in lethal side effect of NSAIDS on β-glucan-administered mice. *FEMS* 31: 1–14.

Takahashi, Y., Nagoya, T., Watanabe, M. et al. (1993a). A new method of counting airborne Japanese cedar (*Cryptomeria japonica*) pollen allergens by immunoblotting. *Allergy* 48: 94–98.

Takahashi, Y., Sakaguchi, M., Inouye, S. et al. (1993b). Airborne grass pollen antigens in a grassland as studied by immunoblotting with anti-lol p I antibody. *Grana* 32: 302–307.

Takahashi, Y., Sakaguchi, M., Inouye, S. et al. (1993c). ConÞrmation of the airborne occurrence of micron-size airborne pollen antigen carrying particles by immunoblotting. *Allergie et Immunologie (Paris)* 25: 132–136.

Tauber, H. (1967). Investigations of the mode of pollen transfer in forested areas. *Review of Palaeobotany and Palynology* 3 (1–4): 277–286.

Thorne, P.S., Perry, S.S., Saito, R. et al. (2010). Evaluation of the Limulus amebocyte lysate and recombinant factor C assays for assessment of airborne endotoxin. *Applied and Environmental Microbiology* 76 (15): 4988–4995.

Traidl-Hoffmann, C., Kasche, A., Menzel, A. et al. (2003). Impact of pollen on human health: more than allergen carriers. *International Archives of Allergy and Immunology* 131 (1): 1–13.

Troko, J., Myles, P., Gibson, J. et al. (2011). Is public transport a risk factor for acute respiratory infection? *BMC Infectious Diseases* 11 (1): 1–6.

Vanhaecke, E., Pijck, J., and Vuye, A. (1987). Endotoxin testing. *Journal of Clinical Pharmacy and Therapeutics* 12 (4): 223–235.

Van Cleemput, J., Poelaert, K.C.K., Laval, K. et al. (2019). Pollens destroy respiratory epithelial cell anchors and drive alphaherpesvirus infection. *Scientific Reports* 9: 1–15.

Vassallo, R., Standing, J.E., and Limper, A.H. (2000). Isolated *Pneumocystis carinii* cell wall glucan provokes lower respiratory tract inflammatory responses. *Journal of Immunology* 164: 3755–3763.

Vinhas, R., Cortes, L., Cardoso, I. et al. (2011). Pollen proteases compromise the airway epithelial barrier through degradation of transmembrane adhesion proteins and lung bioactive peptides. *Allergy* 66: 1088–1098.

Von Essen, S. (1997). The role of endotoxin in grain dust exposure and airway obstruction. *Current Opinion in Pulmonary Medicine* 3 (3): 198–202.

Wan, H., Winton, H.L., Soeller, C. et al. (1999). Derp 1 facilitates transepithelial allergen delivery by disruption of tight junctions. *The Journal of Clinical Investigation* 104 (1): 123–133.

Warren-Gash, C., Fragaszy, E., and Hayward, A.C. (2013). Hand hygiene to reduce community transmission of influenza and acute respiratory tract infection: a systematic review. *Influenza and Other Respiratory Viruses* 7 (5): 738–749.

Werner, J.L., Metz, A.E., Horn, D. et al. (2009). Requisite role for the Dectin-1 beta-glucan receptor in pulmonary defense against Aspergillus fumigates. *Journal of Immunology* 182: 4938–4946.

Whitehead, D.R. (1983). Wind pollination: some ecological and evolutionary perspectives. *Pollination Biology* 97 (8): 97–108.

Whittemore, A.S. (1981). Air pollution and respiratory disease. *Annual Review of Public Health* 2: 397–429.

World Health Organization. (2002). Implementation of the WHO Strategy for Prevention and Control of Chronic Respiratory Diseases: Meeting Report,11–12 February2002 (No. WHO/MNC/CRA/02.2). World Health Organization.

Widmer, F., Hayes, P.J., Whittaker, R.G., and Kumar, R.K. (2000). Substrate preference profiles of proteases released by allergenic pollens. *Clinical and Experimental Allergy* 30: 571–576.

Williams, D.L. (1997). Overview of (1 → 3)-β-d-glucan immunobiology. *Mediators of Inflammation* 6 (4): 247–250. https://doi.org/10.1080/09629359791550.

Williams, D.L., Sherwood, E.R., McNamee, R.B. et al. (1987). Chemo immunotherapy of experimental hepatic metastases. *Hepatology* 7 (6): 1296–1304.

Williams, D.L., Yaeger, R.G., Pretus, H.A. et al. (1989). Immunization against *Trypanosomacruzi*: adjuvant effect of glucan. *International Journal of Immunopharmacology* 11 (4): 403–410.

Williams, D.L., Sherwood, E.R., McNamee, R.B. et al. (1985). Therapeutic efficacy of glucan in a murine model of hepatic metastatic disease. *Hepatology* 5: 198–206.

Williams, D.L., Mueller, A., and Browder, W. (1995). Preclinical and clinical evaluation of carbohydrate immunopharmaceuticals in the prevention of sepsis and septic sequelae. *Journal of Endotoxin Research 2* (3): 203–208.

Williams, D.L., Pretus, H.A., McNamee, R.B. et al. (1991). Development, physicochemical characterization and preclinical efficacy evaluation of a water soluble glucan sulfate derived from Saccharomyces cerevisiae. *Immunopharmacology* 22 (3): 139–156.

Willment, J.A., Gordon, S., and Brown, G.D. (2001). Characterization of the human β-glucan receptor and its alternatively spliced isoforms. *Journal of Biological Chemistry* 276 (47): 43818–43823.

Williams, D. and Browder, W. (1994). Development of natural product (1 → 3)-β-d-glucan polymers as immune stimulating pharmaceuticals. *Polymers for Advanced Technologies* 5: 529–534.

Williams, D.L., Sherwood, E.R., Browder, I.W. et al. (1988). Effect of glucan on neutrophil dynamics and immune function in Escherichia coli peritonitis. *Journal of Surgical Research* 1988 (44): 54–61.

Williams, J.D., Topley, N., Alobaidi, H.M., and Harber, M.J. (1986). Activation of human poly-morpho nuclear leucocytes by particulate zymosan is related to both its major carbohydrate components: glucan and mannan. *Immunology* 58: 117–124.

Xia, Y., Veticka, V., Yan, Y. et al. (1999). The β-glucan-binding lectin site of mouse CR3. (CD11b/CD18) and its function in generating a primed state of the receptor that mediates cytotoxic activation in response to iC3b-opsonized target cells. *Journal of Immunology* 162: 2281–2290.

Yli-Panula, E. and Rantio-Lehtimäki, A. (1994). Antigenic activity in settled dust outdoors; a study of birch pollen antigens. *Grana* 33 (3): 177–180.

11

Problems of Drug Resistance in Indoor Biological Contaminants

Kalpana Sagar and K. Priti

Department of Microbiology, Gurukula Kangri (Deemed to be University), Haridwar, Uttarakhand, India

11.1 Introduction

Indoor biological contaminants are contaminants of biological origin, including bacteria, fungi, viruses, algae, insects, mites, and their by-products such as endotoxins, mycotoxins, and volatile organic compounds. Biological contaminants are present in indoor environments including houses, schools, university rooms, shops, cars, buses, planes, workplaces, and even in highly controlled environments. These contaminants originate from building materials, carpets, tissues (mites), plants (pollen, allergens, odors), and pets (dogs, cats, birds, etc.), and their effects can be exasperated by the heating modes (temperature, humidity), the degree of ventilation. People spend more than 90% time in indoor areas. Therefore, these contaminants affect their health. Indoor areas are never free from microorganisms. The major sources of propagation and spread of microorganisms in indoor areas are fans, coolers, air conditioners, and humidifiers are generally colonized by fungi such as *Aspergillus, Geotrichium, Penicillium,* and *Phialophora* (Dubey and Maheshwari 2014). Although the multiplication and survival of biological contaminants are not dependent on the air only, abiotic factors such as moisture, temperature, relative humidity, air circulation equipment, insulation, and maintenance of ducts regulate the indoor air environment. The number and types of bacteria may be higher in indoor environment than in outdoor environment as ultraviolet light is fatal for airborne in open or outdoor environment. The presence of *Arthrobacter* FB24, *Aspergillus clavatus, Brevibacillus brevis,* and *Trichoderma reesei* in indoor air of residential houses has been reported (Joshi and Srivastava 2013). Common airborne microorganisms such as *Bacillus,*

Airborne Biocontaminants and Their Impact on Human Health, First Edition.
Edited by Rajeev Singh and Anamika Singh.

Staphylococcus, Streptococcus, Micrococcus, Neisseria, Aspergillus, Alternaria, Cladosporium, and *Penicillium* have been found in university libraries (Kausar et al. 2016). Hospitals are the major indoor environment reservoirs responsible for pathogenic microorganisms, which later on transfer to other individuals, e.g. doctors, patients, workers, and visitors through talking, sneezing, coughing, and other human activities (Nevalainen and Seuri 2005). Major airborne transmitted pathogens in hospitals are *Staphylococcus aureus, Mycobacterium tuberculosis, Influenza virus, Aspergillus flavus, A. fumigates, Candida albicans* (Kausar et al. 2016). *Aspergillus* sp., *Alternaria* sp., *Penicillium* sp., *Trichoderma, Talaromyces* are predominantly present in the hematological hospital environment (Cho et al. 2018). *Aspergillus, Penicillium,* and *Cladosporium* may present at higher concentrations in ICU (Gonçalves et al. 2017). Mostly bacteria associated with buildings are saprophytic and are a part of the normal microflora found in mouth, ear, and nose, originating from outdoor air (Moldoveanu 2015). *Mycobacteria, Legionella* are heterotrophic bacteria, associated with biofilm formation in water pipes and water reservoirs of cooling systems. Few species of bacteria such as *Actinobacteria* and *Bacillus* are present in moist areas of buildings. Pets are an important source of bacteria and may be related to levels of endotoxins sourced in indoor environments (Ownby et al. 2013). The storage of organic waste in indoor areas increases the chances of biological contamination. Biological contaminants primarily induce immediate hypersensitivity (IgE) reactions and cause allergic rhinitis, lower and upper respiratory diseases, pneumonitis, measles, influenza, and chickenpox, and other types of immunologic responses, or infection (Chirca 2019). They may act as potential irritants and even toxins that also can contact or irritant dermatitis, and cause mycotoxin-induced flu-like symptoms, diarrhea, and cancer. The major symptoms caused by biological contaminants include fever, coughing, sneezing, watery eyes, shortness of breath, dizziness, lethargy, and digestive problems (Figures 11.1 and 11.2 and Table 11.1).

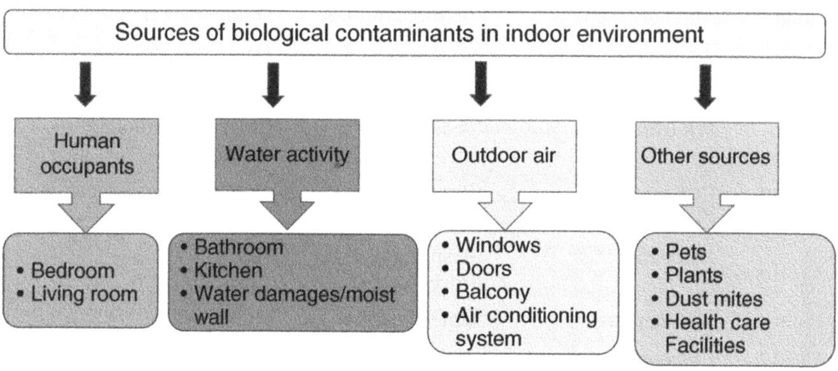

Figure 11.1 Sources of biological contaminants in indoor environment.

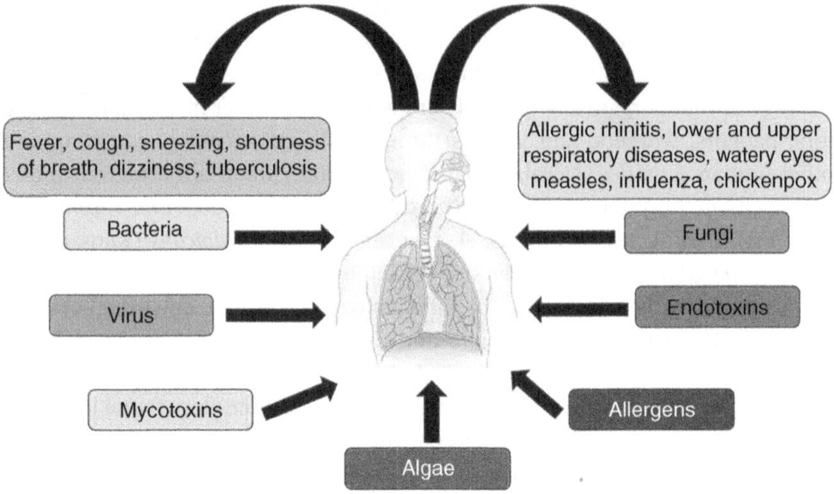

Figure 11.2 Effects of indoor biological contaminants on human health.

Table 11.1 Major indoor biological contaminants, their sources, and health effects (Kumar et al. 2021).

Biological contaminants		Disease cause	Sources in indoor area
Bacteria	*B. anthracis*	Anthrax	Animal handlers, veterinarians
	M. tuberculosis	Tuberculosis (TB)	Human
	Legionella spp.	Pneumonia	Hot water sources
	Y. pestis	Plague	Infected fleas
Fungi	*Alternaria* spp.	Asthma, rhinitis	Outdoor air, damp surfaces
	Aspergillus (aflatoxin)	Cancer	Damp surfaces
	Histoplasma spp.	Histoplasmosis	Bird droppings
	Penicillium spp.	Penicilliosis	Mold-contaminated building
Viruses	Coronavirus (SARS-CoV-2)	Respiratory infection	Human
	Influenza virus	Respiratory infection	Human
	Variola virus	Chickenpox	Human
Algae	*Chlorococcus* spp.	Asthma, rhinitis	Outdoor air

11.2 Drug Resistance: A Major Problem in Indoor Biological Contaminants

Drug resistance is the reduction in effectiveness of a medication such as antibiotics and other antimicrobial medicines and therefore infections become increasingly difficult or harder to treat and increasing the risk of disease spread, severe illness, and death. Antibiotic resistance is a serious problem threatening human health, agriculture, animal, and environment.

It is the emergence and rapid spread of resistant bacteria and genes among humans, animals and the environment, a resistant microbe is one that is not killed with the aid of using an antimicrobial agent after a preferred course of treatment. The use of antimicrobials for any infection, actual or feared, in any dose over a length of time, forces microbes to adapt or die. Microbes, which survive, carry drug-resistance genes and these genes may be transferred to other strains of same genus and species or unrelated species.

Drug resistance in bacteria may be intrinsic or acquired. Intrinsic resistance is naturally occurring resistance that does not require any mutation in gene of an organism, e.g. Vancomycin resistance in *Escherichia coli*, (Hawkey and Munday 2004) acquired resistance occurs when a bacterium, which was previously sensitive to antibiotics develops resistance while acquired antimicrobial resistance results by evolutionary process including several mechanisms such as HGT, drug inactivation, limiting drug uptake, and alteration of drug target by mutations.

Resistance genes produced withinside the bacteria are replicated and transferred through plasmids and transposons resulting in the emergence of multidrug resistant tuberculosis, which has now been identified in more than 100 countries (Vachon 1999). In addition to mutation, microorganisms have advanced a numerous array of biochemical and genetic structures to ensure the evolution and dissemination of antibiotic resistance. These consist of modification of antibiotics; therefore, modified antibiotics no longer react with the target site as occurs in beta-lactamases, which enzymatically cleave four-membered beta-lactam ring, rendering the antibiotic inactive.

Sometimes, antibiotic-resistant bacteria may protect the target of antibiotic action by reducing antibiotic uptake and/or a quick efflux of it, as happens between beta-lactam antibiotics and Gram-negative bacteria (Bush and Bradford 2016). Furthermore, the target antibiotic action may be altered, thus rendering the antibiotic ineffective, e.g. the resistance of *Enterococci* to cephalosporins (Darby et al. 2023). Bacteria may protect themselves from antibiotic action by the production of an alternative target usually an enzyme that is resistant to inhibition by the antibiotic. This mechanism allows bacteria to live on withinside the face of choice because the opportunity enzyme bypasses the impact of the antibiotic. The best-known example is the alternative penicillin-building proteins by methicillin-resistant *Staphylococcus aureus* (MRSA). It is, however, not uncommon to find a bacterium exhibiting more than one of these mechanisms (Chew et al. 2019).

11.3 Genetic Basis of Antimicrobial Resistance

The rapid distribution of antimicrobial resistance among several bacteria greatly results through the spread of mobile genetic elements (MGEs) such as plasmids and transposons. Antimicrobial resistance genes acquire on transposons ensuing the switch of multidrug resistance phenotypes to a susceptible recipient. Bacteria have a great genetic plasticity that permits them to survive in presence of antibiotic molecules. Therefore, bacteria share the same ecological niche with antimicrobial agent-producing organisms. The intrinsic resistance of bacteria allows them to thrive in presence of antimicrobial agents.

Bacteria use two major genetic approaches to survive in presence of antibiotics:

1) mutations in gene(s) often associated with the mechanism of action of the compound, and
2) acquisition of foreign DNA for antibiotic resistance through HGT.

11.4 Mutational Resistance

Bacterial cells that derived from a susceptible population may develop mutations in genes that result in the survival of cells in presence of drug molecules. Once the resistant mutant emerges, the resistant bacteria predominate and the drug eliminates the susceptible population.

Antimicrobial resistance resulting from mutation alters the action of antibiotics using one of the following mechanisms:

1) modifications of the drug target
2) decrease in the drug uptake,
3) activation of efflux mechanisms to eliminate harmful molecules,
4) worldwide modifications in critical metabolic pathways through modulation of regulatory networks

Thus, acquired drug resistance arising due to mutational changes is diverse and varies in complexity.

11.5 Horizontal Gene Transfer

HGT is the most important driver of bacterial evolution and is frequently responsible for the development of antimicrobial resistance. Most of the antimicrobial drugs used in clinical practice are found naturally (mostly in soil) in the environment. Bacteria sharing the environment with those molecules harbor intrinsic drug resistance suggesting that such "environmental resistome" is an abundant source for the accession of antibiotic resistance genes in clinically applicable bacteria. Furthermore, this genetic alternation has been implicated in the dissemination of resistance to many frequently used antibiotics.

Bacteria use three main strategies to acquire external genetic material:

1) Transformation
2) Conjugation
3) Transduction

Transformation is possibly the simplest type of HGT. However, only a handful of clinically applicable bacterial species are able to develop resistance by incorporating naked DNA. Conjugation is a very efficient method of gene transfer that involves cell-to-cell interaction resulting in the emergence of resistance in the hospital environment and probably occurs at high rates in the gastrointestinal tract of people under antibiotic treatment. In general, during conjugation, MGEs are used as vehicles to share valuable genetic information (Read and Woods 2014).

Transposons and plasmids are the most important MGEs, which play an important role to develop and disseminate antimicrobial resistance. Site-specific recombination by integrons, capable of recruiting mobile gene cassettes to incorporate open reading frames is one of the most efficient mechanisms for accumulating drug resistance. Integrons add new genes into bacterial chromosomes, a robust method of genetic exchange and one of the most important drivers of bacterial evolution (Nature 2013) (Tables 11.2 and 11.3).

11.6 Mechanisms of Resistance

Antimicrobial resistance follows four main mechanisms: (i) inactivating a drug; (ii) limiting uptake of a drug; (iii) modifying a drug target; (iv) active drug efflux. Intrinsic drug resistance can also additionally employ drug inactivation, limiting uptake of drug, and drug efflux while acquired drug resistance mechanisms may use the modification in drug target, drug efflux, and inactivation of drug. Types of mechanisms for drug resistance may differ in Gram-positive and Gram-negative bacteria due to differences in structure. Gram-negative bacteria make use of all four main drug resistance mechanisms, whereas Gram-positive

Table 11.2 Antimicrobial groups and their mechanism of action (Sandiumenge et al. 2006; Pakyz et al. 2008).

Mechanism of action	Antimicrobial groups
Inhibit cell wall synthesis	β-lactams
	Carbapenems
	Cephalosporins
	Monobactams
	Penicillins
	Glycopeptides
Depolarize cell membrane	Lipopeptides
Inhibit protein synthesis	Bind to 30S ribosomal subunit
	Aminoglycosides
	Tetracyclines
	Bind to 50S ribosomal subunit
	Chloramphenicol
	Lincosamides
	Macrolides
	Oxazolidinones
	Streptogramins
Inhibit nucleic acid synthesis	Quinolones
	Fluoroquinolones
Inhibit metabolic pathways	Sulfonamides
	Trimethoprim

bacteria less commonly use limiting the uptake of a drug as they do not have lipopolysaccharide outer membrane, and do not have the capacity for drug efflux (Figures 11.3 and 11.4).

11.6.1 Drug Inactivation

Drug resistance in bacteria inactivates drugs using two main ways; by degradation of the drug, or by transfer of a chemical group to the drug. A large group of drug-hydrolyzing enzymes are β-lactamases. Hydrolyzation is tetracycline, via the tetX gene also results in drug resistance. Mostly drug inactivation takes place by the transfer of acetyl, phosphoryl, and adenyl chemical groups to the drug. Transferases are identified to be involved in drug resistance. Adenylation and phosphorylation

Table 11.3 Bacteria with intrinsic resistance (Fajardo et al. 2008; Cox and Wright 2013; Reygaert 2018).

Bacteria	Intrinsic resistance
Acinetobacter spp.	Ampicillin, glycopeptides
All Gram-positives	Aztreonam
All Gram-negatives	Glycopeptides, lipopeptides
Bacteroides (anerobes)	Aminoglycosides, many β-lactams, quinolones
Enterococci	Aminoglycosides, cephalosporins, lincosamides
Escherichia coli	Macrolides
Klebsiella spp.	Ampicillin
Listeria monocytogenes	Cephalosporins
Pseudomonas aeruginosa	Sulfonamides, ampicillin, first- and second-generation cephalosporins, chloramphenicol, tetracycline
Serratia marcescens	Macrolides
Stenotrophomonas maltophilia	Aminoglycosides, β-lactams, carbapenems, quinolones

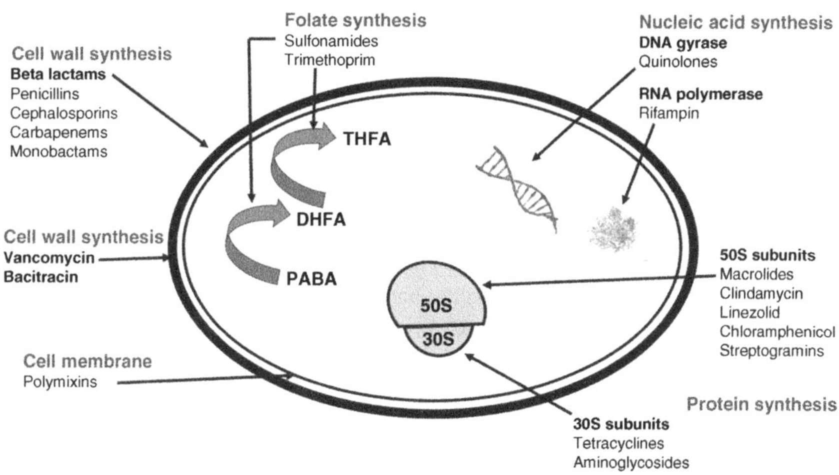

Figure 11.3 Different targets for antibiotics in normal cells.

are known to be used against aminoglycosides. Resistance to aminoglycosides, chloramphenicol, fluoroquinolones, and streptogramins is caused by the most diversely used mechanism acetylation.

β-lactam drugs are broadly used group of antimicrobial agents. The members of β-lactam group consist of a four-sided β-lactam ring. Three general mechanisms

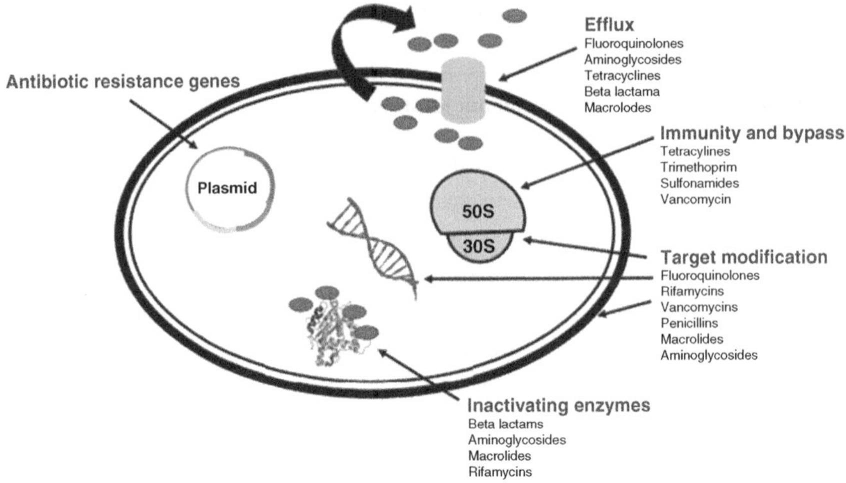

Figure 11.4 General antimicrobial resistance mechanism for different drugs.

follow for the resistance to the β-lactam drugs, which are: (i) modification in the drug target site to prevent the interaction between the drug and target PBP; (ii) extrude β-lactam drugs through efflux pumps; (iii) β-lactamase enzymes also called penicillinases and cephalosporinases, play an important role in the hydrolysis of the drug (Pfeifer et al. 2010; Bush and Bradford 2016). The β-lactamases cause the ring to open by hydrolyzing a specific site in the β-lactam ring structure to inactivate β-lactam drugs. The open β-lactam ring drugs are not capable of binding to their target PBP proteins. Gram-negative bacteria use the most common drug resistance mechanism by producing the β-lactamases enzymes against penicillin and cephalosporin (Bush and Jacoby 2010; Kumar et al. 2013).

The classification of β-lactamase enzymes is based on their molecular structure and functional characterization. On the basis of structure, β-lactamases are categorized into four categories (A, B, C, or D) and three functional groups based on the substrate specificity: serine β-lactamases, metallo (zinc-dependent) β-lactamases, and cephalosporinases. The production of β-lactamases from Gram-negative bacteria from all four structural groups, however, Gram-positive bacteria produce β-lactamases, which are mainly from group A and group B (Schultsz and Geerlings 2012; Bush 2013; Reygaert 2013; Toth et al. 2016). The gene for β-lactamase may be present in chromosomal DNA or may be acquired via plasmid DNA such as *Staphylococcus aureus*, *Enterococcus faecalis*, and *Enterococcus faecium* (Chancey et al. 2012; Schultsz and Geerlings 2012). The Gram-negative bacteria of *Enterobacteriaceae* family possess β-lactamase genes in chromosomal DNA including *Aeromonas* spp., *Acinetobacter* spp., and *Pseudomonas* spp. The first β-lactamase was identified and characterized from *E. coli* chromosomal

DNA. This β-lactamase results in ampicillin resistance encoded by the *ampC* gene and mutations in this may result in overexpression of the gene. β-lactamases encoded by AmpC are most effective against some first-generation cephalosporins and penicillins. Plasmid-borne β-lactamases carry *bla* gene for antibiotic drug resistance. It has been reported that β-lactamases are active against the carbapenems and are found primarily in the *Enterobacteriaceae*. There are two types of carbapenemases: *Klebsiella pneumoniae* carbapenemases (KPCs) and carbapenem-resistant enterobacteriaceae (CRE). The KPCs belong to the serine Class A β-lactamases, and are resistant to all β-lactam drugs, but may still be affected by β-lactamase inhibitors.

11.6.2 Limiting Drug Uptake

The structure of the Lipopolysaccharide layer in Gram-negative bacteria provides a barrier to certain types of molecules resulting in the bacterial cell's uptake of the antimicrobial agents. This mechanism provides bacteria with innate resistance to certain groups of antimicrobial drugs. The outer membrane of mycobacteria is rich in lipid content. Therefore, hydrophobic drugs such as rifampicin and fluoroquinolones can be more easily uptaken than hydrophilic drugs (Lambert 2002; Kumar and Schweizer 2005). Gram-positive bacteria do not contain an outer membrane and generally drug resistance is not as prevalent. *Staphylococcus aureus*, a Gram-positive bacterium, shows resistance to vancomycin. The mechanism for drug resistance to vancomycin by *S. aureus* is that vancomycin allows the bacteria to produce a thickened cell wall, which interferes drug entering the cell and provides resistance to vancomycin. In the enterococci, polar molecules cannot penetrate the cell wall easily, giving intrinsic resistance to some antimicrobial drugs including aminoglycosides.

Bacteria with large outer membranes allow substances to enter the cell through some channels called porin channels. Gram-negative bacteria allow hydrophilic molecules to enter the cell (Gill et al. 1998; Blair et al. 2014). Bacteria use two main ways to limit drug uptake via porin channels: a decrease in the number of porins to pass drug molecules, and a change in selectivity of the porin channel by mutations, e.g. drug resistance for imipenem and cephalosporins by *E. aerogenes* and drug resistance to β-lactams and tetracycline by *Neisseria gonorrhoeae* (Kumar and Schweizer 2005). Members of the Enterobacteriaceae cause drug resistance by reducing or stopping the production of the number of porins for carbapenems (Chow and Shlaes 1991; Cornaglia et al. 1996). The drugs including β-lactams and glycopeptides that target the cell wall of bacteria cannot affect the bacteria that lack the cell wall, such as Mycoplasma and such organisms are intrinsically resistant to these drugs.

Bacterial colonization to form biofilm by a bacterial community such as Pseudomonas aeruginosa is an important factor in drug resistance. Forming biofilm in pathogenic organisms protects the bacteria from antimicrobial drugs and attack by the host immune system. Biofilm is sticky and thick, which contains polysaccharides, DNA, and proteins from the resident bacteria and makes it difficult for antimicrobial drugs to react with bacteria.

11.6.3 Modification of Drug Targets

Antimicrobial drugs target various components in the bacterial cell and these targets may be modified by the bacteria to enable drug resistance. Gram-positive bacteria exclusively create resistance to the β-lactam drugs via alterations in the number and/or structure of penicillin-binding proteins (PBPs). PBPs are transpeptidases and important components involved in the construction of peptidoglycan in the cell wall. A change in the number of PBPs affects the binding of the drug to the target. An increase in PBPs results in a decrease in drug-binding ability. In *S. aureus*, a change in the structure of PBP2a by the acquisition of the mecA gene may totally inhibit the ability of the drug to bind (Reygaert 2009; Beceiro et al. 2013).

Vancomycin prevents the cell wall synthesis and daptomycin depolarizes the cell membrane. However, Gram-negative bacteria show intrinsic resistance to Vancomycin and daptomycin (Randall et al. 2013). *Enterococci* and *Staphylococcus aureus* have major issues of drug resistance to vancomycin. The resistance to vancomycin is mediated through the acquisition of van genes, which enable changes in the structure of peptidoglycan precursors. This results in a decrease in the binding ability of vancomycin to peptidoglycan (Beceiro et al. 2013). Daptomycin binds in the presence of calcium ions. Mutations in mprF genes alter the charge of the cell membrane surface to positive, therefore, inhibiting the binding of calcium.

Drug resistance for aminoglycosides and oxazolidinones via mutation in ribosomal subunits inhibits the binding of drugs that target the ribosomal subunits. Resistance to aminoglycosides, macrolides, oxazolidinones, streptogramins, and tetracyclines via ribosomal subunit methylation inhibits the drug from binding to the ribosome (Eliopoulos et al. 2003; Roberts 2004; Kumar et al. 2013). Resistance to fluoroquinolones via modifications in topoisomerase IV (Gram-positive bacteria, e.g. grlA) or DNA gyrase (Gram-negative bacteria – e.g. gyrA) by mutation decreases the ability of the drug to bind to these targets (Hawkey 2003; Redgrave et al. 2014).

Resistance is via mutations in enzymes involved in different metabolic pathways such as dihydropteroate synthase (DHPS) and dihydrofolate reductase (DHFR) involved in the folate biosynthesis pathway. Sulfonamides bind to DHPS and trimethoprim binds to DHFR due to their being structural analogs of the natural substrate. Mutation in these enzymes prevents the binding of drugs to them, however allowing the binding of natural substrate (Huovinen et al. 1995).

11.6.4 Drug Efflux

Drug efflux is key mechanism of drug resistance in Gram-negative bacteria. Drug efflux system pumps drugs (solute) out of cells and allows the microorganisms to regulate their internal environment by removing toxic substances including antimicrobial drugs, metabolite, and quorum-sensing molecules. Bacteria own chromosomally encoded genes for efflux pumps. Some of these genes are constitutively expressed, and others are overexpressed in high-level resistance. Mutation that modifies the transport channels resulting in drug resistance via drug efflux (Blair et al. 2014; Collu and Cascella 2013; Darby et al. 2023).

Bacterial drug efflux pumps have been classified into six families by the number of transmembrane-spanning regions, the number of components, types of molecules that the pump exports and the energy source used by the pump.

1) The ATP-binding cassette (ABC) superfamily (Martinez 2014)
2) The major facilitator superfamily (MFS) (Cox and Wright 2013)
3) The multidrug and toxic compound extrusion (MATE) (Fajardo et al. 2008)
4) The small multidrug resistance (SMR) family (Davies and Davies 2010)
5) The resistance-nodulation-division (RND) superfamily (Reygaert 2009)
6) The drug metabolite transporter (DMT) superfamily (Martinez 2014)

Most of the efflux pump families transport substrates across the cytoplasmic membrane, are single-component pumps. The RND family exclusively found in Gram-negative bacteria, are multi-component pumps that function in association with an outer membrane protein (OMP-porin) and a periplasmic membrane fusion protein (MFP) to efflux substrate across the cell (Kumar and Schweizer 2005; Villagra et al. 2012; Blair et al. 2014). In Gram-negative bacteria, one member of the ABC family, MacB, works as a tripartite pump (MacAB-TolC) to extrude macrolide drugs. In *E. coli*, nalidixic acid extrudes via members of MFS, EmrB as a tripartite pump (EmrAB-TolC) (Tanabe et al. 2009; Jo et al. 2017). Gram-negative bacteria possess efflux pumps that may come from all the families. However, efflux pump of RND family is the most significant pump (Kourtesi et al. 2013; Blair et al. 2014). In Gram-positive bacteria, efflux pump may confer intrinsic resistance due to being encoded on the chromosome via MATE and MFS families and efflux fluoroquinolones. Efflux pump in Gram-positive bacteria is known to be carried on a plasmid and comes under MFS family (Costa et al. 2013; Kourtesi et al. 2013).

The ABC efflux family may perform both uptake and efflux of a drug. The members of this family use energy derived from ATP hydrolysis. These pumps transport drugs, sugars, amino acids, ions, proteins, and polysaccharides. ABC transporters consisting of α-helices, function in the membrane in pairs, either as homodimers or heterodimers, usually are made up of six transmembrane segments (TMS) and work in conjunction with cytoplasmic ATPases. These pumps are utilized in clinically significant bacteria and have fairly specific substrates. *Vibrio cholerae* possess ABC pump (VcaM), which is capable of transporting tetracycline and fluoroquinolones (Putman et al. 2000; Kumar and Schweizer 2005; Lubelski et al. 2007).

The MFS efflux family functions via cation/solute (H^+ or Na^+) symport or solute/H^+ antiport. They are involved in the transport of drugs (e.g. tetracycline and macrolides), anions, sugars, and metabolites (e.g. bile salts). The MFS pumps have a great diversity of substrates, e.g. *Acinetobacter baumannii* utilizes MFS pumps (SmvA) for erythromycin and (CraA and CmlA) for chloramphenicol, and *Escherichia coli* contains MFS pumps (MefB) for macrolides, (QepA), for fluoroquinolones and (Fsr) for trimethoprim.

MFS pumps with a slightly broader substrate specificity include, the NorA pump in Staphylococcus aureus used to transport chloramphenicol and fluoroquinolones, and LmrS pump in S. aureus, which is used to transport chloramphenicol, erythromycin, trimethoprim and linezolid. Bacterial chromosomes possess most of the MFS pumps. In *E.coli*, nearly 50% of the efflux pumps are MFS pumps (Kumar and Schweizer 2005; Lubelski et al. 2007; Kumar et al. 2013; Blair et al. 2014).

The MATE efflux family efflux fluoroquinolone drugs, cationic dyes and aminoglycosides. The members of MATE efflux family use Na^+ gradient as the energy source. These pumps are found in Gram-negative bacteria. NorM was the first pump characterized in *Vibrio parahaemolyticus* from chromosomal DNA. *Neisseria gonorrhoeae* and *Neisseria meningitidis* also contain NorM pump (Rouquette-Loughlin et al. 2003; Piddock 2006; Kuroda and Tsuchiya 2009).

The SMR efflux family are hydrophobic, mainly efflux lipophilic cations and utilized energy by the proton-motive force (H^+) may have a very narrow substrate range. The genes for SMR efflux pumps are present in transposons, plasmids and chromosomal DNA. These pumps commonly confer resistance to aminoglycosides and β-lactams. *Staphylococcus epidermidis* efflux tetracycline, erythromycin and ampicillin via SMR pumps and *Escherichia coli* efflux tetracycline, erythromycin and vancomycin via EmeR pump (Yerushalmi et al. 1995; Kumar and Schweizer 2005; Blair et al. 2014).

The RND efflux family are found in Gram-negative bacteria, which catalyze substrate efflux via H^+ antiport mechanism. They are involved in the efflux of heavy metals, dyes, detergents, solvents and multidrug transporters (antibiotics). Some of these pumps may be specific for a class of drugs such as Mef pump for macrolides and Tet pump for tetracycline). A wide range of drugs including chloramphenicol, β-lactams, sulfamethoxazole, tetracycline, fluoroquinolones and trimethoprim in *Pseudomonas aeruginosa* confers intrinsic resistance via MexAB-OprM pump. These pumps are multi-component, complex pumps and made up of twelve

TMS. These pumps share a high degree of homology among the RND members. The genes for the RND efflux pumps are generally organized as an operon. AcrAB-TolC pump in *Escherichia coli* confers resistance to tetracycline, penicillins, chloramphenicol, fluoroquinolones and macrolides (Putman et al. 2000; Kumar and Schweizer 2005; Piddock 2006; Poole 2007; Martinez et al. 2009; Kourtesi et al. 2013; Blair et al. 2014; Martinez 2014; Blair et al. 2015).

11.7 Conclusion

A variety of biological contaminants are found in indoor as well as outdoor environment. These contaminants include bacteria, fungi, viruses, algae, insects, and their by-products such as endotoxins and mycotoxins. The presence of biological contaminants is associated with factors such as temperature, available nutrients, ventilation facility, and moisture. The qualitative and quantitative variation in biological contaminants is also affected by various physiological parameters such as temperature, moisture, ventilation, and availability of nutrients, are exposed to indoor biological contaminants and may suffer from various health problems including allergy, infectious diseases, respiratory infection, and other health issues. Millions of microbial cells are inhaled with the respiration of air.

Acquisition of drug resistance is a challenging problem in indoor biological contaminants. Antimicrobial resistance follows four main mechanisms: (i) inactivating a drug; (ii) limiting uptake of a drug; (iii) modifying a drug target; (iv) active drug efflux. Antibiotic resistance determinants may be transferred to other strains by a variety of HGT mechanisms, including transformation of naturally competent bacteria and phages, and the use of conjugative plasmids, transposons, and integrons. Despite barriers to the exchange of genetic information between different genera of bacteria, widespread transfer of resistance genes from chromosomes of environmental and soil bacteria to the mobilizable elements in bacterial isolates seems to have occurred. The prevalence of multidrug-resistant foodborne pathogens is a very high percentage, in many cases, even 100% of multi-drug-resistant foodborne pathogens is a disturbing phenomenon. Drug resistance mediated by HGT of resistance gene by plasmid in various species including *Streptococcus* spp., *Styaphylococcus* spp., *Enterococcus* spp., and *Listeria* spp. indicates a significant global risk of multi-drug resistance among bacteria.

There is an urgent need to monitor these bacteria for drug resistance, which may allow control of the spread of drug resistance among bacteria. Therefore, there is a necessity to develop novel strategies for control of drug resistance in indoor biological contaminants. Several new therapies such as bacteriophages, combinations of two or more antibiotics, and antimicrobial peptides, have been tested. However, it is necessary to take preventive measures such as maintaining hygiene, improving sanitation, ensuring the safety of food and water supplies, rapid identification for drug-resistant microorganisms and treatment, and the development of new-generation vaccines.

Moreover, better algorithms and the use of bioinformatics approaches for determining relationships between resistance determinants of different environmental niches will be highly beneficial. Additional genome sequencing data will also help fill the gaps in our knowledge of intermediate stages and carriers for mobilization.

References

Beceiro, A., Tomás, M., and Bou, G. (2013). Antimicrobial resistance and virulence: a successful or deleterious association in the bacterial world? *Clinical Microbiology Reviews* 26 (2): 185–230.

Blair, J.M., Richmond, G.E., and Piddock, L.J. (2014). Multidrug efflux pumps in Gram-negative bacteria and their role in antibiotic resistance. *Future Microbiology* 9 (10): 1165–1177.

Blair, J.M., Webber, M.A., Baylay, A.J. et al. (2015). Molecular mechanisms of antibiotic resistance. *Nature Reviews Microbiology* 13: 42–51.

Bush, K. (2013). Proliferation and significance of clinically relevant β-lactamases. *Annals of the New York Academy of Sciences* 1277 (1): 84–90.

Bush, K. and Bradford, P.A. (2016). β-Lactams and β-lactamase inhibitors: an overview. *Cold Spring Harbor Perspectives in Medicine* 6 (8): a025247.

Bush, K. and Jacoby, G.A. (2010). Updated functional classification of β-lactamases. *Antimicrobial Agents and Chemotherapy* 54 (3): 969–976.

Chancey, S.T., Zähner, D., and Stephens, D.S. (2012). Acquired inducible antimicrobial resistance in gram-positive bacteria. *Future Microbiology* 7 (8): 959–978.

Chew, J., Peh, S. and Sin Yeang, T., 2019. Non-microbial natural products that inhibit drug-resistant *Staphylococcus aureus*. *Staphylococcus aureus*, pp. 1–30.

Chirca, I. (2019). The hospital environment and its microbial burden: challenges and solutions. *Future Microbiology* 14 (12): 1007–1010.

Cho, S.Y., Myong, J.P., Kim, W.B. et al. (2018). Profiles of environmental mold: indoor and outdoor air sampling in a hematology hospital in Seoul, South Korea. *International Journal of Environmental Research and Public Health* 15 (11): 2560.

Chow, J.W. and Shlaes, D.M. (1991). Imipenem resistance associated with the loss of a 40 kDa outer membrane protein in *Enterobacter aerogenes*. *Journal of Antimicrobial Chemotherapy* 28 (4): 499–504.

Collu, F. and Cascella, M. (2013). Multidrug resistance and efflux pumps: insights from molecular dynamics simulations. *Current Topics in Medicinal Chemistry* 13 (24): 3165–3183.

Cornaglia, G., Mazzariol, A., Fontana, R., and Satta, G. (1996). Diffusion of carbapenems through the outer membrane of enterobacteriaceae and correlation of their activities with their periplasmic concentrations. *Microbial Drug Resistance* 2 (2): 273–276.

Costa, S.S., Viveiros, M., Amaral, L., and Couto, I. (2013). Multidrug efflux pumps in *Staphylococcus aureus*: an update. *The Open Microbiology Journal* 7: 59.

Cox, G. and Wright, G.D. (2013). Intrinsic antibiotic resistance: mechanisms, origins, challenges and solutions. *International Journal of Medical Microbiology* 303 (6–7): 287–292.

Darby, E.M., Trampari, E., Siasat, P. et al. (2023). Molecular mechanisms of antibiotic resistance revisited. *Nature Reviews Microbiology* 21 (5): 280–295.

Davies, J. and Davies, D. (2010). Origins and evolution of antibiotic resistance. *Microbiology and Molecular Biology Reviews* 74 (3): 417–433.

Dubey, R.C. and Maheshwari, D.K. (2014). *Textbook of Microbiology*, 5e. New Delhi: S. Chand and Co., Limited.

Eliopoulos, G.M., Eliopoulos, G.M., and Roberts, M.C. (2003). Tetracycline therapy: update. *Clinical Infectious Diseases* 36 (4): 462–467.

Fajardo, A., Martínez-Martín, N., Mercadillo, M. et al. (2008). The neglected intrinsic resistome of bacterial pathogens. *PLoS One* 3 (2): e1619.

Gill, M.J., Simjee, S., Al-Hattawi, K. et al. (1998). Gonococcal resistance to β-lactams and tetracycline involves mutation in loop 3 of the porin encoded at the penB locus. *Antimicrobial Agents and Chemotherapy* 42 (11): 2799–2803.

Gonçalves, C.L., Mota, F.V., Ferreira, G.F. et al. (2017). Airborne fungi in an intensive care unit. *Brazilian Journal of Biology* 78: 265–270.

Hawkey, P.M. (2003). Mechanisms of quinolone action and microbial response. *Journal of Antimicrobial Chemotherapy* 51 (suppl_1): 29–35.

Hawkey, P.M. and Munday, C.J. (2004). Multiple resistance in gram-negative bacteria. *Reviews and Research in Medical Microbiology* 15 (2): 51–61.

Huovinen, P., Sundström, L., Swedberg, G., and Sköld, O.J.A.A. (1995). Trimethoprim and sulfonamide resistance. *Antimicrobial Agents and Chemotherapy* 39 (2): 279–289.

Jo, I., Hong, S., Lee, M. et al. (2017). Stoichiometry and mechanistic implications of the MacAB-TolC tripartite efflux pump. *Biochemical and Biophysical Research Communications* 494 (3-4): 668–673.

Joshi, M. and Srivastava, R.K. (2013). Identification of indoor airborne microorganisms in residential rural houses of Uttarakhand, India. *International Journal of Current Microbiology and Applied Sciences* 2: 146–152.

Kausar, M.A., Arif, J.M., Alanazi, S.M. et al. (2016). Assessment of microbial load in indoor environment of university and hospitals of hail, ksas. *Biochemical and Cellular Archives* 16 (1): 177–183.

Kourtesi, C., Ball, A.R., Huang, Y.Y. et al. (2013). Suppl 1: microbial efflux systems and inhibitors: approaches to drug discovery and the challenge of clinical implementation. *The Open Microbiology Journal* 7: 34.

Kumar, A. and Schweizer, H.P. (2005). Bacterial resistance to antibiotics: active efflux and reduced uptake. *Advanced Drug Delivery Reviews* 57 (10): 1486–1513.

Kumar, S., Mukherjee, M.M., and Varela, M.F. (2013). Modulation of bacterial multidrug resistance efflux pumps of the major facilitator superfamily. *International Journal of Bacteriology* 2013: 1–15.

Kumar, P., Kausar, M.A., Singh, A.B., and Singh, R. (2021). Biological contaminants in the indoor air environment and their impacts on human health. *Air Quality, Atmosphere and Health* 14 (11): 1723–1736.

Kuroda, T. and Tsuchiya, T. (2009). Multidrug efflux transporters in the MATE family. *Biochimica et Biophysica Acta (BBA)-Proteins and Proteomics* 1794 (5): 763–768.

Lambert, P.A. (2002). Cellular impermeability and uptake of biocides and antibiotics in Gram-positive bacteria and mycobacteria. *Journal of Applied Microbiology* 92 (s1): 46S–54S.

Lubelski, J., Konings, W.N., and Driessen, A.J. (2007). Distribution and physiology of ABC-type transporters contributing to multidrug resistance in bacteria. *Microbiology and Molecular Biology Reviews* 71 (3): 463–476.

Martinez, J.L. (2014). General principles of antibiotic resistance in bacteria. *Drug Discovery Today: Technologies* 11: 33–39.

Martinez, J.L., Sánchez, M.B., Martínez-Solano, L. et al. (2009). Functional role of bacterial multidrug efflux pumps in microbial natural ecosystems. *FEMS Microbiology Reviews* 33 (2): 430–449.

Moldoveanu, A.M. (2015). Biological contamination of air in indoor spaces. In: *Current Air Quality Issues* (ed. F. Nejadkoorki), 489–514. IntechOpen.

Nature, E. (2013). The antibiotic alarm. *Nature* 495 (7440): 141.

Nevalainen, A. and Seuri, M. (2005). Of microbes and men. *Indoor Air* 15: 58–64.

Ownby, D.R., Peterson, E.L., Wegienka, G. et al. (2013). Are cats and dogs the major source of endotoxin in homes? *Indoor Air* 23 (3): 219–226.

Pakyz, A., Powell, J.P., Harpe, S.E. et al. (2008). Diversity of antimicrobial use and resistance in 42 hospitals in the United States. *Pharmacotherapy: The Journal of Human Pharmacology and Drug Therapy* 28 (7): 906–912.

Pfeifer, Y., Cullik, A., and Witte, W. (2010). Resistance to cephalosporins and carbapenems in gram-negative bacterial pathogens. *International Journal of Medical Microbiology* 300 (6): 371–379.

Piddock, L.J. (2006). Clinically relevant chromosomally encoded multidrug resistance efflux pumps in bacteria. *Clinical Microbiology Reviews* 19 (2): 382–402.

Poole, K. (2007). Efflux pumps as antimicrobial resistance mechanisms. *Annals of Medicine* 39 (3): 162–176.

Putman, M., van Veen, H.W., and Konings, W.N. (2000). Molecular properties of bacterial multidrug transporters. *Microbiology and Molecular Biology Reviews* 64 (4): 672–693.

Randall, C.P., Mariner, K.R., Chopra, I., and O'Neill, A.J. (2013). The target of daptomycin is absent from *Escherichia coli* and other gram-negative pathogens. *Antimicrobial Agents and Chemotherapy* 57 (1): 637–639.

Read, A.F. and Woods, R.J. (2014). Antibiotic resistance management. *Evolution, Medicine, and Public Health* 2014 (1): 147.

Redgrave, L.S., Sutton, S.B., Webber, M.A., and Piddock, L.J. (2014). Fluoroquinolone resistance: mechanisms, impact on bacteria, and role in evolutionary success. *Trends in Microbiology* 22 (8): 438–445.

Reygaert, W. (2009). Methicillin-resistant *Staphylococcus aureus* (MRSA): molecular aspects of antimicrobial resistance and virulence. *Clinical Laboratory Science* 22(2): 115.

Reygaert, W.C. (2013). Antimicrobial resistance mechanisms of *Staphylococcus aureus. Microbial Pathogens and Strategies for Combating Them: Science, Technology and Education* 1 (1): 297–305.

Reygaert, W.C. (2018). An overview of the antimicrobial resistance mechanisms of bacteria. *AIMS Microbiology* 4 (3): 482.

Roberts, M.C. (2004). Resistance to macrolide, lincosamide, streptogramin, ketolide, and oxazolidinone antibiotics. *Molecular Biotechnology* 28: 47–62.

Rouquette-Loughlin, C., Dunham, S.A., Kuhn, M. et al. (2003). The NorM efflux pump of *Neisseria gonorrhoeae* and *Neisseria meningitidis* recognizes antimicrobial cationic compounds. *Journal of Bacteriology* 185 (3): 1101–1106.

Sandiumenge, A., Diaz, E., Rodriguez, A. et al. (2006). Impact of diversity of antibiotic use on the development of antimicrobial resistance. *Journal of Antimicrobial Chemotherapy* 57 (6): 1197–1204.

Schultsz, C. and Geerlings, S. (2012). Plasmid-mediated resistance in Enterobacteriaceae: changing landscape and implications for therapy. *Drugs* 72: 1–16.

Tanabe, M., Szakonyi, G., Brown, K.A. et al. (2009). The multidrug resistance efflux complex, EmrAB from *Escherichia coli* forms a dimer in vitro. *Biochemical and Biophysical Research Communications* 380 (2): 338–342.

Toth, M., Antunes, N.T., Stewart, N.K. et al. (2016). Class D β-lactamases do exist in gram-positive bacteria. *Nature Chemical Biology* 12 (1): 9–14.

Vachon, M. (1999). A report presented by George Soros Open Society Institute, New York.

Villagra, N.A., Fuentes, J.A., Jofré, M.R. et al. (2012). The carbon source influences the efflux pump-mediated antimicrobial resistance in clinically important Gram-negative bacteria. *Journal of Antimicrobial Chemotherapy* 67 (4): 921–927.

Yerushalmi, H., Lebendiker, M., and Schuldiner, S. (1995). EmrE, an *Escherichia coli* 12-kDa multidrug transporter, exchanges toxic cations and H+ and is soluble in organic solvents. *Journal of Biological Chemistry* 270 (12): 6856–6863.

12

Methicillin Resistant *Staphylococcus aureus* (MRSA)

An Underestimate Threat

Sudeshna Mandal[1]*, Anirban Kundu*[1]*, Dhitri Borah*[2]*, Mou Singha Ray*[1]*, and Chandrani Mondal*[1]

[1] *Department of Zoology, Laboratory of Helminth Parasitology, Visva-Bharati University, Shantiniketan, West Bengal, India*
[2] *Department of Zoology, Biswanath College, Biswanath Charali, Assam, India*

12.1 Introduction

Staphylococcus aureus (*S. aureus*) was formally isolated by Rosenbach not long after Ogston first discovered it in purulent pus from a leg abscess. It is highly suited to its human host and the healthcare setting (Newsom 2008). Coccoid bacteria belonging to the Firmicutes phylum, *S. aureus* is a Gram-positive, non-motile, coagulase-positive organism. *S. aureus* is by far the most clinically significant of the 52 species and 28 subspecies that make up the *Staphylococcus* genus. 20–40% of the general population has *S. aureus* in their nasal mucosal commensal microbiota (Wertheim et al. 2005; Becker et al. 2017). *S. aureus* is a common commensal and a major contributor to bacteremia, endocarditis, osteomyelitis, and skin and soft tissue infections (SSTIs). *S. aureus* can enter underlying tissues or the circulation and spread infection when the cutaneous and mucosal barriers are compromised, such as due to chronic skin diseases, wounds, or surgical intervention. *S. aureus* infection is more likely to occur in people with invasive medical devices (such as peripheral and central venous catheters) or weakened immune systems (Lowy 1998). With the expansion of hospital-based treatment, *S. aureus* soon rose to the top of the list of illnesses that are related to healthcare. Penicillin provided temporary relief, but in the 1940s, resistance developed through the lactamase gene blaZ. Around 1960, the creation of the first semi-synthetic anti-staphylococcal penicillins, and within

Airborne Biocontaminants and Their Impact on Human Health, First Edition.
Edited by Rajeev Singh and Anamika Singh.

a year after its initial clinical application, methicillin-resistant *S. aureus* (MRSA) was discovered. In fact, according to genetic research, methicillin resistance may have already existed before the first therapeutic application of anti-staphylococcal penicillins (Harkins et al. 2017).

MRSA is to blame for a number of diseases that individuals of all ages frequently contract (Frank et al. 1999; Chambers 2001). These organisms are often acquired in certain environments (healthcare facilities like hospitals and long-term care facilities) and among specific patient populations (patients with prolonged hospitalization, past antimicrobial use, indwelling catheters, bed sore, postoperative surgical wounds, and use of intravenous drugs or treatment with enteral feedings or dialysis) (Karchmer 2000; Graffunder and Venezia 2002; Lodise Jr. et al. 2003). Due to the restricted therapy choices and poor dosage that increases mortality and duration of hospital stay, MRSA infections provide a significant challenge to doctors (Soriano et al. 2000; Graffunder and Venezia 2002; Lodise et al. 2003). Methicillin resistance is mediated by the mobile genetic element mecA, which is transferred horizontally and known as the staphylococcal cassette chromosome mec (SCCmec) (Katayama et al. 2000). Penicillin-binding protein 2a (PBP2a), an enzyme that crosslinks the peptidoglycans in the bacterial cell wall, is produced by the mecA gene. Due to PBP2a's poor affinity for β-lactams, the whole spectrum of antibiotics is resistant to it (Hartman and Tomasz 1984).

Although the primary mechanism of resistance to β-lactam antibiotics is alteration of the target penicillin-binding proteins, MRSA strains have developed multiple mechanisms of resistance to a variety of antimicrobial classes over time, including macrolides, aminoglycosides, fluoroquinolones, tetracyclines, and lincosamide antibiotics like clindamycin. Vancomycin and other glycopeptide antibiotics have long been regarded as the sole medications to which MRSAs have not evolved resistance. Sadly, because glycopeptide antibiotics have been overused, MRSAs have developed with decreased sensitivity to these medicines (Rybak and Akins 2001; Howden et al. 2004).

MRSA was the cause of hospital epidemics (healthcare-associated MRSA, or HA MRSA) in various regions of the world during the decade after its original description (Chambers and Deleo 2009). A substantial change in MRSA epidemiology was observed when it was detected in individuals without previous healthcare contact (referred to as community-associated MRSA [CA-MRSA]), notably among indigenous populations in Australia in the 1980s (Faoagaliet al. 1992) and otherwise healthy persons, including children, in the United States in the 1990s (Fridkin et al. 2005). By the middle of the 2000s, it has also been linked to exposure to livestock (livestock-associated MRSA, or LA MRSA) (Voss et al. 2005). There is still no clear explanation for why some strain types repeatedly increase and decline.

12.2 Epidemiology

Some of the most significant developments in the epidemiology of infectious illnesses may be seen in the emergence and global spread of MRSA. Despite the fact that 209 early MRSA samples were sequenced, whole-genome sequencing (WGS) of these isolates implies that MRSA initially appeared in the mid-1940s, which is significantly earlier than the invention of methicillin (Harkins et al. 2017) though in the early 1960s, first MRSA was reported (Jevons 1961). In fact, rather than the introduction of methicillin, it has been speculated that the widespread use of penicillin was what sparked the growth of MRSA (Harkins et al. 2017). Children, the elderly, those who exercise frequently, those who use drugs intravenously, those from indigenous backgrounds or who live in underserved urban areas, those with HIV or cystic fibrosis, those who frequently interact with healthcare providers, and those in institutionalized populations, such as prisoners, are at higher risk of contracting MRSA (Moran et al. 2006; Hota et al. 2007; Boucher and Corey 2008; David et al. 2008). MRSA infection rates spiked during the 1990s and the beginning of the 2000s. Parallel declines in MRSA infections have been observed in several US and European populations since 2005, particularly in infections of the bloodstream and soft tissues (Kallen et al. 2010; Landrum et al. 2012; Dantes et al. 2013; Klein et al. 2017).

MRSA prevalence varies significantly by region, with Scandinavia having a low frequency and areas of America and Asia having the greatest incidence (Diekema et al. 2001). Spread of already-existing resistance clones and acquisition of SCCmec by a methicillin-sensitive *S. aureus* (MSSA) strain appear to be at least two of the methods by which MRSA spreads. Although the exact method of SCCmec's horizontal transmission is not fully known, epidemiological data indicates that this resistance mechanism has spread to the majority of *S. aureus* clones, including both human and animal pathogenic strains (Enright et al. 2000; Monecke et al. 2011). By the 1990s, incidences of MRSA infection began to occur in people who had never been hospitalized before, which led to the creation of distinct classifications for HA-MRSA and CA-MRSA. Lower rates of clindamycin resistance (especially in the United States), a higher chance of Panton-Valentine leucocidin (PVL) expression, a majority of SCCmec type IV or type V, and a predominance of strain types ST5 or ST8 were first used to identify CA-MRSA isolates (Tenover et al. 2006; Popovich et al. 2017). Nevertheless, genotypic variations by site of acquisition have started to homogenize during the 1990s, showing that CA-MRSA and HA-MRSA may both encroach on one another's niche.

According to surveillance data from European nations, the prevalence of MRSA is generally increasing from the north to the south of the continent. In northern Europe (such as the Netherlands, Norway, Sweden, and Denmark), only 5% of *S. aureus* isolates from invasive infections are methicillin-resistant, compared to 25–50% in southern Europe (for example, Portugal, Spain, Italy, and Greece) (Struyf and Mertens 2017). Different infection control procedures and the use of antibiotics are believed to be responsible for the discrepancies that have been reported (Köck et al. 2010). Since the early 2000s, after years of rising MRSA incidence, constant or declining prevalence has been seen in a number of nations (Struyf and Mertens 2017). Improvements in national control initiatives have been linked to this drop (Nielsen et al. 2012). With the declines largely occurring in HA-MRSA, there has been increasing recognition of animal reservoirs for human MRSA in Europe, particularly from food-producing animals such as pigs, cattle, and poultry (Garcia-Alvarez et al. 2012). The majority of this LA MRSA, which is clonal complex 398 (CC398), has infected people who deal with animals, particularly in the Netherlands, Spain (Lozano et al. 2012), and northern Germany (van Alen et al. 2017), although LA MRSA infections have also been noted in the general population (van Cleef et al. 2011; Larsen et al. 2015). Nonetheless, prolonged person-to-person transmission of LA MRSA appears to be rare at the moment (Hetem et al. 2013; Larsen et al. 2015).

Due in large part to the introduction of CA MRSA, MRSA has also been identified as the most frequent cause of SSTIs presenting to US hospital emergency rooms (particularly the USA300 clone in the early 2000s) (Moran et al. 2006; Talan et al. 2011). The USA300 Latin-American variation (USA300 LV), a parallel outbreak of CA MRSA closely linked to USA300, was originally identified in Colombia in 2005 and has since become the most common CA MRSA clone in northern South America (Planet et al. 2015). Genomic examination of MRSA samples from Switzerland has also revealed the geographic expansion of USA300 clones from South America to Europe (Von Dach et al. 2016). Since 2005, there have been declining trends in HA MRSA incidence in the United States, with hospital-onset HA MRSA infections falling by 54%. This is consistent with findings in Europe (Dantes et al. 2013).

Most hospitals in Asia have an endemic MRSA infection, and certain Asian nations have among the highest MRSA prevalence rates worldwide. Unfortunately, the majority of information is from high-income countries, such as Japan, South Korea, and Singapore, with little data from other countries. MRSA accounts for up to 50% of *S. aureus* bloodstream infections in some regions of Asia, despite regional variations (Chen and Huang 2014). MRSA prevalence is notably high in Japan and South Korea, with >70% of clinical isolates in South Korea having MRSA based on local surveillance data (Mendes et al. 2013; Chen and Huang 2014). The extensive use of inappropriate antibiotics (such as self-medication and over-the-counter usage) and high population densities that promote the fast transmission of multidrug-resistant organisms are considered to be associated with high methicillin resistance rates (Chen and Huang 2014). Since 2001, MRSA-associated

S. aureus isolates in Australia have made up 20–33% of the total (Bell et al. 2020). Australia, like Europe and the United States, has put in place a variety of regional and national initiatives that have been linked to a decline in HA MRSA bloodstream infections since 2002 (Mitchell et al. 2014). Sequence type 93 (ST93; Queensland clone), ST30 (Southwest Pacific clone) (Williamson et al. 2014), and other virulent clones have been discovered as circulating in the area since the first reports of CA MRSA in isolated indigenous tribes in Western Australia in the late 1980s (Faoagali et al. 1992).

The quality and breadth of MRSA prevalence data from Africa vary. There are few published statistics from other countries compared to South Africa, Nigeria, and Mediterranean basin countries. The majority of data come from single-center research as well, and there is a shortage of information from larger monitoring systems. Moreover, the majority of research have relied on phenotypic methods to identify MRSA, and depending on the antibiotic used to detect MRSA, these tests may be less accurate than genotypic approaches (Falagas et al. 2013).

12.2.1 Pathophysiology of Bacterial Colonization and Transmission

In most instances, *S. aureus* colonization occurs before the onset of infection (Klein et al. 2017). Less often, infection may happen without established *S. aureus* colonization, for instance, as a consequence of contamination of catheters or wounds due to subpar infection control procedures by healthcare professionals. *S. aureus* mostly colonizes the nose, although it may also be found in other places, including the throat and perineum (Sim et al. 2013). Three temporal patterns of *S. aureus* colonization – including both MSSA and MRSA have been discovered by longitudinal research. *S. aureus* was never discovered in 15% of people, continuously colonized in 15% of people (known as persistent carriers), intermittently in 70% of people (meaning that most people may frequently acquire *S. aureus* and spontaneously clear it), and never in any people (referred to as noncarriers) (Eriksen et al. 1995). Research examining certain host polymorphisms in genes responsible for the inflammatory response suggests that the host has additional variables that affect the carriage status (Emonts et al. 2008). It is not yet well known what these fundamental components exactly are. The length of colonization, particularly for MRSA, varies, and quoted estimates may be inaccurate due to antibiotic use, which may reduce the length of colonization. Together with host characteristics, the nasal microbiota and variables related to the disease itself might affect host carrier status (Lucet et al. 2009) (Figure 12.1).

Athletes, people in prison, military recruits, children, people in urban or underserved areas, people with an indigenous background, people who own pets or work with livestock, people who have had MRSA infection in the past, people

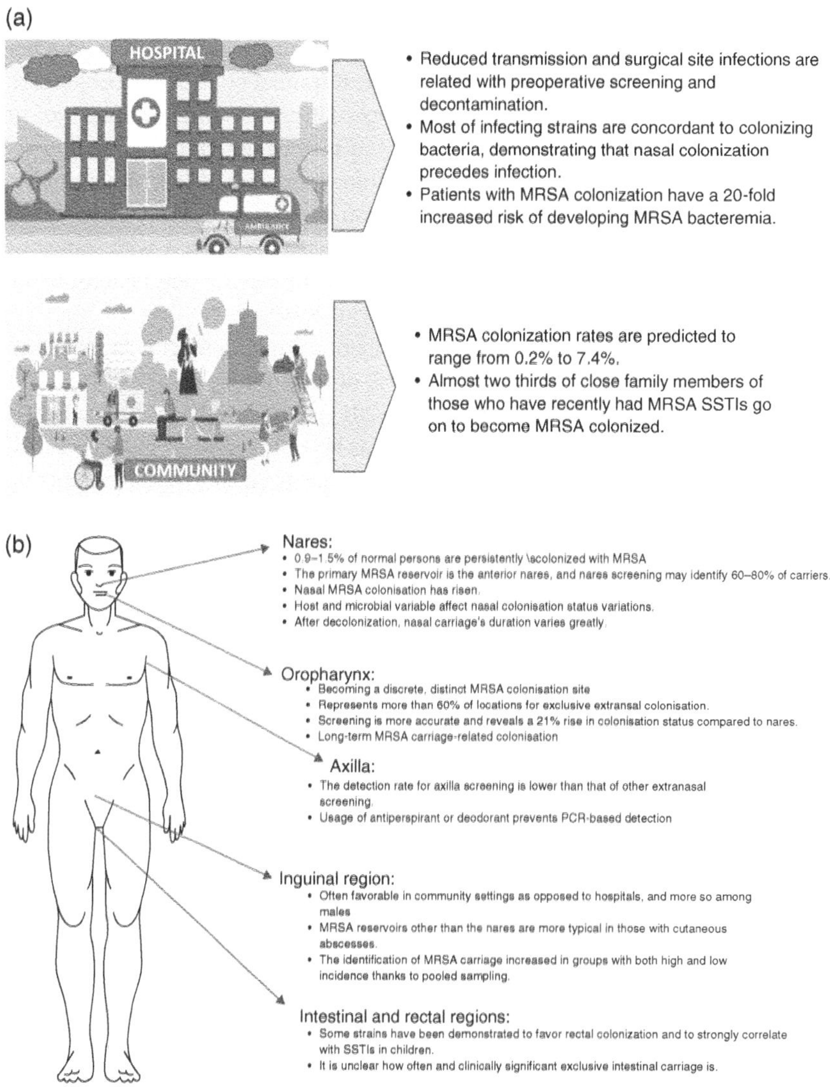

Figure 12.1 *Staphylococcus aureus* colonization with methicillin resistance. (a) Effects of colonization with methicillin-resistant *Staphylococcus aureus* (MRSA) on community transmission and hospital-acquired infections. (b) MRSA anatomical site screening. The most common and extensively used approach for identifying MRSA carriers is nares swab culture, although new findings have indicated that extranasal infection is common. A single evaluation of MRSA nasal carriage is insufficient, as shown by the one-third increase in MRSA detection from extranasal MRSA screening above that obtained from MRSA nares screening alone. Skin and soft tissue infection; SSTI, critical care unit. *Source:* Adapted from Turner et al. (2019)/Springer Nature.

with HIV or cystic fibrosis, and people who frequently interact with healthcare providers are all at increased risk of MRSA colonization (Hota et al. 2007; Boucher and Corey 2008; Schechter-Perkins et al. 2011; Karanika et al. 2016; Lekkerkerk et al. 2017). Recent antibiotic use has also been linked to a higher incidence of MRSA transmission (Lekkerkerk et al. 2017). It may be difficult to pinpoint precisely how long colonization lasts. However, some have seen MRSA remaining for more than six months after the first infection or exposure to MRSA (Cluzet et al. 2015). Moreover, MRSA has been easily recovered from a wide range of fomites, such as domestic settings, phones, pagers, notebooks, ties, pencils, white coats, gloves, and isolation gowns (Mollema et al. 2010).

In up to 50–80% of cases, colonizing and infecting strains are genetically identical, and MRSA colonization may raise the risk of infection by up to 25% (von Eiff et al. 2001; Huang et al. 2008; Clarridge et al. 2013). Particularly, in those where it has colonized, staphylococcal enterotoxin P has been linked to an increased incidence of bacteremia (Calderwood et al. 2014). It has been shown that colonizing strains gradually transition between methicillin-resistant and susceptible phenotypes (Patel et al. 2013; Azarian et al. 2016). Teichoic acid on the cell wall is involved in the early bacterial adhesion to the host's epithelial cells during *S. aureus* colonization, while microbial surface components that recognize sticky matrix molecules are involved in the later stages of nasal colonization (Weidenmaier et al. 2008; Burian et al. 2010). *S. aureus* clumping factor B (ClfB), one of these components, has been researched in vitro (Wertheim et al. 2008). The nose was injected with both the wild-type strain and its single locus clfB knockout variation; the knockout version cleared much more quickly than the wild-type strain. ClfB-deficient strains, however, may still interact with nasal cells, indicating that colonization is influenced by a number of different independent microbial surface components (O'Brien et al. 2002).

In addition to host and pathogen factors, *S. aureus* interacts with other species that colonize the nostrils, such as *Corynebacterium* spp., *Propionibacterium acnes*, and *Staphylococcus epidermidis* (*S. epidermidis*). According to research on the nasal microbiota, some species (such as *S. epidermidis*, which has been positively connected with *S. aureus* occurrence) are related to the presence or absence of *S. aureus* (Lemon et al. 2010; Liu et al. 2015). The nasal microbiota's organisms compete with one another in a variety of ways. For instance, they fight for nutrition and adhesion sites: The human nose contains very little nutrition. *S. aureus* is more suited to the human nose because it can live in conditions with fewer resources than coagulase-negative staphylococci can, probably due to changes in metabolism. Yet, there has been no discernible change in nutritional levels between carriers and noncarriers (Krismer et al. 2014). Moreover, certain microbiota species engage in antibiosis, or the production of chemicals that prevent their rivals from growing. For instance, *Staphylococcus lugdunensis* produces lugdunin, an antibiotic substance that inhibits and kills *S. aureus* (including MRSA) in a mouse model and in vitro, probably by causing a fast breakdown of bacterial energy supplies (Zipperer et al. 2016). Nasal colonization with *S. lugdunensis* in humans has been linked to a six-fold reduced risk of *S. aureus* colonization (Kaspar et al. 2016; Zipperer et al. 2016). Last but not least, *S. aureus* also engages in competitive host defenses by inducing the synthesis of host antimicrobial proteins that are less toxic to *S. aureus* than to other commensal bacteria (Krismer et al. 2017).

12.2.2 Virulence Factors and Toxins

The toxic shock syndrome toxin (TSST-1), enterotoxins, and exotoxins are just a few of the distinctive toxins and virulence factors that *S. aureus* produces (Boyle-Vavra et al. 2015; Thammavongsa et al. 2015) (Table 12.1). These toxins can cause serious clinical syndromes like septic shock, necrotizing pneumonia, and complicated skin and soft tissue infections (Boubaker et al. 2004). The virulence genes of CA-MRSA seem to be very varied and distinctive, clearly setting those apart from conventional HA-MRSA strains. According to genomic research, CA-MRSA strains possess a variety of virulence genes that are unique from those present in other *S. aureus* strains (Fey et al. 2003). *S. aureus* SSTIs are often started by bacterial transmission from the nose, which serves as the body's main reservoir, to open skin lesions and wounds (likely by hand touch) (Kluytmans et al. 1995; von Eiff et al. 2001) (Figure 12.2). *S. aureus* surface proteins that bind to extracellular matrix proteins, such as fibronectin-binding protein A (FnBPA), FnBPB, clumping factor A (ClfA), ClfB, and collagen adhesin (Cna), allow the germs to adhere to and grow on damaged tissues (Foster et al. 2014). The super antigen enterotoxins H and O-encoding genes seh and seo were discovered near the SCCmec complex. Only in this CA-MRSA sample were enterotoxins H and O described, which have very high binding affinities for the molecules of the major histocompatibility complex type II. It is noteworthy that enterotoxin H contributes to acute toxic shock-like syndromes and is generated in excessive numbers compared to other superantigens (Ren et al. 1994).

The *S. aureus* coagulase proteins induce a fibrin pseudo-capsule to develop around the bacteria and polymorphonuclear neutrophils (PMNs) that have been penetrated, halting additional leukocyte influx (Cheng et al. 2011). For example, *S. aureus* may prevent opsonization by producing a polysaccharide microcapsule (Thammavongsa et al. 2015) and blocking the complement cascade (Laarman et al. 2010). Bacteria that are phagocytosed by PMNs can survive not only by counteracting PMN-killing mechanisms (Peschel and Sahl 2006; Spaan et al. 2013; Stapels et al. 2014) but also by gradually

Table 12.1 Major *Staphylococcus aureus* virulence factors and corresponding host ligands.

Type of virulence factors	Virulence factors	Corresponding host ligands
Adhesins		
Surface proteins	• Collagen adhesin (Cna) • Fibronectin-binding protein A (FnBPA) and FnBPB • Iron-regulated surface determinant protein A (IsdA)	• Collagen • Fibronectin • Cytokeratin 10 (also known as KRT10)
Glycopolymers	• Wall teichoic acid (WTA)	• Scavenger receptors
Evasins		
MAMP receptor inhibitors	• Chemotaxis inhibitory protein of S. aureus (CHIPS) • Staphylococcal superantigen-like protein 3 (SSL3) • FPR-like 1 (FPRL1) inhibitory protein (FLIPr) • FLIPr-like	• FPR1 • TLR2 • C5a anaphylatoxin chemotactic receptor (C5aR1) • FPR2
Coagulation factors	• Secreted von Willebrand factor binding protein (vWbp) • Coagulase (Coa) • ClfB • Clumping factor A (ClfA)	• Fibrinogen • Prothrombin
Anticoagulants	• Staphylokinase	• Plasmin
Chemokine receptor inhibitors	SSL5 SSL10	• Several chemokine receptors
Opsonophagocytosis inhibitors	Immunoglobulin-binding protein Sbi Staphylococcus protein A (SpA) Microcapsule FLIPr	• Immunoglobulin-γ receptor (FcγR) • Immunoglobulin G (IgG)

(*Continued*)

Table 12.1 (Continued)

Type of virulence factors	Virulence factors	Corresponding host ligands
Complement inhibitors	Zinc metalloproteinase aureolysin Staphylococcal complement inhibitor (SCIN) Immunoglobulin-binding protein Sbi SSL7 Extracellular complement-binding protein (Ecb, also known as extracellular fibrinogen-binding protein) Fibrinogen-binding protein (Efb)	• Complement proteins C3, C3b, C3bBb, and C5a • Complement factor H
MAMPs		
Chemotactic MAMPs	• Phenol-soluble modulins (PSMs) • Formylated peptides	• N-Formyl-peptide receptor (FPR) 1 and FPR2
Nonchemotactic MAMPs	Peptidoglycan Lipoproteins DNA	Toll-like receptor (TLR) 2 and TLR9 Nucleotide-binding oligomerization domain-containing protein 2 (NOD2)
Toxins		
Superantigen toxins	Toxic shock syndrome toxin 1 (TSST1) Staphylococcal enterotoxin-like X (SEIX) Enterotoxins types (SE) A-Q	Major histocompatibility complex (MHC) class II T cell receptor
Pore-forming peptide toxins	• PSMα1–PSMα4 • δ-Toxin (also known as δ-haemolysin, Hld) • PSMβ1 and PSMβ2 • SCCmec-encoded PSM (PSMmec)	• Host cell membranes
Pore-forming protein toxins	α-Toxin Bi-component HlgCB Bi-component γ-Haemolysin (Hlg) AB Panton–Valentine leukocidin (PVL) Leukocidin (Luc) ED LucAB	Disintegrin and metalloproteinase domain-containing protein 10 (ADAM10) Several chemokine receptors C5aR1 Duffy antigen/chemokine receptor (DARC, also known as atypical chemokine receptor 1)

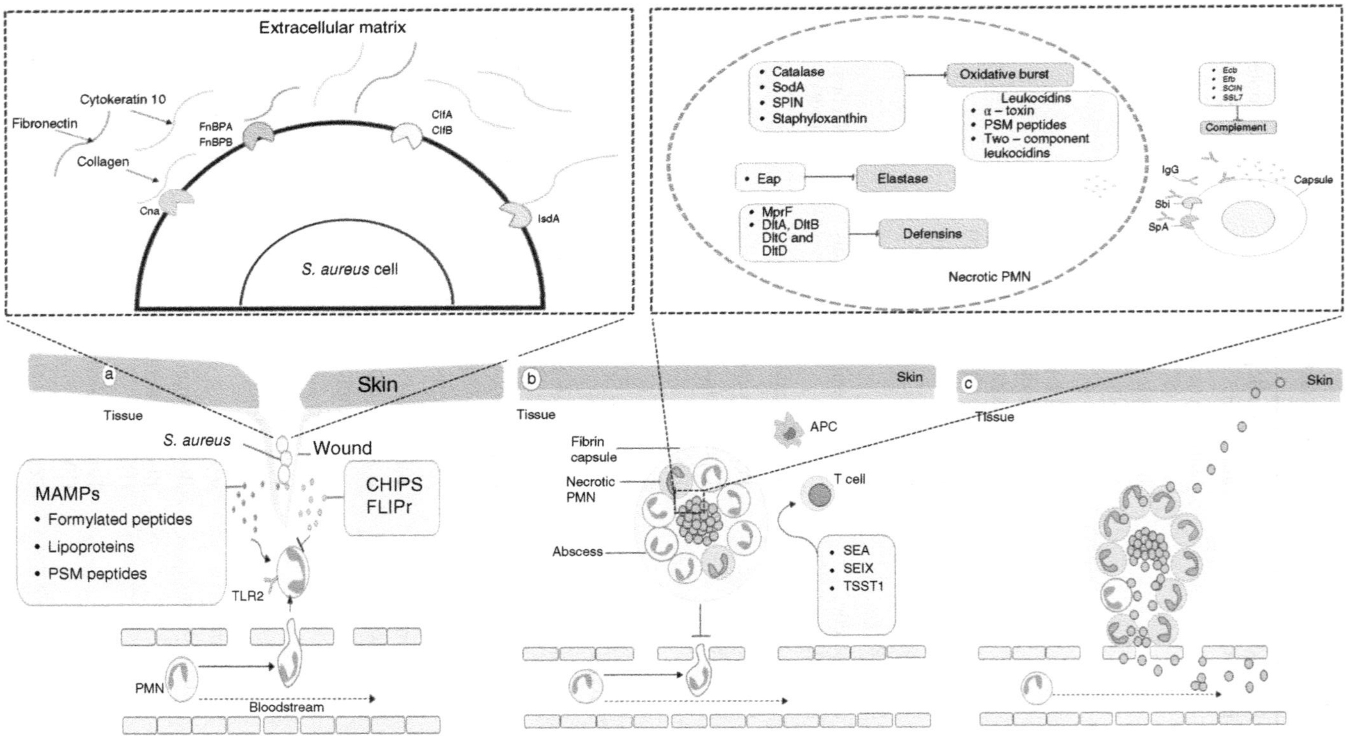

Extracellular matrix
Fibronectin
Cytokeratin 10
Collagen
FnBPA
FnBPB
ClfA
ClfB
Cna
IsdA
S. aureus cell
Catalase
SodA
SPIN
Staphyloxanthin
Oxidative burst
Leukocidins
α – toxin
PSM peptides
Two – component leukocidins
Eap
Elastase
MprF
DltA, DltB
DltC and
DltD
Defensins
Necrotic PMN
Ecb
Efb
SCIN
SSL7
Complement
IgG
Sbi
SpA
Capsule
a
Skin
Tissue
S. aureus
Wound
MAMPs
Formylated peptides
Lipoproteins
PSM peptides
CHIPS
FLIPr
TLR2
PMN
Bloodstream
b
Skin
Tissue
Fibrin capsule
Necrotic PMN
Abscess
APC
T cell
SEA
SEIX
TSST1
c
Skin
Tissue

Figure 12.2 (a) In order to specifically attach to extracellular matrix proteins like fibronectin, cytokeratin 10, and collagen, among others, bacteria use adhesin proteins such as fibronectin-binding protein A (FnBPA), FnBPB, iron-regulated surface determinant protein A (IsdA), clumping factor A (ClfA), ClfB, and collagen adhesin (Cna). *Staphylococcus aureus* may also partially control the entry of polymorphonuclear leukocytes (PMNs) by the use of inhibitors and activators of PMN chemotaxis, such as the chemotaxis inhibitory protein of *S. aureus* (CHIPS) and the FPRL1 inhibitory protein (FLIPr) (Spaan et al. 2013). Moreover, PSM peptides encourage the production of pro-inflammatory lipoproteins, the main *S. aureus* MAMP molecules, which activate Toll-like receptor 2 (TLR2) and cause local inflammation (Hanzelmann et al. 2016). (b) To polymerize fibrin and create an encapsulated abscess surrounding the infection site, *S. aureus* generates coagulases. Leukocidins and virulence factors that prevent opsonophagocytosis and PMN death reduce the ability of PMNs, which are abundantly present in an abscess, to kill *S. aureus*. Using a polysaccharide microcapsule and surface proteins (Staphylococcus protein A [SpA] and immunoglobulin-binding protein Sbi) that ineffectively bind immunoglobulin G (IgG) through the crystallizable fragment (Fc) domain, *S. aureus* may hinder successful opsonization by antibodies (Thammavongsa et al. 2015). Small secreted inhibitors like the staphylococcal complement inhibitor (SCIN), fibrinogen-binding protein (Efb), extracellular complement-binding protein (Ecb), and staphylococcal superantigen-like protein 7 (SSL7), among others, can also prevent the bacteria from activating the complement signalling pathway. Catalase, superoxide dismutase [Mn] 1 (SodA), staphylococcal peroxidase inhibitor (SPIN), staphyloxanthin (against the bactericidal oxidative burst generated by the PMNs), and extracellular adherence protein (Eap) (against elastase) are all produced by phagocytosed bacteria to help them survive inside PMNs (Spaan et al. 2013; Stapels et al. 2014). Multiple peptide resistance factor (MprF) and the d-alanine transfer proteins DltA, DltB, DltC, and DltD mediate the cell envelope alterations that defend against defensins. Furthermore, *S. aureus* secretes cytotoxic chemicals able to eradicate PMNs; Large pore-forming proteins (toxin and a number of two-component leukocidins, such as Panton-Valentine leukocidin [PVL]) (Spaan et al. 2017) and tiny peptides (PSM peptides) (Peschel and Otto 2013) are examples of *S. aureus* leukocidins. Superantigen toxins, including enterotoxin type A (SEA), staphylococcal enterotoxin-like X (SElX), and others, cause nonspecific T cell activation, which results in excessive inflammation. (c) At a later stage, abscesses may discharge live bacteria into the skin's surface or into the circulation; this is perhaps aided by the plasminogen-activating protein staphylokinase. Antigen-presenting cell, or APC. *Source:* Adapted from Turner et al. (2019)/Springer Nature.

destroying them with the help of cytolytic toxins. For instance, several CA MRSA clones generate the protein toxins (also known as α hemolysin) (Chambers and Deleo 2009; Löffler et al. 2010) and pore-forming peptides known as phenol soluble modulins (PSMs) and PVL toxin is another virulence factor unique to CA-MRSA strains. Less than 5% of *S. aureus* isolates were previously documented to generate this bicomponent cytotoxin. It is important to note that the term "toxin" refers to a substance that is present in certain amounts in certain foods. It may kill human leukocytes and cause significant tissue damage. In both children and adults, it has been linked to severe necrotizing pneumonia and necrotic skin sores (Lina et al. 1999; Baggett et al. 2004), consequently raising the pathogenicity of the bacterium. *S. aureus* superantigen toxins, which bind to MHC class II antigen-presenting cells and activate a significant portion of T cells nonspecifically, exacerbate the severe inflammation induced by active or necrotic PMNs. This systemic hyper-inflammation is known as "cytokine storms" (Spaulding et al. 2013).

Abscesses may rupture later on, releasing pus and live bacteria either onto the skin surface to aid in the spread of pathogens or into the circulation to result in bacteremia. *S. aureus* that is endovascular may cling to platelets and endothelial surfaces (Chavakis et al. 2005; Weidenmaier et al. 2005) and this adhesion may cause endocarditis, aid in the development of metastatic abscesses, or cause the absorption of bacteria into endothelial cells, where they are difficult for antibiotics and host defense chemicals to reach (Löffler et al. 2014). If the endovascular spread of the bacteria is not controlled, the agglutinating activity of coagulases is thought to contribute to systemic blood coagulation, and massive release of molecular pattern molecules associated with microorganisms along with superantigen toxin-induced cytokine storms cause sepsis and multi-organ failure (Thomer et al. 2016).

12.2.3 Mechanisms of Drug Resistance

The independent acquisition of the SCCmec complex by multiple multidrug-resistant strains in the early 1960s was a significant development in the evolution of *S. aureus*. Penicillin (blaZ), trimethoprim (dfrA and dfrK), erythromycin (ermC), clindamycin (ermC constitutively expressed), and tetracycline (tetK and tetL) resistance have all been found in MRSA and MSSA on insertion sequences, transposons, and sometimes plasmids (Malachowa and DeLeo 2010). Antibiotic resistance is often genetically related to disinfectant or heavy metal resistance (for example, quaternary ammonia compounds, mercury, or cadmium) among HA-MRSA bacteria, perhaps reflecting the high selection pressures within the hospital environment (Sidhu et al. 2002).

Similar to how nonmulti-drug-resistant, oxacillin-resistant *S. aureus* (NORSA) isolates were often sensitive to a range of non-β-lactams, the majority of CA-MRSA strains were but not all of them. Research demonstrated that CA-MRSA bacteria might develop non-β-lactam resistance via exposure. Additionally, it was discovered that the CA-MRSA strains were less resistant to oxacillin and imipenem-cilastatin (minimum inhibitory concentrations of 8–32 g/ml), suggesting that, in contrast to the HA-MRSA strains, the CA-MRSA strains were not selected by exposure to the potent β-lactam agents typically used in hospital settings. Comparing CA-MRSA with HA-MRSA, doubling times were substantially lower for CA-MRSA. Hence, by giving CA-MRSA the ability to outcompete other bacterial species that are typically a component of the commensal flora, this fast growth rate may aid in its effective colonization. A worldwide collection of 359 MRSA isolates was subjected to multilocus sequence typing and evolutionary mathematical models and the results showed that all MRSAs have a similar genetic history and can be traced to a single MRSA clone. The ST-250 strain of *S. aureus* is thought to have originated from an isolate of MSSA, which later gained the mec gene from an unidentified source. The horizontal transmission of genetic elements that insert into the bacterial genome is more important than the variety of MRSA strains. Recent MRSA isolates are either offspring of pre-existing clones or were produced by horizontal transmission of successful MSSA (Enright et al. 2002). β-lactam resistance in staphylococci is caused by the mecA gene. A penicillin-binding protein with a modest affinity for β-lactam-type antibiotics is encoded by this gene. The SCC is a particular integrative genetic component that harbors the mecA gene complex (SCC). The SCCmec element on the staphylococcal chromosome is integrated and excised by means of these mobile cassettes, which are made up of the mec complex and cassette recombinase genes (Figure 12.3) (Mongkolrattanothai et al. 2003).

The cassette chromosome recombinase (ccr) complex type and the class of the mec complex are used to categorize the twelve known SCCmec kinds (I–XII). For *S. aureus*, five SCCmec kinds have mostly been correctly identified (Table 12.2). The size, makeup, and expression of the related antimicrobial resistance varied across the gene components (Daum et al. 2002; Ma et al. 2002; Ito et al. 2004). Large SCCmec elements with Types I, II, and III are mainly seen in HA-MRSA and include genes that give resistance to multiple antibiotic classes (IWG-SCC 2009). These isolates have many transposable genetic elements and inserted plasmids downstream of the mecA complex. These strains associated with the healthcare industry often exhibit numerous non-β-lactam antibiotic resistance because of the SCCmec types II and III. SCCmec type IV is commonly discovered in isolates of NORSA and CA-MRSA. This SCCmec type is smaller and lacks additional genes for multidrug resistance. Similar to type IV, SCCmec type V is tiny, lacks antimicrobial resistance genes other than mecA, and is mostly present in

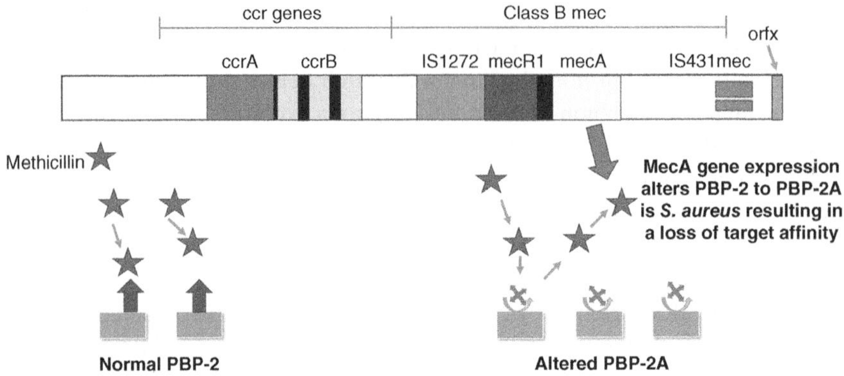

Figure 12.3 Recombinase genes complex located upstream of the *Staphylococcus* cassette chromosome, or mec, in its structure. The mecA gene, which causes *Staphylococcus aureus* to be resistant to β-lactam antibiotics, is part of the mec complex. IS1272 stands for insertion sequence-like element, ccrA and ccrB for cassette chromosome recombinase A and B, mecR1 for mec sensor transducer and repressor genes, IS431 for integrated plasmid encoding tetracyline resistance, and orfx for open reading frame, which is where the mobile elements (*Staphylococcus* cassette chromosome) are located.

Table 12.2 The five SCCmec types in *Staphylococcus aureus* strains.

SCCmec type	Size (kb)	ccr gene complexes	mec gene complexes	High-prevalence setting	Features
I	34.3	1 (A1B1)	B	HA-MRSA	Lacks other resistance genes
II	53.0	2 (A2B2)	A	HA-MRSA	Associated with multiple drug resistance
III	66.9	3 (A3B3)	A	HA-MRSA	Associated with multiple drug resistance
IV	20.9–24.3	2 (A2B2)	B	CA-MRSA and HA-MRSA	Resistant to β-lactam antibiotics
V	28.0	5 (C1)	C2	CA-MRSA and HA-MRSA	Lacks antibiotic resistance genes other than mecA

Source: Adapted from Daum et al. 2002/International Working Group on the Staphylococcal Cassette Chromosome elements.

CA-MRSA and NORSA strains (Ito et al. 2004). MecA, which encodes penicillin-binding protein 2a, is present in all SCCmec types (with the exception of type XI, which has the homologue mecC) (PBP2a) (Hartman and Tomasz 1984), a peptidoglycan transpeptidase. In the presence of lactam antibiotics that block the activity of the four native *S. aureus* penicillin-binding proteins (PBP1, PBP2, PPB3, and PBP4), PBP2a may take over the transpeptidase function of peptidoglycan production despite having a very low affinity for most lactam antibiotics (Figure 12.4). Several *S. aureus* clones from animal and human isolates (García-Álvarez et al. 2011) were found to have the mecA variation named mecC; PBP2aLGA, which is encoded by mecC and is called after the MRSA strain LGA251 from which it was first found (Kim et al. 2012).

The regulators encoded by mecI, mecR1, and mecR2 are primarily responsible for controlling the expression of mecA (Arêde et al. 2012) and on the factors that control how the genes blaZ, blaI, and blaRI express themselves (Zhang et al. 2001). A surprising number of auxiliary or accessory genes also have a significant impact on the resistance phenotype (De Lencastre et al. 1999). The degree of methicillin resistance is not correlated with the amount of mecA transcription, according to three lines of data. First, the antibiotic mupirocin induces the bacterial response to various stress conditions, such as amino acid, fatty acid, and iron restriction as well as heat shock, which results in an increase in PBP2a activity without affecting the transcription of mecA (Kim et al. 2013). Second, inactivating vraS (a component of the VraS-VraR two-component regulatory system, which controls the cell wall peptidoglycan biosynthesis) increased mecA transcription but did not enhance PBP2a activation (Boyle-Vavra et al. 2006). Finally, without changing the transcription of mecA, the chaperone foldase protein PrsA modifies the amount of correctly folded PBP2a in the membrane and, therefore, methicillin resistance (Jousselin et al. 2015). Given the widespread use of this antibiotic to treat MRSA infections, the emergence of vancomycin resistance is the most concerning genetic adaptation in *S. aureus* to date. This illustrates how both core and accessory genomic components individually influence the development of antibiotic resistance. There are two types of vancomycin resistance in *S. aureus*. Instances of vancomycin therapy that are extended or repeated often result in the emergence of vancomycin-intermediate *S. aureus* (VISA) strains. Heteroresistance is a property that refers to the fact that various mutations within a population provide varying degrees of vancomycin resistance. Time and persistent selection pressure, such as extended vancomycin treatment, favor strains with a progressively higher minimum inhibitory dose of vancomycin and an accumulation of numerous mutations (MICs). The recurrent WGS of isolates with continuously increasing vancomycin MICs, some of which had amassed over 30 distinct mutations, provided evidence for this process (Mwangi et al. 2007). The majority of VISA isolate mutations change the key genes responsible for cell wall production and autolysis.

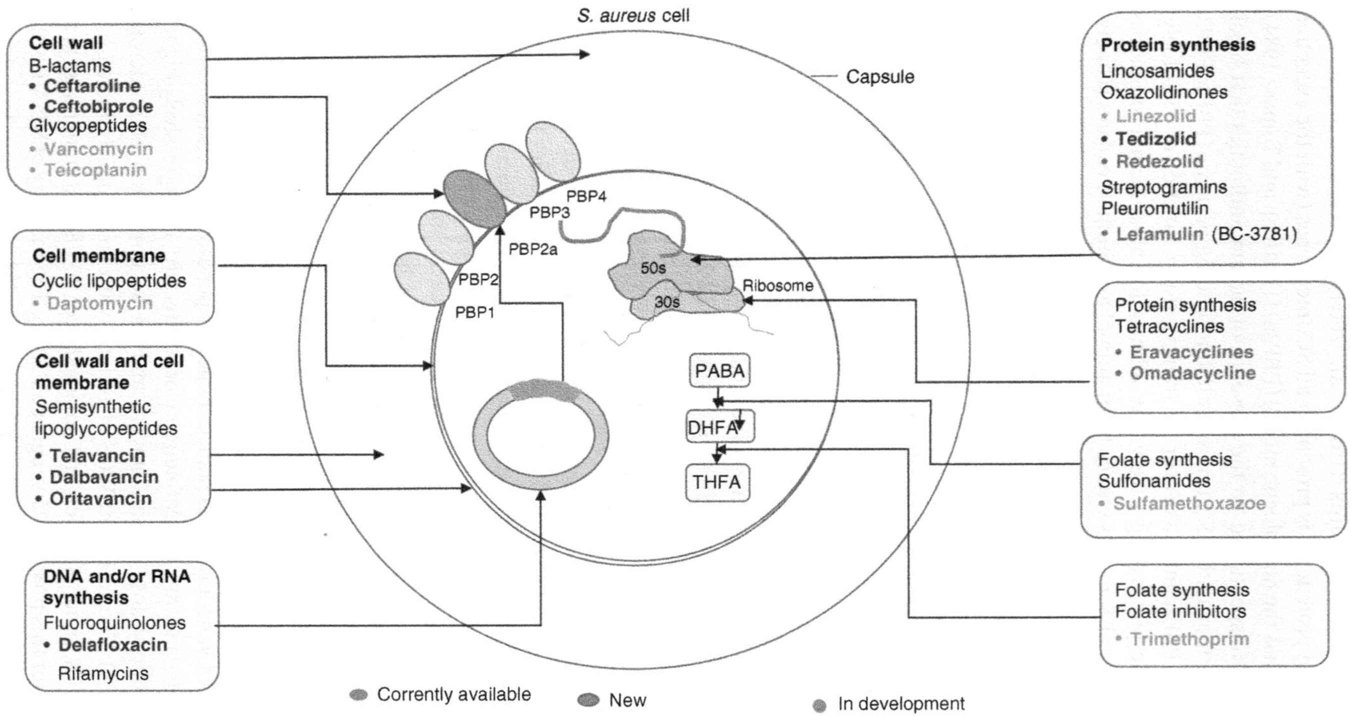

Figure 12.4 Medicines that are effective against MRSA's bacterial targets. Antibiotics target various bacterial structures or metabolic pathways via a variety of distinct modes of action. The colors for currently available antibiotic choices are green, currently authorized and marketed medications are blue, and antibiotics under development are orange. Dihydrofolic acid (DHFA), para-aminobenzoic acid (PABA), penicillin-binding protein (PBP), *Staphylococcus aureus* (*S. aureus*), and tetrahydrofolic acid (THFA). *Source:* Wright (2010)/Licensed Under CC BY 2.0.

The mutations in yycH, mprF, and dltA, among others, also provide cross-resistance to daptomycin (Mwangi et al. 2007; Chen et al. 2015). In contrast to VISA, it has been shown that vanA operon transfer from vancomycin-resistant *Enterococcus faecalis* results in high-level vancomycin-resistant *S. aureus* (Weigel et al. 2003). Since conjugative plasmids are uncommon and MRSA is not natively competent, the majority of plasmid transfer in MRSA happens by transduction. Luckily, the few instances of vancomycin-resistant *S. aureus* that have been recorded thus far suggest that this is an uncommon occurrence (Rossi et al. 2014).

12.2.4 Genomic Regulation

One intriguing aspect of MRSA epidemiology is that even while there is a lot of variation overall, just a few strains predominate. The reconstruction of the transmission of MRSA across healthcare systems and communities has been made possible by WGS. The WGS data of CA-MRSA isolates' phylogenetic reconstructions show that USA300 evolved via fast clonal expansion rather than convergent evolution (Kennedy et al. 2008). Even the co-occurrence of numerous instances of MRSA infection inside one family home, for example, has allowed researchers using WGS to conclude that household clustering likely had a significant role in the spread of USA300 across the population (Uhlemann et al. 2014; Alam et al. 2015). Similar to this, thorough phylogenetic reconstructions indicate that people who had been exposed to circulating community strains later brought USA300 into hospitals, causing CA-MRSA and HA-MRSA strains to eventually combine (Popovich et al. 2017).

The quorum-sensing system of the accessory gene regulator (Agr), as well as other regulatory networks, differentially control the majority of the *S. aureus* virulence factors (Le and Otto 2015). Many CA MRSA clones, such as USA300, have very active Agr systems, which results in copious toxin expression and translates to a high potential to induce SSTIs and invasive infections even in healthy people (Cheung et al. 2011). Arginine deiminase, a crucial enzyme in the arginine catabolism pathway, suppresses innate and adaptive immune responses and enhances pathogen survival. The arginine-catabolic mobile element (ACME) was suggested to improve the fitness of USA300 in comparison to other *S. aureus* strains by upregulating the expression of several genes within this pathway (Diep et al. 2006).

In contrast, many HA MRSA clones include an extra phenol-soluble modulin (PSM; PSMmec) gene that is transcribed by SCCmec and whose mRNA inhibits the production of Agr (Kaito et al. 2013). As a result, Agr is not highly active in many HA MRSA clones, which generate more adhesins than toxins and often lead to bacteremia via contaminated catheters or implanted medical equipment. Many isolates from bloodstream infections have been reported to have Agr-inactivating point mutations, suggesting that high virulence may even be harmful to *S. aureus* in bacteremia (Painter et al. 2014).

12.2.5 Pathogenesis

The skin and subcutaneous tissues are the most often affected organs by MRSA, followed by invasive infections such as osteomyelitis, meningitis, pneumonia, lung abscess, and empyema. MRSA-induced infectious endocarditis is connected to intravenous drug misuse and has a higher morbidity and fatality rate than that produced by any other pathogen.

12.2.5.1 Pneumonia

Prior to the development of antibiotics, staphylococcal pneumonia – previously known as post-influenza pneumonia was a separate clinical entity characterized by a sudden onset of respiratory symptoms and fatality rates between 80% and 90%. It had distinct clinical characteristics such as pulmonary bleeding and the production of microabscesses, as well as radiological characteristics like cavitary lesions, empyema, and pyopneumothorax. The course has been less violent, not necessarily linked to viral influenza, linked to other risk factors for *S. aureus* infections, and has a mortality rate of between 30% and 40% throughout the post-antibiotic period. Yet, CA-MRSA has lately been linked to life-threatening necrotizing pneumonia in otherwise healthy people throughout the United States. Severe respiratory symptoms, high fevers, hemoptysis, and hypotension are its defining features. Rapidly develops leukopenia, high C-reactive protein (more than 350 mg/dl), and sepsis and septic shock. In this clinical setting, multilobar cavitating alveolar infiltrates are consistent with CA-MRSA infection.

Moreover, MRSA is a major factor in both ventilator- and hospital-acquired pneumonia. Nosocomial pneumonia, also known as hospital-acquired pneumonia (HAP), is defined as pneumonia that appears 48 hours or more after being admitted to the hospital and is not incubating at the time of admission. In order to be considered ventilator-associated pneumonia (VAP), pneumonia must appear 48 hours or more following the use of endotracheal intubation and mechanical ventilation. These two diseases have a similar microbiological genesis and have a dire prognosis linked to poor overall results.

12.2.5.2 Endocarditis

MRSA is a significant contributor to bacterial endocarditis, which may be fatal in 30–37% of people who get it. Common risk factors for right-sided MRSA endocarditis include intravenous drug usage and intravenous catheters. Septic pulmonary emboli may cause nodular infiltrates and cavitating lesions in the lungs in people with tricuspid valve vegetations. In the same way, those who have mitral and aortic valve involvement may get secondary infections in distant sites such as bones and joints, kidneys, the brain, and other organs. It is crucial to get a patient's history, conduct a thorough examination, and order any required lab and radiological testing.

12.2.5.3 Bacteremia

According to reports, *S. aureus*-related bacteremia has a death rate range of 15% to 60%. Patients in critical care units with central line insertions often develop MRSA bacteremia. Every patient who has MRSA in their bloodstream should have infectious endocarditis, which is related to MRSA bacteremia, checked out. Because of these individuals' reduced response to vancomycin, the results associated with MRSA bacteremia are worse than those associated with other MRSA infections (Baddour et al. 2015).

12.2.5.4 Skin and Soft Tissue Infections

A common pathogen linked to SSTIs such as cellulitis, necrotizing fasciitis, and a diabetic foot ulcer is CA-MRSA. Moreover, it is more often linked to invasive diseases than those brought on by non-MRSA. These infections are typically multidrug-resistant, which increases recurrence, hospitalization, and death (Clebak and Malone 2018; Khan et al. 2018).

12.2.5.5 Bone and Joint Infections

The most typical cause of bone and joint infections is staphylococci. With these patients, oxacillin resistance is more prevalent. By spreading a local infection from a lesion or as a component of a hematogenous infection, MRSA may cause osteomyelitis of the spine and long bones of the upper and lower limbs. MRSA may also lead to septic arthritis in natural and artificial joints (Baddour et al. 2015).

12.2.6 Diagnosis, Screening, and Prevention

Many infections, including SSTIs, pneumonia, osteoarticular infections, TSS, and bacteremia, which may be worsened by endocarditis or severe sepsis, may be brought on by MRSA (Lowy 1998). The clinical manifestations and infection risk factors for the HA, CA, and LA MRSA strains differ.

12.2.7 HA-MRSA

In hospitalized patients, HA-MRSA is a prevalent cause of bacteraemia, pneumonia, and, less often, SSTIs (Klevens et al. 2007). The organism probably owing to its capacity to form and survive in biofilms (Vanhommerig et al. 2014). Long-term hospitalization, ICU admission, nursing home residency, antibiotic exposure (especially to cephalosporins and fluoroquinolones, which led to antibiotic selection pressure), surgery, hemodialysis, chronic wounds, or indwelling invasive devices are all associated with an increased risk of HA-MRSA infection (Epstein et al. 2016).

12.2.8 CA-MRSA

SSTI is the most frequent clinical manifestation of CA-MRSA, accounting for 90% of cases and often accompanied by pus or abscesses (Moran et al. 2006). Very aggressive infections may be caused by CA-MRSA. There have been reports of CA-MRSA strains causing fulminant infections, including necrotizing fasciitis and pneumonia (Francis et al. 2005; Miller et al. 2005). Very active Agr systems and PVL production have been suggested as reasons for the higher pathogenicity seen with CA-MRSA strains. PVL is present in certain strains, but not others, indicating that additional virulence factors play a role (Chambers and Deleo 2009). Those who have CA-MRSA infections often do not have the established risk factors for HA-MRSA strains. Sports teams, military people, and prisons are some of the groups or places where epidemics of CA-MRSA infection have been documented (Kazakova et al. 2005; Aiello et al. 2006).

12.2.9 LA-MRSA

Localized infections like SSTIs and otitis as well as serious and invasive infections like bacteremia, pneumonia, osteoarticular infections, and endocarditis have all been linked to LA-MRSA (van Alen et al. 2017). Through transmission among household members, LA-MRSA primarily colonizes and infects people who have direct contact with livestock (particularly pigs, but also horses, chickens, and turkeys) (Larsen et al. 2015).

12.2.10 Microbiological Diagnosis

MRSA-producing microbiological samples can be broadly divided into clinical and screening samples. Clinical samples are taken from people who have symptoms or signs of an infection to look for an active infection, whereas screening samples are taken to look for asymptomatic colonization. Clinical samples include specimens of purulent discharge, deep tissues, sputum, and blood. MRSA may be found in clinical or screening samples directly using a variety of phenotypic and nonphenotypic techniques, or it can be found in presumed staphylococcal colonies isolated from clinical samples. In most cases, phenotypic techniques are chosen for clinical diagnoses.

12.2.10.1 Phenotypic Methods

The disk-diffusion technique may be used to test for methicillin resistance in pure *S. aureus* cultures that were produced by plating clinical samples on appropriate culture medium. This procedure calls for the use of a cefoxitin disk on Mueller-Hinton agar or the addition of 4% NaCl and 6 μg of oxacillin per milliliter to Mueller-Hinton agar (CLSI guidelines). Originally, oxacillin was used as the marker antibiotic to identify MRSA; however, CLSI now advises cefoxitin since it induces mecA and mecC more effectively than oxacillin and produces a phenotype that is easily recognized (CLSI 2013). To avoid misleading negative findings, the disk-diffusion technique must strictly adhere to temperature (35 °C) and time (reading after 24 hours). This is because the resistant isolates grow more slowly because the mecA-encoded PBP2a is less effective at crosslinking the pentapeptide chains of the cell wall peptidoglycan during cell wall synthesis (Ryffel et al. 1994). Rarely, MRSA may exhibit phenotypic sensitivity to cefoxitin (and oxacillin) before developing resistance after spending the night in an environment with low levels of cefoxitin (Penn et al. 2013). The existence of inducible mecA in this situation should be taken into account. An antigen-antibody-based latex agglutination test that identifies PBP2a by employing an anti-PBP2a antibody may also be used to identify methicillin resistance in *S. aureus* colonies and cultures (Winstanley and Courvalin 2011).

The development of bacteriophage-based assays for the direct phenotypic identification of MRSA from positive blood cultures has received considerable attention. A US FDA-approved, nongenotypic, fast test for the identification of *S. aureus* and the detection of methicillin resistance directly from positive blood cultures is the Key Path MRSA/MSSA blood culture test (MicroPhage Inc, Longmont, Colorado, United States). The test has a five-hour turnaround time and may identify the amplification of bacteriophages unique to *S. aureus* when methicillin is present (Bhowmick et al. 2013).

12.2.10.2 Non-Phenotypic Methods

Matrix-assisted laser desorption-ionization time-of-flight mass spectrometry (MALDI-TOF MS) is one of the most promising nongenotypic methods for identifying pathogens directly from positive blood cultures. Nevertheless, since MALDI-TOF MS performance greatly relies on the purity and quantity of a bacterium, enrichment and purification techniques for bacteria are needed from positive blood cultures, which include significant amounts of interfering nonmicrobial material (Opota et al. 2015).

Multiplex real-time PCR techniques to identify *S. aureus* and the presence of mecA are currently used DNA-based approaches for directly detecting MRSA from clinical samples (Malhotra-Kumar et al. 2010). The time it takes to get results is around 1.5 hours. The Film Array (Idaho Technology, Salt Lake City, Utah, United States) is a multiplex PCR-based technology created to identify the presence of mecA, 25 microorganisms (90–95% of the pathogens implicated in blood cultures), and the vancomycin (vanA and vanB) and carbapenem resistance genes (blaKPC) (Blaschke et al. 2012). With an average turnaround time of 2.5 hours, this technique offers greater sensitivity than MALDI-TOF MS in detecting bacteria from blood culture bottles before positive (Almuhayawi et al. 2014).

The single most significant development in diagnostic microbiology and surveillance since in vitro culture was the use of WGS to bacterial pathogens. Nevertheless, there are still few direct uses of WGS in diagnostic microbiology due to technical limitations in collecting findings quickly enough to affect patient care, as well as the need for established methods and automated data interpretation (Cao et al. 2016).

The specificity and turnaround time have been significantly enhanced by the coupling of chromogenic medium with MALDI-TOF MS, which allows the species identification of numerous colonies in less than one hour. Moreover, efforts are being made to create automated colony scoring that would improve specificity and speed up turnaround (Faron et al. 2016). For MRSA screening from nasal swabs, real-time PCR-based assays may reduce turnaround time to 1–2 hours, while the findings of chromogenic media-based tests might require at least 14–18 hours without confirmatory testing and, thus, might not always be appropriate to guide therapeutic choices.

12.2.11 Prevention

Interventions to combat MRSA have been extensively used in healthcare institutions. These measures attempt to regulate the reservoir of patients who are carriers, prevent MRSA transmission between patients, facilitate prudent use of antimicrobial drugs (including prescribing limits), prevent the development of infection in carriers, and control the spread of MRSA. Usually, a number of precautions are necessary to properly stop MRSA transmission and infection (Calfee et al. 2014). The Management section discusses decolonization, a crucial control measure for which there is accumulating evidence.

12.2.11.1 Hand Hygiene

Health care personnel may get MRSA on their hands via handling equipment that has been contaminated with MRSA or from coming into touch with patients who have the infection. MRSA can then be spread from one patient to another through this method (Lotfinejad et al. 2021). With soap and water or alcohol-based hand rub, good hand hygiene seeks to reduce the transmission of MRSA via this pathway.

12.2.11.2 Contact Precautions and Isolation

While caring for patients who have MRSA colonization, many hospitals have health care staff utilize contact precautions (use of disposable gowns and gloves) to limit the risk of MRSA transmission due to contamination of hands and clothes.

12.2.11.3 Active Surveillance

As most patients with MRSA colonization are asymptomatic, depending only on the culture of clinical samples, which are only taken when a person exhibits symptoms or indications of infection, may result in up to 85% of carriers of MRSA being undetected (Salgado and Farr 2006). Active surveillance programs may locate this sizable pool of asymptomatic carriers using screening techniques and target therapies (such as topical decontamination) to lower the risk of infection or transmission.

12.2.12 Management

Depending on the incidence of MRSA in a given area and the accessibility of antibiotics, especially newer medicines, the treatment strategy for MRSA differs across various geographic locations.

12.2.12.1 Decolonization of Carriers

MRSA colonization raises the risk of infection and aids in the spread of the disease. An approximately 10-fold greater risk of recurring infection is linked to both MRSA colonization upon admission and acquisition during hospitalization (Davis et al. 2004). Decolonization may thus aid in the management of MRSA by lowering the risk of infection and dissemination. The majority of decolonization techniques apply topical treatments to the nostrils, the primary colonization site (Kluytmans et al. 1997). The primary drug is mupirocin (pseudomonic acid A), which inhibits bacterial isoleucyl tRNA synthetase and prevents protein synthesis. Chlorhexidine bathing is often used in conjunction with mupirocin (Loeb et al. 2003).

12.2.12.2 Short-Term Decolonization

Decolonization is most often utilized as a preventative measure during very brief intervals of heightened infection risk, such as during perioperative times or ICU stays. Before surgery for known *S. aureus* carriers, topical mupirocin to the nares and chlorhexidine body washing dramatically decreased the chance of post-surgical *S. aureus* infection (Bode et al. 2010).

12.2.12.3 Permanent Decolonization

Permanent MRSA eradication is sometimes sought. In nations with a low frequency of MRSA, the "seek and destroy approach" includes permanent decolonization (Ammerlaan et al. 2011b). Because treatment failure is three times more likely in people with throat colonization than in people without throat colonization, the Dutch protocol, for instance, makes a distinction between uncomplicated and complicated carriers based on the MRSA strain, host characteristics, and colonization site (Ammerlaan et al. 2011a). In real-world settings, several more clinical studies investigating permanent decolonization techniques have been carried out, but the findings have been relatively underwhelming due to a high incidence of endogenous recolonization (Loeb et al. 2003).

12.2.13 Treatment

Based on the kind of illness, local *S. aureus* resistance patterns, medication availability, side effect profile, and patient characteristics, empiric antibiotic therapy is chosen to treat MRSA infections (Kavanagh et al. 2018; Lewis et al. 2018).

12.2.13.1 Bacteremia

Source control and empiric MRSA coverage are both important components of the therapy for MRSA bacteremia until the susceptibility data are known (Eisenschenk 2018; Sirijatuphat et al. 2018).

According to the 2011 Infectious Diseases Society of America (IDSA) recommendations, vancomycin and daptomycin are deemed appropriate forms of empiric treatment. Vancomycin may not be effective against bloodstream MRSA isolates having vancomycin MIC levels higher than or equal to 2 μg/ml. Daptomycin is thus a superior choice in these circumstances. Telavancin and ceftaroline are also used as alternate antibiotics for treating bacteremia if either vancomycin or daptomycin could not be administered because of allergy, toxicity, or resistance. Combinations of parenteral antibiotic medication regimens, such as daptomycin plus ceftaroline, vancomycin plus ceftaroline, daptomycin plus trimethoprim-sulfamethoxazole, and ceftaroline plus trimethoprim-sulfamethoxazole, have also been utilized in difficult-to-treat patients.

Teicoplanin is a bacteriostatic glycopeptide that is more tolerable than vancomycin and has a comparable range of action and effectiveness. But, because of its rarity, it is utilized less often. Teicoplanin is unavailable in the United States and has not received FDA approval. For the treatment of *S. aureus* bacteremia, the use of linezolid, quinupristin-dalfopristin, tigecycline, or fluoroquinolones is not recommended. Clinical investigations have shown that linezolid is ineffective when compared to vancomycin. Repeated follow-up cultures are necessary to prove that the infection has been removed from the circulation. After 48 hours of therapy, persistent, positive cultures should trigger further testing for medication susceptibility and source management.

12.2.13.2 Endocarditis

The American Heart Association's (AHA's) 2015 recommendations suggest using intravenous vancomycin as the first endocarditis therapy. Those who are unable to tolerate vancomycin should get intravenous daptomycin. Six weeks are the suggested amount of time to treat native valve endocarditis. For the first two weeks of treatment, gentamicin should be administered. Prosthetic valve endocarditis owing to MRSA should be treated with intravenous vancomycin and rifampin for six weeks. For native valve endocarditis, however, there is no proof that combining rifampin or gentamicin with vancomycin or daptomycin would be beneficial. In a small number of patients, telavancin, ceftaroline, and quinupristin-dalfopristin have been used successfully as salvage therapies. Due to their inferior effectiveness when compared to cell walls, active medications like vancomycin, daptomycin, clindamycin, and linezolid should not be used to treat endocarditis (Baddour et al. 2015).

12.2.13.3 Skin and Soft Tissue Infection

Purulent acute bacterial skin and skin structure infections (ABSSSIs) should be treated as soon as feasible by incision and drainage. After incision and drainage, antibiotic therapy lowers the risk of recurring abscesses or treatment failure, according to a new large placebo-controlled experiment (Daum et al. 2017). With the recent approval of delafloxacin and omadacycline, two studies proving the effectiveness of iclaprim, and ongoing trials evaluating afabicin (Debio 1450), the number of antimicrobials licensed to treat MRSA ABSSSIs is rising (Holland et al. 2018; Huang et al. 2018; O'Riordan et al. 2018). Oritavancin and Dalbavancin, two long-acting, one-dose injectable drugs, have also shown their superiority over vancomycin (Boucher et al. 2014; Corey et al. 2014; Dunne et al. 2016).

12.2.13.4 Osteomyelitis

The tendency of *S. aureus* to develop biofilms makes the treatment of infections associated with bones, joints, and prosthetics more challenging. In spite of inadequate bone penetration and failure rates as high as 35–46%, vancomycin remains the first-line treatment (Dombrowski and Winston 2008). Some retrospective investigations support the use of dapoxetine as an alternate medication (Gallagher et al. 2012; Moenster et al. 2012; Seaton et al. 2013). While linezolid has a nearly 100% absorption rate and high bone penetration, chronic usage is linked to neuropathy and myelosuppression (Senneville et al. 2006). The investigational drug tedizolid may have fewer side effects than linezolid (US National Library of Medicine 2018).

It is commonly acceptable for pediatric patients to go from first intravenous medication to oral clindamycin (Feigin et al. 1975). Initial MRSA osteomyelitis therapy in adults is often given intravenously for at least the first two weeks (Daver et al. 2007). While the ideal treatment period is still debatable, shorter treatments, such as those for MRSA vertebral osteomyelitis, may increase the chance of recurrence (Park et al. 2016).

12.2.13.5 Pneumonia

Vancomycin, linezolid, or clindamycin are suggested as treatments for MRSA pneumonia under current American Thoracic Society and IDSA recommendations. The prescription for clindamycin is mostly based on a few small observational studies in kids and the assumption that it would be beneficial since it will cause PVL-positive bacteria to produce fewer toxins (Martínez-Aguilar et al. 2003). While linezolid has good pulmonary pharmacokinetics and probably improves clinical cure rates, it has not consistently reduced mortality (Conte et al. 2002; Kalil et al. 2010; Wunderink et al. 2012). There is little published data to support the use of ceftaroline in MRSA pneumonia since few recruited patients had MRSA in their sputum despite the fact that it is licensed for the treatment of community-acquired pneumonia (Casapao et al. 2014).

12.2.13.6 Vaccine Development

MRSA vaccine research attempts have so far proven unsuccessful. Phase IIb/III studies have been advanced for three candidates. Durable immunity was not induced by StaphVAX, a bivalent conjugate vaccination that targets capsular polysaccharides types 5 and 8 (Shinefield et al. 2002). The experiment was prematurely terminated because V710, a monovalent vaccination that targets the iron salvage protein IsdB, was actually linked to higher mortality (Giersing et al. 2016). The phase IIb trial for Pfizer's multi-antigen vaccine (PF-06290510) was terminated for futility in December 2018 (https://www.pfizer.com/news/press-).

References

Aiello, A.E., Lowy, F.D., Wright, L.N., and Larson, E.L. (2006). Meticillin-resistant *Staphylococcus aureus* among US prisoners and military personnel: review and recommendations for future studies. *The Lancet Infectious Diseases* 6 (6): 335–341. https://doi.org/10.1016/s1473-3099(06)70491-1.

Alam, M.T., Read, T.D., Petit, R.A. 3rd et al. (2015). Transmission and microevolution of USA300 MRSA in U.S. households: evidence from whole-genome sequencing. *MBio* 6 (2): e00054. https://doi.org/10.1128/mBio.00054-15.

van Alen, S., Ballhausen, B., Peters, G. et al. (2017). In the centre of an epidemic: fifteen years of LA-MRSA CC398 at the University Hospital Münster. *Veterinary Microbiology* 200: 19–24. https://doi.org/10.1016/j.vetmic.2016.01.021.

Almuhayawi, M., Altun, O., Strålin, K., and Ozenci, V. (2014). Identification of microorganisms by FilmArray and matrix-assisted laser desorption ionization-time of flight mass spectrometry prior to positivity in the blood culture system. *Journal of Clinical Microbiology* 52 (9): 3230–3236. https://doi.org/10.1128/jcm.01084-14.

Ammerlaan, H.S., Kluytmans, J.A., Berkhout, H. et al. (2011a). Eradication of carriage with methicillin-resistant *Staphylococcus aureus*: determinants of treatment failure. *The Journal of Antimicrobial Chemotherapy* 66 (10): 2418–2424. https://doi.org/10.1093/jac/dkr250.

Ammerlaan, H. S., Kluytmans, J. A., Berkhout, H., Buiting, A., de Brauwer, E. I., van den Broek, P. J., ... Bonten, M. J. (2011b). Eradication of carriage with methicillin-resistant *Staphylococcus aureus*: effectiveness of a national guideline. *The Journal of Antimicrobial Chemotherapy*, 66 (10), 2409–2417. doi:https://doi.org/10.1093/jac/dkr243

Arêde, P., Milheiriço, C., de Lencastre, H., and Oliveira, D.C. (2012). The anti-repressor MecR2 promotes the proteolysis of the mecA repressor and enables optimal expression of β-lactam resistance in MRSA. *PLoS Pathogens* 8 (7): e1002816. https://doi.org/10.1371/journal.ppat.1002816.

Azarian, T., Daum, R.S., Petty, L.A. et al. (2016). Intrahost evolution of methicillin-resistant *Staphylococcus aureus* USA300 among individuals with reoccurring skin and soft-tissue infections. *The Journal of Infectious Diseases* 214 (6): 895–905. https://doi.org/10.1093/infdis/jiw242.

Baddour, L.M., Wilson, W.R., Bayer, A.S. et al. (2015). Infective endocarditis in adults: diagnosis, antimicrobial therapy, and management of complications: a scientific statement for healthcare professionals from the American Heart Association. *Circulation* 132 (15): 1435–1486. https://doi.org/10.1161/cir.0000000000000296.

Baggett, H.C., Hennessy, T.W., Rudolph, K. et al. (2004). Community-onset methicillin-resistant *Staphylococcus aureus* associated with antibiotic use and the cytotoxin Panton-Valentine leukocidin during a furunculosis outbreak in rural Alaska. *The Journal of Infectious Diseases* 189 (9): 1565–1573. https://doi.org/10.1086/383247.

Becker, K., Schaumburg, F., Fegeler, C. et al. (2017). *Staphylococcus aureus* from the German general population is highly diverse. *International Journal of Medical Microbiology* 307 (1): 21–27. https://doi.org/10.1016/j.ijmm.2016.11.007.

Bell, J.M., Lubian, A.F., Partridge, S. et al. (2020). Australian Group on Antimicrobial Resistance (AGAR) Australian Gram-negative Sepsis Outcome Programme (GNSOP) Annual Report, p. 44.

Bhowmick, T., Mirrett, S., Reller, L.B. et al. (2013). Controlled multicenter evaluation of a bacteriophage-based method for rapid detection of *Staphylococcus aureus* in positive blood cultures. *Journal of Clinical Microbiology* 51 (4): 1226–1230. https://doi.org/10.1128/jcm.02967-12.

Blaschke, A.J., Heyrend, C., Byington, C.L. et al. (2012). Rapid identification of pathogens from positive blood cultures by multiplex polymerase chain reaction using the Film Array system. *Diagnostic Microbiology and Infectious Disease* 74 (4): 349–355. https://doi.org/10.1016/j.diagmicrobio.2012.08.013.

Bode, L.G., Kluytmans, J.A., Wertheim, H.F. et al. (2010). Preventing surgical-site infections in nasal carriers of *Staphylococcus aureus*. *The New England Journal of Medicine* 362 (1): 9–17. https://doi.org/10.1056/NEJMoa0808939.

Boubaker, K., Diebold, P., Blanc, D.S. et al. (2004). Panton-valentine leukocidin and staphyloccoccal skin infections in schoolchildren. *Emerging Infectious Diseases* 10 (1): 121–124. https://doi.org/10.3201/eid1001.030144.

Boucher, H.W. and Corey, G.R. (2008). Epidemiology of methicillin-resistant *Staphylococcus aureus*. *Clinical Infectious Diseases* 46 (Suppl 5): S344–S349. https://doi.org/10.1086/533590.

Boucher, H.W., Wilcox, M., Talbot, G.H. et al. (2014). Once-weekly dalbavancin versus daily conventional therapy for skin infection. *The New England Journal of Medicine* 370 (23): 2169–2179. https://doi.org/10.1056/NEJMoa1310480.

Boyle-Vavra, S., Yin, S., and Daum, R.S. (2006). The VraS/VraR two-component regulatory system required for oxacillin resistance in community-acquired methicillin-resistant *Staphylococcus aureus*. *FEMS Microbiology Letters* 262 (2): 163–171. https://doi.org/10.1111/j.1574-6968.2006.00384.x.

Boyle-Vavra, S., Li, X., Alam, M.T. et al. (2015). USA300 and USA500 clonal lineages of *Staphylococcus aureus* do not produce a capsular polysaccharide due to conserved mutations in the cap5 locus. *mBio* 6 (2): https://doi.org/10.1128/mBio.02585-14.

Burian, M., Rautenberg, M., Kohler, T. et al. (2010). Temporal expression of adhesion factors and activity of global regulators during establishment of *Staphylococcus aureus* nasal colonization. *The Journal of Infectious Diseases* 201 (9): 1414–1421. https://doi.org/10.1086/651619.

Calderwood, M.S., Desjardins, C.A., Sakoulas, G. et al. (2014). Staphylococcal enterotoxin P predicts bacteremia in hospitalized patients colonized with methicillin-resistant *Staphylococcus aureus*. *The Journal of Infectious Diseases* 209 (4): 571–577. https://doi.org/10.1093/infdis/jit501.

Calfee, D.P., Salgado, C.D., Milstone, A.M. et al. (2014). Strategies to prevent methicillin-resistant *Staphylococcus aureus* transmission and infection in acute care hospitals: 2014 update. *Infection Control and Hospital Epidemiology* 35 (7): 772–796. https://doi.org/10.1086/676534.

Cao, M.D., Ganesamoorthy, D., Elliott, A.G. et al. (2016). Streaming algorithms for identification of pathogens and antibiotic resistance potential from real-time MinION(TM) sequencing. *Gigascience* 5 (1): 32. https://doi.org/10.1186/s13742-016-0137-2.

Casapao, A.M., Davis, S.L., Barr, V.O. et al. (2014). Large retrospective evaluation of the effectiveness and safety of ceftaroline fosamil therapy. *Antimicrobial Agents and Chemotherapy* 58 (5): 2541–2546. https://doi.org/10.1128/aac.02371-13.

Chambers, H.F. (2001). The changing epidemiology of *Staphylococcus aureus*? *Emerging Infectious Diseases* 7 (2): 178–182. https://doi.org/10.3201/eid0702.010204.

Chambers, H.F. and Deleo, F.R. (2009). Waves of resistance: *Staphylococcus aureus* in the antibiotic era. *Nature Reviews. Microbiology* 7 (9): 629–641. https://doi.org/10.1038/nrmicro2200.

Chavakis, T., Wiechmann, K., Preissner, K.T., and Herrmann, M. (2005). *Staphylococcus aureus* interactions with the endothelium: the role of bacterial "secretable expanded repertoire adhesive molecules" (SERAM) in disturbing host defense systems. *Thrombosis and Haemostasis* 94 (2): 278–285. https://doi.org/10.1160/th05-05-0306.

Chen, C.J. and Huang, Y.C. (2014). New epidemiology of *Staphylococcus aureus* infection in Asia. *Clinical Microbiology and Infection* 20 (7): 605–623. https://doi.org/10.1111/1469-0691.12705.

Chen, C.J., Huang, Y.C., and Chiu, C.H. (2015). Multiple pathways of cross-resistance to glycopeptides and daptomycin in persistent MRSA bacteraemia. *The Journal of Antimicrobial Chemotherapy* 70 (11): 2965–2972. https://doi.org/10.1093/jac/dkv225.

Cheng, A.G., DeDent, A.C., Schneewind, O., and Missiakas, D. (2011). A play in four acts: *Staphylococcus aureus* abscess formation. *Trends in Microbiology* 19 (5): 225–232. https://doi.org/10.1016/j.tim.2011.01.007.

Cheung, G.Y., Wang, R., Khan, B.A. et al. (2011). Role of the accessory gene regulator agr in community-associated methicillin-resistant *Staphylococcus aureus* pathogenesis. *Infection and Immunity* 79 (5): 1927–1935. https://doi.org/10.1128/iai.00046-11.

Clarridge, J.E. 3rd, Harrington, A.T., Roberts, M.C. et al. (2013). Impact of strain typing methods on assessment of relationship between paired nares and wound isolates of methicillin-resistant *Staphylococcus aureus*. *Journal of Clinical Microbiology* 51 (1): 224–231. https://doi.org/10.1128/jcm.02423-12.

Clebak, K.T. and Malone, M.A. (2018). Skin infections. *Primary Care* 45 (3): 433–454. https://doi.org/10.1016/j.pop.2018.05.004.

van Cleef, B.A., Monnet, D.L., Voss, A. et al. (2011). Livestock-associated methicillin-resistant *Staphylococcus aureus* in humans, Europe. *Emerging Infectious Diseases* 17 (3): 502–505. https://doi.org/10.3201/eid1703.101036.

Clinical and Laboratory Standards Institute (2023). M100-ED33. Performance Standards for Antimicrobial Susceptibility Testing; Thirty ThreeInformational Supplement (CLSI).

Cluzet, V.C., Gerber, J.S., Nachamkin, I. et al. (2015). Duration of colonization and determinants of earlier clearance of colonization with methicillin-resistant *Staphylococcus aureus*. *Clinical Infectious Diseases* 60 (10): 1489–1496. https://doi.org/10.1093/cid/civ075.

Conte, J.E. Jr., Golden, J.A., Kipps, J., and Zurlinden, E. (2002). Intrapulmonary pharmacokinetics of linezolid. *Antimicrobial Agents and Chemotherapy* 46 (5): 1475–1480. https://doi.org/10.1128/aac.46.5.1475-1480.2002.

Corey, G.R., Kabler, H., Mehra, P. et al. (2014). Single-dose oritavancin in the treatment of acute bacterial skin infections. *The New England Journal of Medicine* 370 (23): 2180–2190. https://doi.org/10.1056/NEJMoa1310422.

Dantes, R., Mu, Y., Belflower, R. et al. (2013). National burden of invasive methicillin-resistant *Staphylococcus aureus* infections, United States, 2011. *JAMA Internal Medicine* 173 (21): 1970–1978. https://doi.org/10.1001/jamainternmed.2013.10423.

Daum, R.S., Ito, T., Hiramatsu, K. et al. (2002). A novel methicillin-resistance cassette in community-acquired methicillin-resistant *Staphylococcus aureus* isolates of diverse genetic backgrounds. *The Journal of Infectious Diseases* 186 (9): 1344–1347. https://doi.org/10.1086/344326.

Daum, R.S., Miller, L.G., Immergluck, L. et al. (2017). A placebo-controlled trial of antibiotics for smaller skin abscesses. *The New England Journal of Medicine* 376 (26): 2545–2555. https://doi.org/10.1056/NEJMoa1607033.

Daver, N.G., Shelburne, S.A., Atmar, R.L. et al. (2007). Oral step-down therapy is comparable to intravenous therapy for *Staphylococcus aureus* osteomyelitis. *The Journal of Infection* 54 (6): 539–544. https://doi.org/10.1016/j.jinf.2006.11.011.

David, M.Z., Mennella, C., Mansour, M. et al. (2008). Predominance of methicillin-resistant *Staphylococcus aureus* among pathogens causing skin and soft tissue infections in a large urban jail: risk factors and recurrence rates. *Journal of Clinical Microbiology* 46 (10): 3222–3227. https://doi.org/10.1128/jcm.01423-08.

Davis, K.A., Stewart, J.J., Crouch, H.K. et al. (2004). Methicillin-resistant *Staphylococcus aureus* (MRSA) nares colonization at hospital admission and its effect on subsequent MRSA infection. *Clinical Infectious Diseases* 39 (6): 776–782. https://doi.org/10.1086/422997.

De Lencastre, H., Wu, S.W., Pinho, M.G. et al. (1999). Antibiotic resistance as a stress response: complete sequencing of a large number of chromosomal loci in *Staphylococcus aureus* strain COL that impact on the expression of resistance to methicillin. *Microbial Drug Resistance* 5 (3): 163–175. https://doi.org/10.1089/mdr.1999.5.163.

Diekema, D.J., Pfaller, M.A., Schmitz, F.J. et al. (2001). Survey of infections due to *Staphylococcus* species: frequency of occurrence and antimicrobial susceptibility of isolates collected in the United States, Canada, Latin America, Europe, and the Western Pacific region for the SENTRY Antimicrobial Surveillance Program, 1997–1999. *Clinical Infectious Diseases* 32 (Suppl 2): S114–S132. https://doi.org/10.1086/320184.

Diep, B.A., Gill, S.R., Chang, R.F. et al. (2006). Complete genome sequence of USA300, an epidemic clone of community-acquired meticillin-resistant

Staphylococcus aureus. *Lancet* 367 (9512): 731–739. https://doi.org/10.1016/s0140-6736(06)68231-7.

Dombrowski, J.C. and Winston, L.G. (2008). Clinical failures of appropriately-treated methicillin-resistant *Staphylococcus aureus* infections. *The Journal of Infection* 57 (2): 110–115. https://doi.org/10.1016/j.jinf.2008.04.003.

Dunne, M.W., Puttagunta, S., Giordano, P. et al. (2016). A randomized clinical trial of single-dose versus weekly dalbavancin for treatment of acute bacterial skin and skin structure infection. *Clinical Infectious Diseases* 62 (5): 545–551. https://doi.org/10.1093/cid/civ982.

von Eiff, C., Becker, K., Machka, K. et al. (2001). Nasal carriage as a source of *Staphylococcus aureus* bacteremia. Study Gropu. *The New England Journal of Medicine* 344 (1): 11–16. https://doi.org/10.1056/nejm200101043440102.

Eisenschenk, M. (2018). A concern with the clinical consensus guidelines on meticillin-resistant staphylococci. *Veterinary Dermatology* 29 (2): 174. https://doi.org/10.1111/vde.12509.

Emonts, M., Uitterlinden, A.G., Nouwen, J.L. et al. (2008). Host polymorphisms in interleukin 4, complement factor H, and C-reactive protein associated with nasal carriage of *Staphylococcus aureus* and occurrence of boils. *The Journal of Infectious Diseases* 197 (9): 1244–1253. https://doi.org/10.1086/533501.

Enright, M.C., Day, N.P., Davies, C.E. et al. (2000). Multilocus sequence typing for characterization of methicillin-resistant and methicillin-susceptible clones of *Staphylococcus aureus*. *Journal of Clinical Microbiology* 38 (3): 1008–1015. https://doi.org/10.1128/jcm.38.3.1008-1015.2000.

Enright, M.C., Robinson, D.A., Randle, G. et al. (2002). The evolutionary history of methicillin-resistant *Staphylococcus aureus* (MRSA). *Proceedings of the National Academy of Sciences of the United States of America* 99 (11): 7687–7692. https://doi.org/10.1073/pnas.122108599.

Epstein, L., Mu, Y., Belflower, R. et al. (2016). Risk factors for invasive methicillin-resistant *Staphylococcus aureus* infection after recent discharge from an acute-care hospitalization, 2011–2013. *Clinical Infectious Diseases* 62 (1): 45–52. https://doi.org/10.1093/cid/civ777.

Eriksen, N.H., Espersen, F., Rosdahl, V.T., and Jensen, K. (1995). Carriage of *Staphylococcus aureus* among 104 healthy persons during a 19-month period. *Epidemiology and Infection* 115 (1): 51–60. https://doi.org/10.1017/s0950268800058118.

Falagas, M.E., Karageorgopoulos, D.E., Leptidis, J., and Korbila, I.P. (2013). MRSA in Africa: filling the global map of antimicrobial resistance. *PLoS One* 8 (7): e68024. https://doi.org/10.1371/journal.pone.0068024.

Faoagali, J.L., Thong, M.L., and Grant, D. (1992). Ten years' experience with methicillin-resistant *Staphylococcus aureus* in a large Australian hospital. *The Journal of Hospital Infection* 20 (2): 113–119. https://doi.org/10.1016/0195-6701(92)90113-z.

Faron, M.L., Buchan, B.W., Vismara, C. et al. (2016). Automated scoring of chromogenic media for detection of methicillin-resistant *Staphylococcus aureus* by use of WASPLab image analysis software. *Journal of Clinical Microbiology* 54 (3): 620–624. https://doi.org/10.1128/jcm.02778-15.

Feigin, R.D., Pickering, L.K., Anderson, D. et al. (1975). Clindamycin treatment of osteomyelitis and septic arthritis in children. *Pediatrics* 55 (2): 213–223.

Fey, P.D., Saïd-Salim, B., Rupp, M.E. et al. (2003). Comparative molecular analysis of community- or hospital-acquired methicillin-resistant *Staphylococcus aureus*. *Antimicrobial Agents and Chemotherapy* 47 (1): 196–203. https://doi.org/10.1128/aac.47.1.196-203.2003.

Foster, T.J., Geoghegan, J.A., Ganesh, V.K., and Höök, M. (2014). Adhesion, invasion and evasion: the many functions of the surface proteins of *Staphylococcus aureus*. *Nature Reviews. Microbiology* 12 (1): 49–62. https://doi.org/10.1038/nrmicro3161.

Francis, J.S., Doherty, M.C., Lopatin, U. et al. (2005). Severe community-onset pneumonia in healthy adults caused by methicillin-resistant *Staphylococcus aureus* carrying the Panton-Valentine leukocidin genes. *Clinical Infectious Diseases* 40 (1): 100–107. https://doi.org/10.1086/427148.

Frank, A.L., Marcinak, J.F., Mangat, P.D., and Schreckenberger, P.C. (1999). Increase in community-acquired methicillin-resistant *Staphylococcus aureus* in children. *Clinical Infectious Diseases* 29 (4): 935–936. https://doi.org/10.1086/520463.

Fridkin, S.K., Hageman, J.C., Morrison, M. et al. (2005). Methicillin-resistant *Staphylococcus aureus* disease in three communities. *The New England Journal of Medicine* 352 (14): 1436–1444. https://doi.org/10.1056/NEJMoa043252.

Gallagher, J.C., Huntington, J.A., Culshaw, D. et al. (2012). Daptomycin therapy for osteomyelitis: a retrospective study. *BMC Infectious Diseases* 12: 133. https://doi.org/10.1186/1471-2334-12-133.

García-Álvarez, L., Holden, M.T., Lindsay, H. et al. (2011). Meticillin-resistant *Staphylococcus aureus* with a novel mecA homologue in human and bovine populations in the UK and Denmark: a descriptive study. *The Lancet Infectious Diseases* 11 (8): 595–603. https://doi.org/10.1016/s1473-3099(11)70126-8.

Garcia-Alvarez, L., Dawson, S., Cookson, B., and Hawkey, P. (2012). Working across the veterinary and human health sectors. *The Journal of Antimicrobial Chemotherapy* 67 (Suppl 1): i37–i49. https://doi.org/10.1093/jac/dks206.

Giersing, B.K., Dastgheyb, S.S., Modjarrad, K., and Moorthy, V. (2016). Status of vaccine research and development of vaccines for *Staphylococcus aureus*. *Vaccine* 34 (26): 2962–2966. https://doi.org/10.1016/j.vaccine.2016.03.110.

Graffunder, E.M. and Venezia, R.A. (2002). Risk factors associated with nosocomial methicillin-resistant *Staphylococcus aureus* (MRSA) infection including previous use of antimicrobials. *The Journal of Antimicrobial Chemotherapy* 49 (6): 999–1005. https://doi.org/10.1093/jac/dkf009.

Hanzelmann, D., Joo, H.S., Franz-Wachtel, M. et al. (2016). Toll-like receptor 2 activation depends on lipopeptide shedding by bacterial surfactants. *Nature Communications* 7: 12304. https://doi.org/10.1038/ncomms12304.

Harkins, C.P., Pichon, B., Doumith, M. et al. (2017). Methicillin-resistant *Staphylococcus aureus* emerged long before the introduction of methicillin into clinical practice. *Genome Biology* 18 (1): 130. https://doi.org/10.1186/s13059-017-1252-9.

Hartman, B.J. and Tomasz, A. (1984). Low-affinity penicillin-binding protein associated with beta-lactam resistance in *Staphylococcus aureus*. *Journal of Bacteriology* 158 (2): 513–516. https://doi.org/10.1128/jb.158.2.513-516.1984.

Hetem, D.J., Bootsma, M.C., Troelstra, A., and Bonten, M.J. (2013). Transmissibility of livestock-associated methicillin-resistant *Staphylococcus aureus*. *Emerging Infectious Diseases* 19 (11): 1797–1802. https://doi.org/10.3201/eid1911.121085.

Holland, T.L., O'Riordan, W., McManus, A. et al. (2018). A phase 3, randomized, double-blind, multicenter study to evaluate the safety and efficacy of intravenous iclaprim versus vancomycin for treatment of acute bacterial skin and skin structure infections suspected or confirmed to be due to gram-positive pathogens (REVIVE-2 Study). *Antimicrobial Agents and Chemotherapy* 62 (5): https://doi.org/10.1128/aac.02580-17.

Hota, B., Ellenbogen, C., Hayden, M.K. et al. (2007). Community-associated methicillin-resistant *Staphylococcus aureus* skin and soft tissue infections at a public hospital: do public housing and incarceration amplify transmission? *Archives of Internal Medicine* 167 (10): 1026–1033. https://doi.org/10.1001/archinte.167.10.1026.

Howden, B.P., Ward, P.B., Charles, P.G. et al. (2004). Treatment outcomes for serious infections caused by methicillin-resistant *Staphylococcus aureus* with reduced vancomycin susceptibility. *Clinical Infectious Diseases* 38 (4): 521–528. https://doi.org/10.1086/381202.

Huang, S.S., Diekema, D.J., Warren, D.K. et al. (2008). Strain-relatedness of methicillin-resistant *Staphylococcus aureus* isolates recovered from patients with repeated infection. *Clinical Infectious Diseases* 46 (8): 1241–1247. https://doi.org/10.1086/529381.

Huang, D.B., O'Riordan, W., Overcash, J.S. et al. (2018). A phase 3, randomized, double-blind, multicenter study to evaluate the safety and efficacy of intravenous iclaprim vs vancomycin for the treatment of acute bacterial skin and skin structure infections suspected or confirmed to be due to gram-positive pathogens: REVIVE-1. *Clinical Infectious Diseases* 66 (8): 1222–1229. https://doi.org/10.1093/cid/cix987.

International Working Group on the Classification of Staphylococcal Cassette Chromosome Elements (IWG-SCC) (2009). Classification of staphylococcal cassettechromosome mec (SCCmec): guidelines for reportingnovel SCCmec elements. *Antimicrobial Agents and Chemotherapy* 53: 4961–4967.

Ito, T., Ma, X.X., Takeuchi, F. et al. (2004). Novel type V staphylococcal cassette chromosome mec driven by a novel cassette chromosome recombinase, ccrC. *Antimicrobial Agents and Chemotherapy* 48 (7): 2637–2651. https://doi.org/10.1128/aac.48.7.2637-2651.2004.

Jevons, M.P.J.B.M.J. (1961). "Celbenin"-resistant staphylococci. *BMJ* 1 (5219): 124.

Jousselin, A., Manzano, C., Biette, A. et al. (2015). The *Staphylococcus aureus* Chaperone PrsA is a new auxiliary factor of oxacillin resistance affecting penicillin-binding protein 2A. *Antimicrobial Agents and Chemotherapy* 60 (3): 1656–1666. https://doi.org/10.1128/aac.02333-15.

Kaito, C., Saito, Y., Ikuo, M. et al. (2013). Mobile genetic element SCCmec-encoded psm-mec RNA suppresses translation of agrA and attenuates MRSA virulence. *PLoS Pathogens* 9 (4): e1003269. https://doi.org/10.1371/journal.ppat.1003269.

Kalil, A.C., Murthy, M.H., Hermsen, E.D. et al. (2010). Linezolid versus vancomycin or teicoplanin for nosocomial pneumonia: a systematic review and meta-analysis. *Critical Care Medicine* 38 (9): 1802–1808. https://doi.org/10.1097/CCM.0b013e3181eb3b96.

Kallen, A.J., Mu, Y., Bulens, S. et al. (2010). Health care-associated invasive MRSA infections, 2005–2008. *JAMA* 304 (6): 641–648. https://doi.org/10.1001/jama.2010.1115.

Karanika, S., Kinamon, T., Grigoras, C., and Mylonakis, E. (2016). Colonization with methicillin-resistant *Staphylococcus aureus* and risk for infection among asymptomatic athletes: a systematic review and metaanalysis. *Clinical Infectious Diseases* 63 (2): 195–204. https://doi.org/10.1093/cid/ciw240.

Karchmer, A.W. (2000). Nosocomial bloodstream infections: organisms, risk factors, and implications. *Clinical Infectious Diseases* 31 (Suppl 4): S139–S143. https://doi.org/10.1086/314078.

Kaspar, U., Kriegeskorte, A., Schubert, T. et al. (2016). The culturome of the human nose habitats reveals individual bacterial fingerprint patterns. *Environmental Microbiology* 18 (7): 2130–2142. https://doi.org/10.1111/1462-2920.12891.

Katayama, Y., Ito, T., and Hiramatsu, K. (2000). A new class of genetic element, staphylococcus cassette chromosome mec, encodes methicillin resistance in *Staphylococcus aureus*. *Antimicrobial Agents and Chemotherapy* 44 (6): 1549–1555. https://doi.org/10.1128/aac.44.6.1549-1555.2000.

Kavanagh, K.T., Abusalem, S., and Calderon, L.E. (2018). View point: gaps in the current guidelines for the prevention of methicillin-resistant *Staphylococcus aureus* surgical site infections. *Antimicrobial Resistance and Infection Control* 7: 112. https://doi.org/10.1186/s13756-018-0407-0.

Kazakova, S.V., Hageman, J.C., Matava, M. et al. (2005). A clone of methicillin-resistant *Staphylococcus aureus* among professional football players. *The New England Journal of Medicine* 352 (5): 468–475. https://doi.org/10.1056/NEJMoa042859.

Kennedy, A.D., Otto, M., Braughton, K.R. et al. (2008). Epidemic community-associated methicillin-resistant *Staphylococcus aureus*: recent clonal expansion and diversification. *Proceedings of the National Academy of Sciences of the United States of America* 105 (4): 1327–1332. https://doi.org/10.1073/pnas.0710217105.

Khan, T.M., Kok, Y.L., Bukhsh, A. et al. (2018). Incidence of methicillin resistant *Staphylococcus aureus* (MRSA) in burn intensive care unit: a systematic review. *Germs* 8 (3): 113–125. https://doi.org/10.18683/germs.2018.1138.

Kim, C., Milheiriço, C., Gardete, S. et al. (2012). Properties of a novel PBP2A protein homolog from *Staphylococcus aureus* strain LGA251 and its contribution to the β-lactam-resistant phenotype. *The Journal of Biological Chemistry* 287 (44): 36854–36863. https://doi.org/10.1074/jbc.M112.395962.

Kim, C., Mwangi, M., Chung, M. et al. (2013). The mechanism of heterogeneous beta-lactam resistance in MRSA: key role of the stringent stress response. *PLoS One* 8 (12): e82814. https://doi.org/10.1371/journal.pone.0082814.

Klein, E.Y., Mojica, N., Jiang, W. et al. (2017). Trends in methicillin-resistant *Staphylococcus aureus* hospitalizations in the United States, 2010–2014. *Clinical Infectious Diseases* 65 (11): 1921–1923. https://doi.org/10.1093/cid/cix640.

Klevens, R.M., Morrison, M.A., Nadle, J. et al. (2007). Invasive methicillin-resistant *Staphylococcus aureus* infections in the United States. *JAMA* 298 (15): 1763–1771. https://doi.org/10.1001/jama.298.15.1763.

Kluytmans, J.A., Mouton, J.W., Ijzerman, E.P. et al. (1995). Nasal carriage of *Staphylococcus aureus* as a major risk factor for wound infections after cardiac surgery. *The Journal of Infectious Diseases* 171 (1): 216–219. https://doi.org/10.1093/infdis/171.1.216.

Kluytmans, J., van Belkum, A., and Verbrugh, H. (1997). Nasal carriage of *Staphylococcus aureus*: epidemiology, underlying mechanisms, and associated risks. *Clinical Microbiology Reviews* 10 (3): 505–520. https://doi.org/10.1128/cmr.10.3.505.

Köck, R., Becker, K., Cookson, B. et al. (2010). Methicillin-resistant *Staphylococcus aureus* (MRSA): burden of disease and control challenges in Europe. *Euro Surveillance* 15 (41): 19688. https://doi.org/10.2807/ese.15.41.19688-en.

Krismer, B., Liebeke, M., Janek, D. et al. (2014). Nutrient limitation governs *Staphylococcus aureus* metabolism and niche adaptation in the human nose. *PLoS Pathogens* 10 (1): e1003862. https://doi.org/10.1371/journal.ppat.1003862.

Krismer, B., Weidenmaier, C., Zipperer, A., and Peschel, A. (2017). The commensal lifestyle of *Staphylococcus aureus* and its interactions with the nasal microbiota.

Nature Reviews. Microbiology 15 (11): 675–687. https://doi.org/10.1038/nrmicro.2017.104.

Laarman, A., Milder, F., van Strijp, J., and Rooijakkers, S. (2010). Complement inhibition by gram-positive pathogens: molecular mechanisms and therapeutic implications. *Journal of Molecular Medicine (Berlin, Germany)* 88 (2): 115–120. https://doi.org/10.1007/s00109-009-0572-y.

Landrum, M.L., Neumann, C., Cook, C. et al. (2012). Epidemiology of *Staphylococcus aureus* blood and skin and soft tissue infections in the US military health system, 2005–2010. *JAMA* 308 (1): 50–59. https://doi.org/10.1001/jama.2012.7139.

Larsen, J., Petersen, A., Sørum, M. et al. (2015). Meticillin-resistant *Staphylococcus aureus* CC398 is an increasing cause of disease in people with no livestock contact in Denmark, 1999 to 2011. *Euro Surveillance* 20 (37): https://doi.org/10.2807/1560-7917.Es.2015.20.37.30021.

Le, K.Y. and Otto, M. (2015). Quorum-sensing regulation in staphylococci – an overview. *Frontiers in Microbiology* 6: 1174. https://doi.org/10.3389/fmicb.2015.01174.

Lekkerkerk, W.S.N., Haenen, A., van der Sande, M.A.B. et al. (2017). Newly identified risk factors for MRSA carriage in The Netherlands. *PLoS One* 12 (11): e0188502. https://doi.org/10.1371/journal.pone.0188502.

Lemon, K.P., Klepac-Ceraj, V., Schiffer, H.K. et al. (2010). Comparative analyses of the bacterial microbiota of the human nostril and oropharynx. *mBio* 1 (3): https://doi.org/10.1128/mBio.00129-10.

Lewis, P.O., Heil, E.L., Covert, K.L., and Cluck, D.B. (2018). Treatment strategies for persistent methicillin-resistant *Staphylococcus aureus* bacteraemia. *Journal of Clinical Pharmacy and Therapeutics* 43 (5): 614–625. https://doi.org/10.1111/jcpt.12743.

Lina, G., Piémont, Y., Godail-Gamot, F. et al. (1999). Involvement of Panton-Valentine leukocidin-producing *Staphylococcus aureus* in primary skin infections and pneumonia. *Clinical Infectious Diseases* 29 (5): 1128–1132. https://doi.org/10.1086/313461.

Liu, C.M., Price, L.B., Hungate, B.A. et al. (2015). *Staphylococcus aureus* and the ecology of the nasal microbiome. *Science Advances* 1 (5): e1400216. https://doi.org/10.1126/sciadv.1400216.

Lodise, T.P. Jr., McKinnon, P.S., and Rybak, M. (2003). Prediction model to identify patients with *Staphylococcus aureus* bacteremia at risk for methicillin resistance. *Infection Control and Hospital Epidemiology* 24 (9): 655–661. https://doi.org/10.1086/502269.

Lodise, T.P., McKinnon, P.S., Swiderski, L., and Rybak, M.J. (2003). Outcomes analysis of delayed antibiotic treatment for hospital-acquired *Staphylococcus aureus* bacteremia. *Clinical Infectious Diseases* 36 (11): 1418–1423. https://doi.org/10.1086/375057.

Loeb, M.B., Main, C., Eady, A., and Walker-Dilks, C. (2003). Antimicrobial drugs for treating methicillin-resistant *Staphylococcus aureus* colonization. *Cochrane Database of Systematic Reviews* 4: Cd003340. https://doi.org/10.1002/14651858.Cd003340.

Löffler, B., Hussain, M., Grundmeier, M. et al. (2010). *Staphylococcus aureus* panton-valentine leukocidin is a very potent cytotoxic factor for human neutrophils. *PLoS Pathogens* 6 (1): e1000715. https://doi.org/10.1371/journal.ppat.1000715.

Löffler, B., Tuchscherr, L., Niemann, S., and Peters, G. (2014). *Staphylococcus aureus* persistence in non-professional phagocytes. *International Journal of Medical Microbiology* 304 (2): 170–176. https://doi.org/10.1016/j.ijmm.2013.11.011.

Lotfinejad, N., Peters, A., Tartari, E. et al. (2021). Hand hygiene in health care: 20 years of ongoing advances and perspectives. *The Lancet infectious diseases* 21 (8): e209–e221.

Lowy, F.D. (1998). *Staphylococcus aureus* infections. *The New England Journal of Medicine* 339 (8): 520–532. https://doi.org/10.1056/nejm199808203390806.

Lozano, C., Rezusta, A., Gómez, P. et al. (2012). High prevalence of spa types associated with the clonal lineage CC398 among tetracycline-resistant methicillin-resistant *Staphylococcus aureus* strains in a Spanish hospital. *The Journal of Antimicrobial Chemotherapy* 67 (2): 330–334. https://doi.org/10.1093/jac/dkr497.

Lucet, J.C., Paoletti, X., Demontpion, C. et al. (2009). Carriage of methicillin-resistant *Staphylococcus aureus* in home care settings: prevalence, duration, and transmission to household members. *Archives of Internal Medicine* 169 (15): 1372–1378. https://doi.org/10.1001/archinternmed.2009.217.

Ma, X.X., Ito, T., Tiensasitorn, C. et al. (2002). Novel type of staphylococcal cassette chromosome mec identified in community-acquired methicillin-resistant *Staphylococcus aureus* strains. *Antimicrobial Agents and Chemotherapy* 46 (4): 1147–1152. https://doi.org/10.1128/aac.46.4.1147-1152.2002.

Malachowa, N. and DeLeo, F.R. (2010). Mobile genetic elements of *Staphylococcus aureus*. *Cellular and Molecular Life Sciences* 67 (18): 3057–3071. https://doi.org/10.1007/s00018-010-0389-4.

Malhotra-Kumar, S., Van Heirstraeten, L., Lee, A. et al. (2010). Evaluation of molecular assays for rapid detection of methicillin-resistant *Staphylococcus aureus*. *Journal of Clinical Microbiology* 48 (12): 4598–4601. https://doi.org/10.1128/jcm.00004-10.

Martínez-Aguilar, G., Hammerman, W.A., Mason, E.O. Jr., and Kaplan, S.L. (2003). Clindamycin treatment of invasive infections caused by community-acquired, methicillin-resistant and methicillin-susceptible *Staphylococcus aureus* in children. *The Pediatric Infectious Disease Journal* 22 (7): 593–598. https://doi.org/10.1097/01.inf.0000073163.37519.ee.

Mendes, R.E., Mendoza, M., Banga Singh, K.K. et al. (2013). Regional resistance surveillance program results for 12 Asia-Pacific nations (2011). *Antimicrobial Agents and Chemotherapy* 57 (11): 5721–5726. https://doi.org/10.1128/aac.01121-13.

Miller, L.G., Perdreau-Remington, F., Rieg, G. et al. (2005). Necrotizing fasciitis caused by community-associated methicillin-resistant *Staphylococcus aureus* in Los Angeles. *The New England Journal of Medicine* 352 (14): 1445–1453. https://doi.org/10.1056/NEJMoa042683.

Mitchell, B.G., Collignon, P.J., McCann, R. et al. (2014). A major reduction in hospital-onset *Staphylococcus aureus* bacteremia in Australia-12 years of progress: an observational study. *Clinical Infectious Diseases* 59 (7): 969–975. https://doi.org/10.1093/cid/ciu508.

Moenster, R.P., Linneman, T.W., Finnegan, P.M., and McDonald, J.R. (2012). Daptomycin compared to vancomycin for the treatment of osteomyelitis: a single-center, retrospective cohort study. *Clinical Therapeutics* 34 (7): 1521–1527. https://doi.org/10.1016/j.clinthera.2012.06.013.

Mollema, F.P., Richardus, J.H., Behrendt, M. et al. (2010). Transmission of methicillin-resistant *Staphylococcus aureus* to household contacts. *Journal of Clinical Microbiology* 48 (1): 202–207. https://doi.org/10.1128/jcm.01499-09.

Monecke, S., Coombs, G., Shore, A.C. et al. (2011). A field guide to pandemic, epidemic and sporadic clones of methicillin-resistant *Staphylococcus aureus*. *PLoS One* 6 (4): e17936. https://doi.org/10.1371/journal.pone.0017936.

Mongkolrattanothai, K., Boyle, S., Kahana, M.D., and Daum, R.S. (2003). Severe *Staphylococcus aureus* infections caused by clonally related community-acquired methicillin-susceptible and methicillin-resistant isolates. *Clinical Infectious Diseases* 37 (8): 1050–1058. https://doi.org/10.1086/378277.

Moran, G.J., Krishnadasan, A., Gorwitz, R.J. et al. (2006). Methicillin-resistant *S. aureus* infections among patients in the emergency department. *The New England Journal of Medicine* 355 (7): 666–674. https://doi.org/10.1056/NEJMoa055356.

Mwangi, M.M., Wu, S.W., Zhou, Y. et al. (2007). Tracking the in vivo evolution of multidrug resistance in *Staphylococcus aureus* by whole-genome sequencing. *Proceedings of the National Academy of Sciences of the United States of America* 104 (22): 9451–9456. https://doi.org/10.1073/pnas.0609839104.

Newsom, S.W. (2008). Ogston's coccus. *The Journal of Hospital Infection* 70 (4): 369–372. https://doi.org/10.1016/j.jhin.2008.10.001.

Nielsen, K.L., Pedersen, T.M., Udekwu, K.I. et al. (2012). Fitness cost: a bacteriological explanation for the demise of the first international methicillin-resistant *Staphylococcus aureus* epidemic. *The Journal of Antimicrobial Chemotherapy* 67 (6): 1325–1332. https://doi.org/10.1093/jac/dks051.

O'Brien, L.M., Walsh, E.J., Massey, R.C. et al. (2002). *Staphylococcus aureus* clumping factor B (ClfB) promotes adherence to human type I cytokeratin 10: implications for nasal colonization. *Cellular Microbiology* 4 (11): 759–770. https://doi.org/10.1046/j.1462-5822.2002.00231.x.

Opota, O., Croxatto, A., Prod'hom, G., and Greub, G. (2015). Blood culture-based diagnosis of bacteraemia: state of the art. *Clinical Microbiology and Infection* 21 (4): 313–322. https://doi.org/10.1016/j.cmi.2015.01.003.

O'Riordan, W., McManus, A., Teras, J. et al. (2018). A comparison of the efficacy and safety of intravenous followed by oral delafloxacin with vancomycin plus aztreonam for the treatment of acute bacterial skin and skin structure infections: a phase 3, multinational, double-blind, randomized study. *Clinical Infectious Diseases* 67 (5): 657–666. https://doi.org/10.1093/cid/ciy165.

Painter, K.L., Krishna, A., Wigneshweraraj, S., and Edwards, A.M. (2014). What role does the quorum-sensing accessory gene regulator system play during *Staphylococcus aureus* bacteremia? *Trends in Microbiology* 22 (12): 676–685. https://doi.org/10.1016/j.tim.2014.09.002.

Park, K.H., Cho, O.H., Lee, J.H. et al. (2016). Optimal duration of antibiotic therapy in patients with hematogenous vertebral osteomyelitis at low risk and high risk of recurrence. *Clinical Infectious Diseases* 62 (10): 1262–1269. https://doi.org/10.1093/cid/ciw098.

Patel, A.B., Hill, E., Simpson, E.L., and Hanifin, J.M. (2013). Reversion of methicillin-resistant *Staphylococcus aureus* skin infections to methicillin-susceptible isolates. *JAMA Dermatology* 149 (10): 1167–1171. https://doi.org/10.1001/jamadermatol.2013.4909.

Penn, C., Moddrell, C., Tickler, I.A. et al. (2013). Wound infections caused by inducible meticillin-resistant *Staphylococcus aureus* strains. *Journal of Global Antimicrobial Resistance* 1 (2): 79–83. https://doi.org/10.1016/j.jgar.2013.03.009.

Peschel, A. and Otto, M. (2013). Phenol-soluble modulins and staphylococcal infection. *Nature Reviews Microbiology* 11 (10): 667–673. https://doi.org/10.1038/nrmicro3110.

Peschel, A. and Sahl, H.G. (2006). The co-evolution of host cationic antimicrobial peptides and microbial resistance. *Nature Reviews. Microbiology 4* (7): 529–536. https://doi.org/10.1038/nrmicro1441.

Planet, P.J., Diaz, L., Kolokotronis, S.O. et al. (2015). Parallel epidemics of community-associated methicillin-resistant *Staphylococcus aureus* USA300 infection in North and South America. *The Journal of Infectious Diseases* 212 (12): 1874–1882. https://doi.org/10.1093/infdis/jiv320.

Popovich, K.J., Snitkin, E.S., Hota, B. et al. (2017). Genomic and epidemiological evidence for community origins of hospital-onset methicillin-resistant *Staphylococcus aureus* bloodstream infections. *The Journal of Infectious Diseases* 215 (11): 1640–1647. https://doi.org/10.1093/infdis/jiw647.

Ren, K., Bannan, J.D., Pancholi, V. et al. (1994). Characterization and biological properties of a new staphylococcal exotoxin. *The Journal of Experimental Medicine* 180 (5): 1675–1683. https://doi.org/10.1084/jem.180.5.1675.

Rossi, F., Diaz, L., Wollam, A. et al. (2014). Transferable vancomycin resistance in a community-associated MRSA lineage. *The New England Journal of Medicine* 370 (16): 1524–1531. https://doi.org/10.1056/NEJMoa1303359.

Rybak, M.J. and Akins, R.L. (2001). Emergence of methicillin-resistant *Staphylococcus aureus* with intermediate glycopeptide resistance: clinical significance and treatment options. *Drugs* 61 (1): 1–7. https://doi.org/10.2165/00003495-200161010-00001.

Ryffel, C., Strässle, A., Kayser, F.H., and Berger-Bächi, B. (1994). Mechanisms of heteroresistance in methicillin-resistant *Staphylococcus aureus*. *Antimicrobial Agents and Chemotherapy* 38 (4): 724–728. https://doi.org/10.1128/aac.38.4.724.

Salgado, C.D. and Farr, B.M. (2006). What proportion of hospital patients colonized with methicillin-resistant *Staphylococcus aureus* are identified by clinical microbiological cultures? *Infection Control and Hospital Epidemiology* 27 (2): 116–121. https://doi.org/10.1086/500624.

Schechter-Perkins, E.M., Mitchell, P.M., Murray, K.A. et al. (2011). Prevalence and predictors of nasal and extranasal staphylococcal colonization in patients presenting to the emergency department. *Annals of Emergency Medicine* 57 (5): 492–499. https://doi.org/10.1016/j.annemergmed.2010.11.024.

Seaton, R.A., Malizos, K.N., Viale, P. et al. (2013). Daptomycin use in patients with osteomyelitis: a preliminary report from the EU-CORE (SM) database. *The Journal of Antimicrobial Chemotherapy* 68 (7): 1642–1649. https://doi.org/10.1093/jac/dkt067.

Senneville, E., Legout, L., Valette, M. et al. (2006). Effectiveness and tolerability of prolonged linezolid treatment for chronic osteomyelitis: a retrospective study. *Clinical Therapeutics* 28 (8): 1155–1163. https://doi.org/10.1016/j.clinthera.2006.08.001.

Shinefield, H., Black, S., Fattom, A. et al. (2002). Use of a *Staphylococcus aureus* conjugate vaccine in patients receiving hemodialysis. *The New England Journal of Medicine* 346 (7): 491–496. https://doi.org/10.1056/NEJMoa011297.

Sidhu, M.S., Heir, E., Leegaard, T. et al. (2002). Frequency of disinfectant resistance genes and genetic linkage with beta-lactamase transposon Tn552 among clinical staphylococci. *Antimicrobial Agents and Chemotherapy* 46 (9): 2797–2803. https://doi.org/10.1128/aac.46.9.2797-2803.2002.

Sim, B.L., McBryde, E., Street, A.C., and Marshall, C. (2013). Multiple site surveillance cultures as a predictor of methicillin-resistant *Staphylococcus aureus* infections. *Infection Control and Hospital Epidemiology* 34 (8): 818–824. https://doi.org/10.1086/671273.

Sirijatuphat, R., Sripanidkulchai, K., Boonyasiri, A. et al. (2018). Implementation of global antimicrobial resistance surveillance system (GLASS) in patients with bacteremia. *PLoS One* 13 (1): e0190132. https://doi.org/10.1371/journal.pone.0190132.

Soriano, A., Martínez, J.A., Mensa, J. et al. (2000). Pathogenic significance of methicillin resistance for patients with *Staphylococcus aureus* bacteremia. *Clinical Infectious Diseases* 30 (2): 368–373. https://doi.org/10.1086/313650.

Spaan, A.N., Surewaard, B.G., Nijland, R., and van Strijp, J.A. (2013). Neutrophils versus *Staphylococcus aureus*: a biological tug of war. *Annual Review of Microbiology* 67: 629–650. https://doi.org/10.1146/annurev-micro-092412-155746.

Spaan, A.N., van Strijp, J.A.G., and Torres, V.J. (2017). Leukocidins: staphylococcal bi-component pore-forming toxins find their receptors. *Nature Reviews Microbiology* 15 (7): 435–447. https://doi.org/10.1038/nrmicro.2017.27.

Spaulding, A.R., Salgado-Pabón, W., Kohler, P.L. et al. (2013). Staphylococcal and streptococcal superantigen exotoxins. *Clinical Microbiology Reviews* 26 (3): 422–447. https://doi.org/10.1128/cmr.00104-12.

Stapels, D.A., Ramyar, K.X., Bischoff, M. et al. (2014). *Staphylococcus aureus* secretes a unique class of neutrophil serine protease inhibitors. *Proceedings of the National Academy of Sciences of the United States of America* 111 (36): 13187–13192. https://doi.org/10.1073/pnas.1407616111.

Struyf, T. and Mertens, K. (2017). European Antimicrobial Resistance Surveillance Network (EARS-Net Belgium) Report 2017.

Talan, D.A., Krishnadasan, A., Gorwitz, R.J. et al. (2011). Comparison of *Staphylococcus aureus* from skin and soft-tissue infections in US emergency department patients, 2004 and 2008. *Clinical Infectious Diseases* 53 (2): 144–149. https://doi.org/10.1093/cid/cir308.

Tenover, F.C., McDougal, L.K., Goering, R.V. et al. (2006). Characterization of a strain of community-associated methicillin-resistant *Staphylococcus aureus* widely disseminated in the United States. *Journal of Clinical Microbiology* 44 (1): 108–118. https://doi.org/10.1128/jcm.44.1.108-118.2006.

Thammavongsa, V., Kim, H.K., Missiakas, D., and Schneewind, O. (2015). Staphylococcal manipulation of host immune responses. *Nature Reviews Microbiology* 13 (9): 529–543. https://doi.org/10.1038/nrmicro3521.

Thomer, L., Schneewind, O., and Missiakas, D. (2016). Pathogenesis of *Staphylococcus aureus* bloodstream infections. *Annual Review of Pathology 11*: 343–364. https://doi.org/10.1146/annurev-pathol-012615-044351.

Turner, N.A., Sharma-Kuinkel, B.K., Maskarinec, S.A. et al. (2019). Methicillin-resistant *Staphylococcus aureus*: an overview of basic and clinical research. *Nature Reviews. Microbiology* 17 (4): 203–218. https://doi.org/10.1038/s41579-018-0147-4.

Uhlemann, A.C., Dordel, J., Knox, J.R. et al. (2014). Molecular tracing of the emergence, diversification, and transmission of *S. aureus* sequence type 8 in a New York community. *Proceedings of the National Academy of Sciences of the United States of America* 111 (18): 6738–6743. https://doi.org/10.1073/pnas.1401006111.

US National Library of Medicine (2018). ClinicalTrials.gov. https://clinicaltrials.gov/ct2/show/NCT03009045 (accessed January 2023).

Vanhommerig, E., Moons, P., Pirici, D. et al. (2014). Comparison of biofilm formation between major clonal lineages of methicillin resistant *Staphylococcus aureus*. *PLoS One* 9 (8): e104561. https://doi.org/10.1371/journal.pone.0104561.

Von Dach, E., Diene, S.M., Fankhauser, C. et al. (2016). Comparative genomics of community-associated methicillin-resistant *Staphylococcus aureus* shows the emergence of clone ST8-USA300 in Geneva, Switzerland. *The Journal of Infectious Diseases* 213 (9): 1370–1379. https://doi.org/10.1093/infdis/jiv489.

Voss, A., Loeffen, F., Bakker, J., Klaassen, C., & Wulf, M. (2005). Methicillin-resistant *Staphylococcus aureus* in pig farming. *Emerging Infectious Diseases*, *11* (12), 1965–1966. doi:https://doi.org/10.3201/eid1112.050428

Weidenmaier, C., Peschel, A., Xiong, Y.Q. et al. (2005). Lack of wall teichoic acids in *Staphylococcus aureus* leads to reduced interactions with endothelial cells and to attenuated virulence in a rabbit model of endocarditis. *The Journal of Infectious Diseases* 191 (10): 1771–1777. https://doi.org/10.1086/429692.

Weidenmaier, C., Kokai-Kun, J.F., Kulauzovic, E. et al. (2008). Differential roles of sortase-anchored surface proteins and wall teichoic acid in *Staphylococcus aureus* nasal colonization. *International Journal of Medical Microbiology* 298 (5–6): 505–513. https://doi.org/10.1016/j.ijmm.2007.11.006.

Weigel, L.M., Clewell, D.B., Gill, S.R. et al. (2003). Genetic analysis of a high-level vancomycin-resistant isolate of *Staphylococcus aureus*. *Science* 302 (5650): 1569–1571. https://doi.org/10.1126/science.1090956.

Wertheim, H.F., Melles, D.C., Vos, M.C. et al. (2005). The role of nasal carriage in *Staphylococcus aureus* infections. *The Lancet Infectious Diseases* 5 (12): 751–762. https://doi.org/10.1016/s1473-3099(05)70295-4.

Wertheim, H.F., Walsh, E., Choudhurry, R. et al. (2008). Key role for clumping factor B in *Staphylococcus aureus* nasal colonization of humans. *PLoS Medicine* 5 (1): e17. https://doi.org/10.1371/journal.pmed.0050017.

Williamson, D.A., Coombs, G.W., and Nimmo, G.R. (2014). *Staphylococcus aureus* 'Down Under': contemporary epidemiology of *S. aureus* in Australia, New Zealand, and the South West Pacific. *Clinical Microbiology and Infection* 20 (7): 597–604. https://doi.org/10.1111/1469-0691.12702.

Winstanley, T. and Courvalin, P. (2011). Expert systems in clinical microbiology. *Clinical Microbiology Reviews* 24 (3): 515–556. https://doi.org/10.1128/cmr.00061-10.

Wright, G.D. (2010). Q&A: antibiotic resistance: where does it come from and what can we do about it? *BMC Biology* 8: 123. https://doi.org/10.1186/1741-7007-8-123.
Wunderink, R.G., Niederman, M.S., Kollef, M.H. et al. (2012). Linezolid in methicillin-resistant *Staphylococcus aureus* nosocomial pneumonia: a randomized, controlled study. *Clinical Infectious Diseases* 54 (5): 621–629. https://doi.org/10.1093/cid/cir895.
Zhang, H.Z., Hackbarth, C.J., Chansky, K.M., and Chambers, H.F. (2001). A proteolytic transmembrane signaling pathway and resistance to beta-lactams in staphylococci. *Science* 291 (5510): 1962–1965. https://doi.org/10.1126/science.1055144.
Zipperer, A., Konnerth, M.C., Laux, C. et al. (2016). Human commensals producing a novel antibiotic impair pathogen colonization. *Nature* 535 (7613): 511–516. https://doi.org/10.1038/nature18634.

13

Immunological and Molecular Mechanism Induced by Indoor Bio-contaminants

Ravi Kumar Goswami[1], *Mohit Kumar*[1], *Atavika Chandan*[1], *Sudeshna Mandal*[2], *Pratibha Kumari*[3], *and Varunendra Singh Rawat*[1]

[1] *Department of Zoology, Hindu College, University of Delhi, Delhi, New Delhi, India*
[2] *Department of Zoology, Laboratory of Helminth Parasitology, Visva-Bharati University, Shantiniketan, West Bengal, India*
[3] *Department of Botany, Daulat Ram College University of Delhi, Delhi, New Delhi, India*

13.1 Introduction

Biological contaminants are pollutants of biological origin of diverse variety, including bacteria, fungi, viruses, algae, insects, and their by-products like endotoxins, mycotoxins, volatile organic compounds (VOCs), etc., which can be found in the indoor environment (Kumar et al. 2021). Allergies, viral disorders, acute and chronic respiratory conditions, illnesses linked to an individual's environment, and even cancer can all be brought on by exposure to biocontaminants and these have been classified into several distinct groups (Nath et al. 2023). Indoor surroundings are considered at least twice as contaminated as outdoor ones. The major sources of the growth and dissemination of microorganisms in indoor structures include air conditioning units, fans, coolers, and humidification devices, which are frequently inhabited by fungi, bacteria, and yeasts (Kraakman et al. 2021). It is plausible to speculate that fungi that contaminate building materials' surfaces and release them into indoor air could potentially have a substantial impact on indoor air quality. Health may be directly affected by the aerosolization of allergic substances or mycotoxins carried by fungi (Al Hallak et al. 2023). The local environment affects the number of fungi present and in addition to that, the season, temperature, humidity, using insecticide, burning incense, heating, building material, ventilation condition, sunlight exposure, and longitude all affect the type and concentration of fungi in a structure (Rogawansamy et al. 2015; Sousa 2022). The molds have

Airborne Biocontaminants and Their Impact on Human Health, First Edition.
Edited by Rajeev Singh and Anamika Singh.

the capacity to release bioaerosols, which are a mixture of biological particles (spores, mycelium) and harmful substances like allergic chemicals and mycotoxins (Al Hallak et al. 2023). These are airborne suspended particles and are volatile chemical compounds (Kumar et al. 2021). Mycotoxins are hazardous secondary metabolites released by fungi that can sicken and kill both people and animals and are believed to cause skin rashes, nausea, disorientation, and immunosuppression in addition to irritating mucous membranes (Rogawansamy et al. 2015).

The level of aerial bacterial concentration depends on a number of factors, including ventilation, layout of buildings, factors like temperature and humidity, the presence of organic matter, the capacity to form biofilms, setting, outdoor environment, human occupants' activities, as well as outdoor air entering indoor spaces (Monteiro et al. 2021). However, only a few heterotrophic bacteria have been linked to the proliferation of biofilms. Bacterial contamination risks are considered to have risen when organic waste is stored indoors (Kumar et al. 2021). VOCs and particulate matter (PM) are one of the most prevalent indoor air contaminants. While sources of coarse particles include dogs, housework, and human activity, sources of PM include smoking, heating, cooking, lit candles, and insecticides. Household goods, cleaning products, adhesives, personal grooming items, supplies for construction, and vehicle exhausts were among the sources of VOCs (Maung et al. 2022). Upper airway difficulties and asthma symptoms have been linked to high VOC levels (Kumar et al. 2021). In addition to these, endotoxins are also a contributing factor in indoor contamination. Observable mold scent, the heating system, the presence of mattress and rug, and endotoxin concentration are all positively correlated (Sousa 2022).

VOCs are chemicals with a low molecular weight and a high vapor pressure. VOC can also be produced as a by-product of microorganism metabolism and is thus referred to as a microbial VOC (mVOC). Even if quantitative measurements are negative, the presence of fungi in indoor air is indicated by the presence of mVOC. Some mVOCs, such as 2octen-1-ol, serve as indicators of biological contamination. Because this compound is strongly emitted during the early stages of *Aspergillus* spp., *Alternaria alternata*, and *Penicillium* spp. growth, it has been linked as a nonspecific indicator of fungal contamination (Flint et al. 2020) (Table 13.1).

13.2 Sources of Biocontaminants in Indoor Environment

Indoor air quality may be at risk from inadequate ventilation in enclosed interior spaces, including contaminated external air, poor air distribution, improper air filtration, and poorly maintained ventilation systems (Mandal and Brandl 2011). The presence and concentration of biocontaminants in indoor environment depends on a variety of factors and activities like building

Table 13.1 List of biocontaminants and their significance, biological properties, and physical properties.

Biocontaminants	Significance	Biological properties	Physical properties	References
Bacteria e.g. *Dienococcus, Methylobacterium*	Acts as a biological marker for polycyclic aromatic hydrocarbons (PAH) pollution, causes various diseases and infections	Single-celled microorganisms, prokaryotes, reproduce through binary fission, and can have diverse shapes	Various sizes, with lengths ranging from 0.2 to 10 μm, discovered in water reservoirs of cooling systems and water pipes, particularly in connection with the production of biofilms	Marzuki et al. (2021), Tortora et al. (2015), Kumar et al. (2021), and Silva et al. (2022)
Fungi e.g. *Aspergillus, Cladosporium, penicillium, Alternaria, Rhodotorula, Sporobolomyces*	Causing many harmful health problems, Fungal bio-contaminants can cause allergies, respiratory problems, and infections	Eukaryotic organisms, have a distinct cell wall made of chitin, reproduce through spores	Are either unicellular or multicellular and have conidia, which are different kinds of spore-forming structures, linked to their lengthy, microscopic hyphal strands	Deacon (2005) and Kumar et al. (2021)
Allergens e.g. Endotoxin, β-glucan	Cause IgE-mediated allergy responses, create type 1 hypersensitivity reactions, which results in a range of respiratory illnesses	Allergens can be derived from various sources such as pollen, dust mites, pet dander, and certain foods	They are glycoproteins or proteins with high molecular weights that range from 10 kDa to more than 100 kDa	Kumar et al. (2021), Adkinson Jr et al. (2013), and Wu et al. (2009)

Table 13.1 (Continued)

Biocontaminants	Significance	Biological properties	Physical properties	References
Viruses e.g. *Influenza virus*, *Variola virus*	Viral infections and illnesses can be brought on by viral bio-contaminants. Through droplets, human activities including coughing, sneezing, chatting, and laughing may transmit millions of airborne viruses inside buildings	Infectious nonliving agents made of protein-coated genetic material (DNA or RNA). Viruses are acellular organisms that can develop on both natural and artificial things, but they need a host in order to replicate and multiply	Extremely small, typically ranging from 20 to 300 nm in size	Kumar et al. (2021), Flint et al. (2020), and Stetzenbach et al. (2004)
Protozoa e.g. A*moeba*, *Giardia*	Are mostly present in water bodies and cause diseases such as malaria and amoebic dysentry	Eukaryotic organisms with a single cell, some of which are motile and contain sophisticated organelles for movement	Present in colonial and solitary ciliated peritrichous forms and vary in size, ranging from 10 to 100 μm	Rivas-Castillo et al. (2022), Murray et al. (2015), Efstratiou et al. (2017), and Dixon (2021)

inhabitants, building systems, furniture, and food (Kausar et al. 2007; Yassin and Almouqatea 2010; Badea et al. 2015; Sheik et al. 2015).

13.2.1 Humans

In the indoor environment, humans are the main sources of biological pollutants' transportation and survival. A healthy individual may contain 10^{12}–10^{14} microbial cells in various body parts (Ananthanaryan and Paniker 2009). In human-occupied areas, compared to uninhabited outdoor areas, there are two times as many germs connected to humans (Prussin and Marr 2015). The bacterial species *Micrococcus*, *Staphylococcus*, *Bacillus*, and *Streptococcus* are some of the predominant species found in inhabited buildings (Charlson et al. 2011). Only a few fungal genera, including *Candida*, *Rhodotorula*, *Trichosporon*, and *Cryptococcus*, are connected to human skin and spread in inhabited buildings as bioaerosols. The microflora of the human male and female bodies are usually varied. In contrast to *Lactobacillus* genera in females, *Corynebacterium* spp., *Roseburia* spp., and *Dermabacter* genera are more prevalent in men (Charlson et al. 2011; Prussin and Marr 2015).

13.2.2 Building Systems

Through the creation of bioaerosols in toilets, bathtubs, sinks, drain traps, aquariums, etc., biological pollutants may spread throughout a facility (World Health Organization 1988). When building materials degrade, they produce organic chemicals that may be particularly favorable for the development of microbes. High temperatures and high relative humidity can encourage the development and spread of indoor bacteria (World Health Organization 1988). Possible biological contaminants may also be present in construction materials and other components. Other significant sources of potential contamination in the interior space include certain natural materials used in carpets, furniture, drapes, ornamental items, plush toys, cushions, and straw or reed baskets (World Health Organization 1988). Unwanted biological contamination in a building may be connected to issues like sick building syndrome (SBS) (World Health Organization 1988; Mandal and Brandl 2011).

13.2.3 Pets

A large number of bacterial genera inhabit the bodies of cats and dogs. *Arthrobacter* spp., *Bacteroides* spp., *Blautia* spp., *Neisseria* spp., *Moraxella* spp. was found predominantly in dogs and *Bifdobacterium* spp., *Sporosarcina* spp., *Jeotgalicoccus* spp., *Moraxella* spp., *Sporosarcina* spp., *Porphyromonas* spp., and *Prevotella* spp. are predominant in cats (Prussin and Marr 2015). However, according to several investigations, exposure to bioaerosols and dog-derived dust is beneficial for children's and babies' health (Fujimura et al. 2010, 2014; Prussin and Marr 2015).

13.2.4 Healthcare Facilities

One of the main places where bacteria propagate indoors is the hospital area. As patients with various illnesses are housed in hospitals, microbes spread through creation of bio-aerosols, ventilation systems, sweeping, dressing, beds, etc. (Knights et al. 2011) (Figure 13.1).

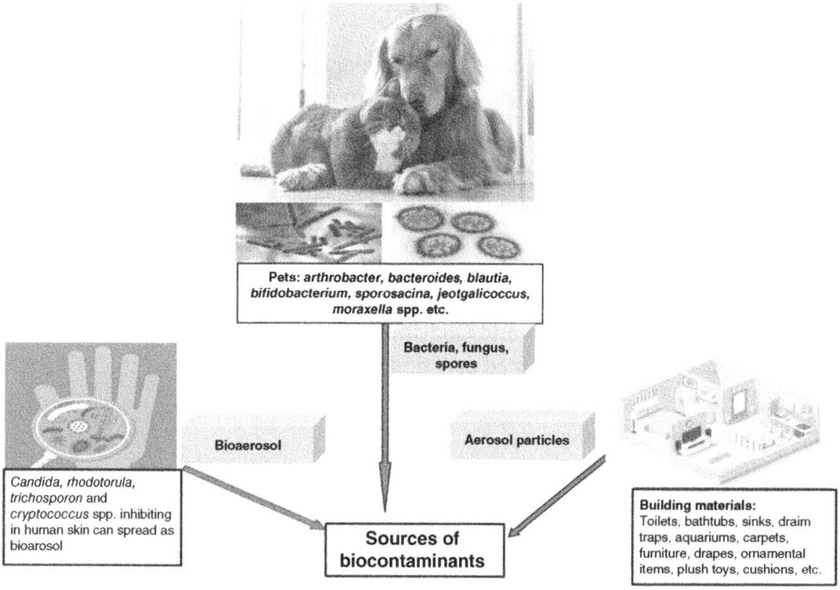

Figure 13.1 Sources of different biocontaminants in indoor environments and their types.

13.3 Detection Techniques for Quantification of Biocontaminants

To measure and characterize the microbial populations in the indoor environment, a variety of sampling techniques are required. However, evaluating the indoor environment to investigate biological contaminants is a difficult task. The selection of appropriate sampling tests, knowing where and when to sample, assessing harmful effects, and interpreting results are the major challenges. Depending on the concentration and kind of biological pollutants, evaluation techniques may vary. Air, dust, and other surfaces may all be examined for microbes (bacteria, fungi, spores, etc.). Since no strategy could possibly address every element of sampling, the techniques must be chosen in accordance with the requirements of the inquiry. First, sample analysis is done to identify any biocontaminant in the air, and then other procedures are used. Depending on the viability of the samples, these techniques may be culture-based or nonculture-based. A good technique for morphologically identifying bacteria and fungi is direct microscopy. The most practical and affordable nonculture approach for most molds is using cello tape by adhering it to the surface, transferring it to a slide, and then staining it using Carbela's solution. Culture techniques depend on the type of biological agents present in an enclosed environment by giving them the right medium and other conditions for in-vitro development.

13.3.1 Sampling Methods

A suitable medium is used to collect viable biological contaminants from samples of the air. Sampling of dust, airborne microorganisms, and other particulates is based on the same principle. There are two primary categories of air sampling: (i) Active sampling and (ii) Passive sampling. Studies that compared active and passive approaches to gauge indoor quality concluded that active methods performed better than passive ones. However, often, quantitative data is unbalanced because of the proliferation of live organisms (Napoli et al. 2012; Moldoveanu 2015). Although sampling time is critical for determining the number of airborne microorganisms, a study by (Godish and Godish 2007) indicated that airborne culturable mold concentrations remained constant for 1–10 minutes.

13.3.1.1 Passive Sampling

The basis for passive monitoring is the exposure of agar plates and glass slides covered with agar to air, which results in the deposition of bacteria and airborne particles. Microorganisms are passively sampled using Koch's sedimentation technique. Growth of viable biological particles was expressed in the form of colony-forming units (CFUs) by the following formula: $CFU/m^3 = n \times 10{,}000/s.t/5$ where "n" is the number of colonies, "t" is exposure time, and "s" is the surface area of the petri plate. However, the sampler's impaction approach is more suited than the Koch's sedimentation method. The inexpensive and straightforward sedimentation technique offers an approximate count of bacteria and fungi in the indoor environment (Srikanth et al. 2008; Moldoveanu 2015).

13.3.1.2 Active Sampling

Active sampling techniques depend on the use of any external force to gather airborne microorganisms on the culture medium. Active sampling techniques need specialized equipment for various air samples. Active sampling techniques based on solid impaction surfaces, an impingement in liquids, filtration, electric precipitation, thermal precipitation, and centrifugation are used to assess airborne microorganisms quantitatively (Kang Frank 1989; Sutton 2004). An essential physical characteristic of a sampler is inlet efficiency, which refers to the form, size, density, and collecting efficiency of the airborne particles (Srikanth et al. 2008).

13.3.1.2.1 Impaction Methods Utilizing air current, impaction techniques capture airborne microorganisms on the surface of agar and sometimes on a dry surface. A common example of an impaction technique using an air-jet driven across a plate for the collision of the particle on a plate is the Anderson multistage sieve sampler (Anderson 1958). The force of the vacuum causes suspended particles and airborne microorganisms to settle down. After sample collection, impaction procedures are more accurate and do not call for any kind of processing (Kang Frank 1989; Sutton 2004; Moldoveanu 2015).

13.3.1.2.2 Filtration Methods Filters are often employed for the collection of microorganisms owing to their great cost-effectiveness and straightforward process. Filter pore size and construction materials may change depending on the application; the most commonly used materials are cellulose, gelatin membrane filter, and glass fiber material (Kang Frank 1989; Sutton 2004).

13.3.1.2.3 Impingement Methods For the capture of airborne microorganisms, impingement liquid medium (salt solution with antifoamants, proteins) is used. Particles in a liquid are dispersed after being drawn into it by air, and this is done quantitatively by serial dilution. The high-velocity All Glass Impinger-30 (AGI-30) sampler is a typical example of this group. These techniques are affordable and very effective for gathering viable cells (Kang Frank 1989; Sutton 2004).

13.4 Biocontaminants and Immunological Responses

A variety of allergic responses, including hypersensitivity, allergic rhinitis, pneumonitis, and various types of asthma, may be brought on by biological pollutants. Sneezing, watery eyes, coughing, shortness of breath, dizziness, fatigue, fever, and stomach issues are among the primary signs of health issues brought on by biological pollutants (Stetzenbach et al. 2004; United States Environmental Protection Agency (EPA) 2020). After exposure to a certain biological allergen repeatedly, different allergic responses could happen (Seltzer 1994; Srikanth et al. 2008; Chirca 2019; Rautiainena et al. 2019). Children, the elderly, and those with breathing issues, allergies, and lung conditions are particularly vulnerable to disease-causing biocontaminants in indoor air. In a sizeable fraction of the population, pollens and molds may induce allergic responses.

13.4.1 Immune Receptors that Recognize Fungal Molecules

Three major mechanisms are responsible for recognition of allergens by the cellular immune system, namely (i) engagement of pattern recognition receptors (PRRs), (ii) molecular mimicry of Toll-like receptor (TLR) signaling complex molecules, and (iii) recognition of the allergen's biological activities, such as proteases (Karp 2010; Wills-Karp et al. 2010). Recognition of fungi and fungal allergens likely occurs by the first and third mechanisms. Several families of PRRs have been shown to be involved in interactions with fungi.

13.4.1.1 Pattern Recognition Receptors

These are the key components of immune system. These are present on the surface of various types of cells like epithelial, endothelial cells, etc., and recognize components termed pathogen associated molecular patterns (PAMPs), which are shared by pathogens but not by host cells (Takeuchi Oand Akira 2010; Janeway 2013). Some of the PAMPs are peptidoglycan, flagellin, and lipopolysaccharides present on the surface of bacteria. They also recognize some components shared by fungi and other parasites and also ssRNA and dsRNA of bacteria (Takeuchi Oand Akira 2010). Often multiple PRRs are activated after recognizing the pathogen, ensuring a robust immune response.

13.4.1.2 Toll-Like Receptors

Pathogen-associated molecular patterns (PAMPs) are recognized by TLRs. Unmethylated CpG-rich DNA, double-stranded RNA, O-linked mannans, and Zymosan are some of the fungal products that interact with TLRs (Williams et al. 2016; Taghavi et al. 2017). β-Glucan from hyphae and fungal conidia are recognized by TLR1/TLR2 heterodimer (Portnoy et al. 2016). TLR2 interacts with unidentified ligands from *Aspergillus fumigatus* (Portnoy et al. 2016). Unmethylated CpG from fungal DNA and dsRNA from conidia activate TLR9 and TLR3, respectively (Portnoy et al. 2016). The presence of viral genetic material is detected by TLR3, 7, and 8. ssRNA is recognized by TLR7/8 and dsRNA by TLR3, which can be an intermediate produced during viral replication or the viral genome (Heil et al. 2004). Viral proteins like F protein of RSV can directly activate TLR4 (Funchal et al. 2015). Surface glycoprotein of EBOV and nonstructural protein (NSP) 1 of DENV also directly activate TLR4 (Modhiran et al. 2015). Depending on the target cell type and the activated TLRs, the outcome could be pro- or anti-inflammatory (Portnoy et al. 2016).

13.4.1.3 C-Type Lectin Receptors

Dectin-1 and Dectin-2, mannose receptor (MR) and dendritic cell-specific intercellular adhesion molecule-3-Grabbing Nonintegrin (DC-SIGN) are the C-type lectin receptors (CLRs) that recognize fungal proteins, lipids, and carbohydrates (Hardison and Brown 2012; Williams et al. 2016). Dectin-2, DC-SIGN, Dectin-2, and MR bind mannose-based molecules while Dectin-1 recognizes β-glucan (Hardison and Brown 2012). Proinflammatory immune responses like phagocytosis and production of cytokines and lipid mediators are triggered by activation of these responses. 60 DG-SIGN is a receptor, which recognizes mannose on Influenza A Virus (IAV) (Yu et al. 2017), while vaccinia virus is recognized by DNGR-1 (Iborra et al. 2012).

13.4.1.4 Recognition of Proteases

Proinflammatory activities of fungal proteolytic and glycolytic enzymes have been recognized for a number of years (Reed and Kita 2004). The disruption of airway epithelial tight junctions by fungal proteases can lead to an increase in recognition of fungal proteins by immune cells present in mucosal tissues (Rodrigues et al. 2016). Second, the cell surface receptors of innate immune cells can directly recognize the fungal proteases. Third, immune response can be indirectly promoted by fungal proteases by facilitating production of molecule(s) within the host, which then activates immune cells.

13.4.2 Innate Immune Responses

13.4.2.1 Epithelial Barrier and Epithelial Cells

The tight junctions formed by epithelial cells prevent invasion of fungal spores and products (Schleimer and Berdnikovs 2017) while the layer of mucus along with cilia provides additional protection by removing them physically (Crameri and Blaser 2002). Cytokines IL-25, IL-33, and TSLP that promote type 2 immune response are released by airway epithelial cells in response to fungi like Alternaria and fungal proteases (Kouzaki et al. 2009, 2011, 2013). Accumulation of extracellular ATP, followed by increase in Ca^{2+}, and autocrine, paracrine activation of P2 purinergic receptors is involved in IL-33 release (Kouzaki et al. 2011; Hristova et al. 2016). Activation of oxidative stress pathways in airway epithelial cells that are exposed to Alternaria stimulates IL-33 release.

13.4.2.2 Group 2 Innate Lymphoid Cells

Group 2 innate lymphoid cells (ILC2s) play an important role in airway immunopathology associated with fungal infection. Studies in mice have shown that on exposure to Alternaria extract, levels of IL-5 and IL-13 increased very early after the exposure (Kouzaki et al. 2011). Studies indicate that the rapid type 2 cytokine response is driven by ILC2 population that is activated by IL-33 (Bartemes et al. 2012; Doherty et al. 2012). IL-33-independent pathway for ILC2 activation also occurs through CysLT1R (Doherty et al. 2013).

13.4.2.3 Natural Killer Cells

The cells infected by virus have activated Natural Killer (NK) cell surface receptors and are a potential target for NK cells. NK cell receptors are of activating and inhibitory types (Bartel et al. 2013; Biassoni and Malnati 2018). During viral infection, NK cells actively remove cells infected by virus. Increased NK numbers correlated to decreased disease severity in individuals infected with IAV (Björkström et al. 2021). These cells have also been found to be infected by viruses, which renders them non-functional (van Erp et al. 2019).

13.4.2.4 Dendritic Cells

Dendritic cells play an important role in providing immunity against fungal infection. Dendritic cells are responsible for taking antigens, undergoing maturation, and migrating to lymph nodes where they stimulate proliferation and differentiation of naïve T cells. Dendritic cells recognize fungal mannans through DC-SIGN, Dectin-2, and MR leading to production of immunoregulatory and proinflammatory cytokines, including IL-23, IL-10, IL-6, IL-2, and IL-1 (Trzeciak-Ryczek et al. 2015). Dendritic cells have been found to produce IL-12, which leads to Th-1-type immune response when exposed to *C. albicans* conidia. However, exposure to hyphae leads to Th2-type response by induction of IL-4 and IL-10 (Bozza et al. 2002). This suggests that there might exist a differential response of dendritic cells to different parts of fungi. Fungal antigens may lead to direct activation of dendritic cells, or in association with ILC2s like immune cells, may mediate the innate immune response to fungi that leads to regulation of adaptive immunity (Halim et al. 2016).

13.4.2.5 Neutrophils and Eosinophils

Neutrophils are the type of white blood cells (WBCs) that are predominantly associated with immune response against bacteria. Antifungal immunity is also considered to be mediated by neutrophils and eosinophils. Neutrophils phagocytose smaller conidia, while hyphae lead to formation of neutrophil extracellular DNA traps (Branzk et al. 2014). Neutrophils also produce and release antimicrobial peptides, granule proteins, and reactive oxygen species (Underhill and Pearlman 2015). The smaller conidia are generally phagocytized by neutrophils while larger hyphae trigger NETs (Branzk et al. 2014). Opsonization of fungi by IgG can lead to enhancement of neutrophil phagocytosis and can promote antibody-directed cell toxicity (Verma and Deepe 2015).

In vitro, it has been observed that in response to Alternaria conidia and hyphae, eosinophils release granule proteins extracellularly, and the process was dependent on adhesion of eosinophil CD11b to fungal β-glucans (Yoon et al. 2008). Alternaria proteases also induce extracellular release of granular contents (Inoue et al. 2005; Matsuwaki et al. 2009). The viscous and sticky mucus observed in patients with allergic bronchopulmonary aspergillosis and many other disorders may be a result of activity of eosinophil (Muniz et al. 2018). More work and information are required to elucidate the roles of neutrophils and eosinophils in antifungal immunity as well as to ascertain the pathology in patients with fungus-associated diseases. Further information will be necessary to better understand the roles of neutrophils and eosinophils in antifungal immunity in humans.

13.4.3 Adaptive Immune Response

T cells and B cells are the adaptive immune cells that express surface receptors that are genetically rearranged and are specific for a particular antigen. Adaptive immune response requires concerted activation of cells residing in tissues, antigen-presenting cells, and antigen-specific T cells and B cells.

13.4.3.1 Th1/Th17 Cell Response

Th1- and Th17-type CD4+ T cells are involved in the clearance of fungal organisms (Hardison and Brown 2012). Th1 cells produce IFNγ that enhances the function of phagocytic cells and also promotes production of opsonizing antifungal bodies by B-cell (Romani 2011). Th 17 cells activate epithelial cells and fibroblasts that produce chemokines, which lead to recruitment of phagocytes to the site of immune response (Verma and Deepe 2015). Exposure of airways to fungal β-glucan enhanced Th-2 type immune response driven by allergens leading to a mixed Th2 and Th 17 response (Zhang et al. 2017). In infection by Aspergillus, factors from hyphae and germinating spores favored Th2-type response while those from dormant conidia promoted Th-1-type response (Bozza et al. 2002).

13.4.3.2 Th2 Cell Response

Exposure of epithelial cells to allergens leads to a release of cytokines like IL-25, IL-33, GM-CSF, and TSLP that promote differentiation of Th-2-type CD4+ T cells. Prostaglandins may also play a role in regulation of Type-2 immunity. Prolonged exposure of mice to fungal antigens led to an increase in ILC2s by about two-fold, whereas the absolute number and proportion of IL-13 or IL-5 producing ILC2s and CD4+ cells showed an increase of about four-folds, suggesting that both Th2-type CD4+ and ILC-2s cells may provide immunity against fungus (Iijima et al. 2014). Chronic airway inflammation induced by fungus may lead to tripartite interactions among airway epithelial cells, ILC-2s and Th2 cells forming a positive feedback loop, that results in continued and persistent immunopathology.

13.5 Potential Role of Biocontaminants in Increasing Risk of Diseases and Disorders

Exposure to indoor biocontaminants is a major factor in the development of many unusual disorders, most notably, allergies. Biological contaminants can cause allergic reactions such as hypersensitivity, allergic rhinitis, pneumonitis, and asthma. Toxins that cause disease are released by mold and mildews

(Chirca 2019). Sneezing, watery eyes, coughing, shortness of breath, dizziness, lethargy, fever, and digestive problems are common symptoms of biological contaminants (United States Environmental Protection Agency (EPA) 2020).

Infectious diseases are caused by bacteria, viruses, fungi, and other microorganisms and can easily spread from person to person. In humans, infectious diseases can be identified by a variety of symptoms such as headache, rhinitis, fever, chills, coughs, shortness of breath, vomiting, and diarrhea. Infectious diseases can be spread through the direct entry of pathogenic biological contaminants (mostly microorganisms in the form of bioaerosol) or through the entry of harmful toxins (e.g. neurotoxin, enterotoxin, etc.) produced by them (Kumar et al. 2021).

The biocontaminants disrupt the skin's and mucosal surfaces' epithelial barriers, and lead to an increase in the prevalence and severity of allergic and inflammatory diseases such as atopic dermatitis, food allergy, allergic rhinitis, chronic rhinosinusitis, eosinophilic esophagitis, and asthma (Zeynep et al. 2022).

13.5.1 Role of Fungi in Diseases

Fungi are common inside homes, particularly in bedroom microenvironments. Fungal exposure has been linked to respiratory and allergic symptoms in epidemiological studies. Infants and young children are a vulnerable population because their respiratory system is still developing and they spend the majority of their day in their rooms (Stamatelopoulou et al. 2019, 2020). *Cladosporium*, *Penicillium*, *Alternaria*, and *Aspergillus* are the most common fungal genera in the indoor environment. Fungal fragments have also recently been identified as a trigger factor for respiratory diseases and atopic dermatitis (Tischer and Heinrich 2013). Exposure to mold leads to asthma and allergies. However, despite substantial evidence linking residential dampness and fungal exposure to the prevalence of respiratory or allergic diseases, the fungi responsible for asthma exacerbation remain inconsistent (Fan et al. 2017; Stamatelopoulou et al. 2019, 2020). (1,3)-D glucans are a class of -D-glucose polysaccharides that are a structural component of fungi and some bacteria and may be present in house dust. In the presence of (1,3)-D glucans in the indoor environment, respiratory infections can occur (Seo et al. 2015).

Mycotoxins are toxic secondary metabolites produced by fungi that have the potential to cause disease and death in humans and animals. These disrupt RNA synthesis and cause DNA damage. Cancer can be caused by aflatoxin produced by *Aspergillus flavus* and *A. parasiticus*. *Stachybotrys chartarum* produces another mycotoxin that has been linked to acute pulmonary hemorrhage (Kumar et al. 2021).

13.5.2 Role of Viruses in Diseases

Viruses are the microscopic entities that cause highly contagious diseases such as SARS, MERS, Ebola, and avian influenza. Numerous viruses, such as the influenza virus (H1N1), intestinal enteric viruses, respiratory syncytial virus (RSV), hantavirus from rodent feces, the severe acute respiratory syndrome (SARS) virus, the measles, mumps, and rubella viruses, spread readily via the air (Diglisic et al. 1999; Yu et al. 2004). One of the most prevalent airborne infections among children is chickenpox. Adults may get this illness, which has a higher pathogenicity in some regions. Recent pandemics, such as SARS-CoV-2, have resulted in mutations that strengthen the virus and allow it to evade treatment (Patel et al. 2023). The virus can be spread through direct contact with bodily fluids, sores on an infected person, or objects that come into contact with bodily fluids or sores, such as clothing or linens. They can also be transmitted through respiratory secretions when there is prolonged face-to-face contact (Patel et al. 2023).

13.5.3 Role of Bacteria in Diseases

Most bacteria associated with buildings are saprophytic, meaning they are part of the normal microflora found in the nose, mouth, ear, and so on, and are derived from outdoor air. Few heterotrophic bacteria, such as *Mycobacteria* and *Legionella*, have been linked to biofilm formation, particularly in cooling systems, water reservoirs and water pipes (Moldoveanu 2015). *Corynebacterium* spp., *Streptococcus* spp., *Staphylococcus* spp., and *Micrococcus* spp. are among the prominent bacterial species identified in indoor areas (Kausar et al. 2016). Bacterial cells, spores, peptidoglycans, microbial VOCs, exotoxins, and other bacteria-growing metabolites are among the elements released into the atmosphere (Moldoveanu 2015). Tuberculosis (TB) is one of the leading bacterial diseases responsible for deaths worldwide. It is a common community-acquired disease caused by inhaling aerosolized bacilli in expectorated sputum-positive TB patients' droplet nuclei. *Mycobacterium tuberculosis* is the causative agent of TB. This is spread to others through coughing, sneezing, and talking by infected people in the form of bio-aerosols (Srikanth et al. 2008). Legionellosis is a hospital-acquired and community-acquired pneumonia caused by *Legionella pneumophila* in humans who work in an industrial or occupational setting. *Legionella* is caused by active aerosolizing processes (aeration of contaminated water). *Legionella* is frequently found in manmade water systems, biofilms in cooling towers, and air conditioning systems (United States Environmental Protection Agency (EPA) 2020).

13.5.4 Allergens

Allergens are substances that can trigger IgE-mediated allergic reactions in humans. They are high molecular weight proteins or glycoproteins ranging in size from 10 kDa to more than 100 kDa. *Aspergillus*, *Penicillium*, *Cladosporium*, and

Alternaria spp. allergens cause type 1 hypersensitivity, which causes a variety of respiratory diseases. Insects and their by-products have been reported to be powerful allergens. According to a clinico-immunological study, mosquito debris could be a significant inhalant allergen in indoor air. *Dermatophagoides pteronyssius* (Der p I and Der p II) and *Dermatophagoides farinae* (Der f I) are two major allergen-producing mite species. Allergic fungi include *Penicillium, Aspergillus, Alternaria, Cladosporium, Curvularia, Epicoccum, Pullularia, Ganoderma, Trichoderma*, and *Stemphylium* (United States Environmental Protection Agency (EPA) 2020).

13.6 Cell Signaling Pathways Alterations and Endocrine Disruption

Biocontaminants are to blame for an increase in the risk of many diseases and disorders in humans. They have the potential to significantly impact body metabolism, signalling pathways, and even cause endocrine disruption.

13.6.1 Changes in Cell Signalling Pathways

Extensive biochemical and structural analyses over the last few decades have greatly advanced our understanding of bacterial protein toxins that modulate G-proteins. Bacterial protein toxins are a powerful tool for modulating host-pathogen interactions. *Pasteurella multocida* (PMT) and Bordetella's dermonecrotic toxin, cytotoxic necrotizing factor (CNF)1–3 of *Escherichia coli*, are some examples of toxins (Wilson and Ho 2010). A number of these toxins are large multipartite enzymes (A–B toxins) or effector proteins (also known as exoenzymes) that catalyze reactions that can interfere with host cell signal transduction and physiological processes (Wilson and Ho 2010).

Recent research suggests that local and distant microbiota play an active role in the initiation, progression, and overall prognosis of breast cancer. A dysbiotic microbiota predisposes the body to cancer by inducing genetic instability, initiating DNA damage and progeny proliferation, metabolic dysregulation, and altered response to therapy (Parida and Sharma 2019). For breast cancer, distinct bacterial, fungal, viral, and parasite signatures have been identified. These microbes can produce a variety of secondary metabolites that can act as signalling mediators in the progression of breast cancer (Vogtmann and Goedert 2016).

Mycobacterium tuberculosis is a well-known pathogen that causes millions of deaths worldwide. The emergence of drug resistance in *Mycobacterium tuberculosis* has posed new challenges to the global TB elimination effort (Naz and Arish 2022). This invasive bacterium actively induces its own uptake via macropinocytosis in order to enter normally nonphagocytic cells such as AECs

(Garcia-Perez et al. 2003). TLR2, TLR4, and C-type lectins have previously been implicated in mycobacteria-induced epithelial signalling and uptake (Alaridah et al. 2017). Mycobacteria have also been shown to manipulate epithelial production of the cytokine CXCL8 via the inhibitory G-protein-coupled receptor (GPCR) kinase 2 (Hakansson et al. 2013). Intracellular bacteria, such as *Shigella*, are known to engage GTPases in actin polymerization to trigger GPCR signal transduction in AECs. These low-molecular-weight proteins, which include Rab and Rho/Rac, belong to the Ras GTPase superfamily and have the ability to act as molecular switches by coupling extracellular signals to different cellular responses, cytoskeletal integrity (Alaridah et al. 2017), intracellular vesicular transport, and protein trafficking. Rac1 inhibition has recently been shown to reverse tumor protein p53 suppression of STAT and NF-B, and Rho is required for the formation and maintenance of tight junctions (Guo et al. 2014).

Mycotoxin contamination is common in food and feed around the world, and it can have serious consequences for both human and animal health. Aflatoxin B1 (AFB1), ochratoxin A (OTA), deoxynivalenol (DON), T-2 toxin (T-2), fumonisin B1 (FB1), and zearalenone (ZEA) are the ones that are legislated by the European Food Safety Authority (EFSA), more controlled in food and feed, and thus routinely determined and studied (Choi et al. 2020). OTA, which is derived from *Aspergillus* and *Penicillium*, is also hepatotoxic, nephrotoxic, and immunotoxic, with the kidney as its target organ (Tao et al. 2018). When immunologic stimulants are present, these six mycotoxins play anti-inflammatory roles, but when they are absent, they play pro-inflammatory roles. Pigs are the most sensitive to mycotoxins, followed by humans and poultry, rodents and marine organisms, and ruminants. Females are more vulnerable to mycotoxins than males. Mycotoxins suppress immune cells through oxidative stress, apoptosis, and autophagy, as well as by inhibiting immunity-related signal pathways; and AFB1, OTA, DON, and T-2 induce immunostimulant by directly activating the TLRs/NF-B pathway and other crossing pathways such as cyclooxygenase-2 (COX-2) and mitogen-activated protein kinase (MAPK) (Sun et al. 2022).

13.6.2 Endocrine Disruption

SARS-CoV-2 microinvasion is now recognized. Sauve et al. (2021) examined SARS-CoV-2 brain entry using animal and cellular models, as well as adult COVID-19 patient and fetal human brains, to see if low testosterone levels observed in men with severe COVID-19 could be of central origin. The majority of hypotestosteronemic patients had hypogonadotropic hypogonadism or dysregulation of the hypothalamic-pituitary-gonadal axis. In all COVID-19 patients' brains, neurons producing gonadotropin-releasing hormone (GnRH), the master molecule controlling fertility, expressed angiotensin-converting enzyme 2 and neuropilin-1, two host-cell factors mediating infection, and were infected and dying. Tanycytes, which are hypothalamic glia that control GnRH secretion, were also infected. Furthermore, both the above host-cell susceptibility factors and formyl peptide receptor 2, a putative vomeronasal receptor implicated in SARS-CoV-2 pathogenesis in humans and mice, were abundantly expressed in human fetal olfactory and vomeronasal epithelia, from which GnRH neurons arise. Finally, a SARS-CoV-2-like pseudovirus could infect a fetal human GnRH cell line expressing all of these receptors. GnRH neurons, which may be involved in brain development and aging as well as reproduction, are especially vulnerable to SARS-CoV-2 in both adults and fetuses/newborns, with potentially devastating long-term consequences (Sauve et al. 2021).

Aflatoxin B1 (AFB1) is a naturally occurring environmental pollutant found in feed and foodstuffs. It is the most toxic of all aflatoxins and can severely impair testicular development and function (Lin et al. 2022). According to previous research, oxidative stress is a master mechanism of AFB1 toxicity (Huang et al. 2019). AFB1 can cause germ cell apoptosis, endocrine disruption, and spermatogenesis obstruction in rats and mice by inducing oxidative stress and significantly impairing the testicular antioxidant defense system (Malekzadeh and Razi 2018).

13.7 Probable Methods of Detoxification

People spend most of their time (90% or more) indoors, in houses, schools, university rooms, shops, cars, planes, and workplaces (Moldoveanu 2015). Indoor air is never free of microorganisms and spores, and even clean rooms contain approximately 25 spores/m^3 (Chirca 2019). According to Chirca (Chirca 2019), biocontaminants are present even in highly controlled indoor environments (e.g. modern hospital operating rooms). Air conditioners, fans, coolers, humidifiers, and other similar devices are major sources of microorganism propagation and spread in indoor buildings. Fungi (e.g. *Aspergillus*, *Penicillium*, *Phialophora*, and *Geotrichium*), bacteria, and yeasts colonize ventilation, fans, and air conditioners (Dubey and Maheshwari 2014). Because the outdoor and indoor environments are interconnected, microorganisms may spread at random. Although air is not an ideal medium for microorganism multiplication and survival, fungal spore concentrations in contaminated farm buildings can reach thousands of spores/m^3.

The prevention of biological contamination in indoor air is primarily based on avoiding conditions that provide a substrate for the growth of viable particles, followed by the containment and removal of such growth (World Health Organization 1988). Using air filters and improving ventilation performance to control bioaerosols may be a viable solution (Vlad et al. 2013). Most of the indoor control strategies are based on controlling the dew point, which reduces the growth of various microorganisms. When contaminated water is aerosolized, it can cause the formation and spread of harmful bioaerosols in both indoor and outdoor environments (World Health Organization 1988). Airborne transmission of infectious disease within a building is affected by occupant density; the higher the density, the greater the risk. However, the presence of vectors such as pets and insects can be blamed for the transmission of nonviable aerosols that cause allergies (World Health Organization 1988).

Controlling or avoiding biocontamination can be accomplished through a variety of methods, including:

13.7.1 Humidity Control

The relative humidity in a building and thus the risk of biological contamination can be reduced by use of air conditioners and humidifiers. Humidity levels below 20% may induce itching and dryness of the skin while levels more than 65% may aggravate the symptoms of people suffering from asthma and allergies (World Health Organization 1988). Use of humidifiers, chillers, and dehumidifiers to control indoor humidity levels can reduce the formation of aerosols and thereby reduce the chances of aerosol formation (Kumar et al. 2021).

13.7.2 Biofumigation

Since ancient times, fumes from herbal plants and their products have been used to clean the air. Biofumigation is considered a holy act in Hindu culture, known as "Agnihotra," in which cow dung, fine cow butter, meditational plant leaves, wood, and other materials are burned and the fumes clean the air. Scientists recorded the data of microflora before and after performing Agnihotra in various locations and found a significant decrease in microbial quantity after Agnihotra (Kumar et al. 2021). Microorganisms are controlled by biofumigation by lowering relative humidity (Pachori et al. 2013). In Haridwar, researchers also used Vedic purification methods. Lad and colleagues prepared Dhoop from herbal products to clean the air, and its effectiveness against indoor microorganisms was also tested (Lad and Palekar 2016).

13.7.3 Hybrid Biomembrane-Functionalized Nanorobots

With the rapid advancement of robotic research, developing biomimetic micro- or nanorobots that translate biological principles into robotic systems is becoming increasingly interesting and important. The design, construction, and testing of a dual-cell membrane-functionalized nanorobot for multipurpose biological threat agent removal, specifically concurrent targeting and neutralization of pathogenic bacteria and toxins (de Ávila et al. 2018). They demonstrated ultrasound-powered biomimetic nanorobots made of gold nanowires covered in a hybrid of red blood cell (RBC) and platelet (PL) membranes. These hybrid cell membranes contain a variety of functional proteins associated with human RBCs and PLs, which provide the nanorobots with a number of appealing biological capabilities, such as adhesion and binding to PL-adhering pathogens (e.g. *Staphylococcus aureus* bacteria) and neutralization of pore-forming toxins. Furthermore, the biomimetic nanorobots demonstrated rapid and efficient prolonged acoustic propulsion in whole blood with no visible biofouling, mimicking the movement of natural motile cells. This propulsion improved the robots' binding and detoxification efficiency against pathogens and toxins. Overall, combining these diverse biological functions of hybrid cell membranes with the nanorobots' fuel-free propulsion resulted in a dynamic robotic system for efficient isolation and simultaneous removal of different biological threats, an important step toward the development of a broad-spectrum detoxification robotic platform (Li et al. 2017; de Ávila et al. 2018).

13.7.4 Silver Nanoparticles

Green, silver nanoparticles (AgNPs) synthesized using *Moringa oleifera* seed (MOS) as a reducing/capping agent kill Gram-positive *Staphylococcus aureus, Escherichia coli*, *Salmonella enterica typhimurium*, and *Pseudomonas aeruginosa* (Baghbani-Arani et al. 2017; Mehwish et al. 2021). The effect of AgNPs mixed with paint in reducing the fungal contamination of air (Rostami et al. 2021).

13.7.5 Air Filters

Air filters are one of the primary methods by which harmful microbes and viruses can be filtered out. For filtering different sizes of aerosols, nonwoven, woven, nanofiber, nanoparticle-coated and nanofiber electret filters are used (Feng and Cao 2019).

13.7.6 Probiotics-Combating Biofilms

Biofilms are aggregates of microorganisms in a sessile state embedded in a self-produced polymeric matrix (Bjarnsholt et al. 2018). Biofilm-related infections, which include chronic infections, device-related infections, and medical device malfunction, have been a major clinical problem. There have been approaches to dealing with biofilms that are largely based on disrupting their adhesion and maturation. Nowadays, there is a growing interest in using probiotics and their derivatives to combat pathogenic biofilms (Barzegari et al. 2020).

Biocontamination of medical devices and implants is a growing problem that leads to medical complications and higher costs. Developing synthetic surfaces that reduce microbe adhesion and provide biocidal activity or combinatory effects has emerged as a major global strategy in the fight against biocontamination. Advances in nanotechnology and biological sciences have enabled the development of smart surfaces for infection control (Warnes et al. 2015). Nonetheless, the clinical performance of these surfaces is heavily dependent on the material used. Surfaces can be coated with antimicrobial coatings, such as cationic polymers, metal coatings, and antifouling micro-/nanostructures. Once applied to biomedical devices and medical workers' protective equipment, the nanotechnology-based strategies may be useful in producing materials that reduce or prevent the transmission of airborne viral droplets.

In the food industry, microbial entities on equipment surfaces can easily contaminate the food being processed, causing significant economic problems and even public health issues. Cleaning processes are commonly used to ensure the hygiene of materials in contact with food. Traditional biofilm removal strategies require a lot of energy and water. They also use a significant amount of chemicals. Despite the strict enforcement of traditional cleaning procedures, sanitary organizations report alarming statistics each year, emphasizing the importance of improving the safety and hygiene of equipment surfaces. As a result, research into strategies for effectively removing biofilm biomass is becoming increasingly important (Bénézech and Faille 2018). In addition to the widely used physical methods (conventional CIP, water jet, and static foam cleaning), other promising methods such as pulsed flow, bubble flow, foam flow, ice pigging, snow CO_2, electromagnetic methods, plasma, and air jet can be used. All of these methods have demonstrated a promising ability to remove biofilm, with some, such as foam flow and pulsed flow, being found to be eco-friendly due to lower environmental impacts based on LCA analyses (Dallagi et al. 2023).

13.8 Conclusion

It has been seen that many individuals who even though spend most of their time indoors are susceptible to many diseases. These diseases are caused by contaminants of biological origin. Though immune response is generated, many a times it does not suffice and individuals fall ill, suffering from allergies, cold, fever, and gastrointestinal diseases. The materials used in building, equipment, and appliances also play a major role in multiplication and spread of these contaminants. To check the spread of these contaminants, it is imperative that detoxification be carried out frequently in homes, schools, colleges and hospitals.

References

Adkinson, N.F. Jr., Bochner, B.S., Burks, A.W. et al. (2013). *Middleton's Allergy E-Book: Principles and Practice*. Elsevier Health Sciences.

Al Hallak, M., Verdier, T., Bertron, A. et al. (2023). Fungal contamination of building materials and the aerosolization of particles and toxins in indoor air and their associated risks to health: a review. *Toxins* 15 (3): 175.

Alaridah, N., Lutay, N., Tenland, E. et al. (2017). Mycobacteria manipulate G-protein-coupled receptors to increase mucosal Rac 1 expression in the lungs. *Journal of Innate Immunity* 9 (3): 318–329.

Ananthanaryan R, Paniker CKJ (2009) Text of Microbiology, 8th Edition, Universities Press

Anderson, A.A. (1958). A new sampler for collection, sizing and enumeration of viable airborne particles. *Journal of Bacteriology* 76: 471–484.

de Ávila, B.E.-F. et al. (2018). Hybrid biomembrane-functionalized nanorobots for concurrent removal of pathogenic bacteria and toxins Sci. *Robotics* 3: eaat0485. https://doi.org/10.1126/scirobotics.aat0485.

Badea, S., Chirita, A.L., Androne, C., and Olaru, B.G. (2015). Indoor air quality assessment through microbiological methods. Scientific papers. *Series Journal of Young Scientist* 3: 12–17.

Baghbani-Arani, F., Movagharnia, R., Sharifian, A. et al. (2017). Photo-catalytic, anti-bacterial, and anti-cancer properties of phyto-mediated synthesis of silver nanoparticles from Artemisia tournefortiana Rchb extract. *Journal of Photochemistry and Photobiology B: Biology* 173: 640–649. 10.1016/j.jphotobiol.2017.07.003.

Bartel, Y., Bauer, B., and Steinle, A. (2013). Modulation of NK cell function by genetically coupled C-type lectin-like receptor/ligand pairs encoded in the human natural killer gene complex. *Frontiers in Immunology* 4: 362.

Bartemes, K.R., Iijima, K., Kobayashi, T. et al. (2012). IL-33-responsive lineage-CD25 + CD44 (hi) lymphoid cells mediate innate type 2 immunity and allergic inflammation in the lungs. *Journal of Immunology* 188: 1503–1513.

Barzegari, A., Kheyrolahzadeh, K., Khatibi, S.M.H. et al. (2020). The battle of probiotics and their derivatives against biofilms. *Infection and Drug Resistance* 13: 659–672. https://doi.org/10.2147/IDR.S232982.

Bénézech, T. and Faille, C. (2018). Two-phase kinetics of biofilm removal during CIP. Respective roles of mechanical and chemical effects on the detachment of single cells vs cell clusters from a *Pseudomonas fluorescens* biofilm. *Journal of Food Engineering* 219: 121–128. https://doi.org/10.1016/j.jfoodeng.2017.09.013.

Biassoni, R. and Malnati, M.S. (2018). Human natural killer receptors, co-receptors, and their ligands. *Current Protocols in Immunology* 121 (1): e47.

Bjarnsholt, T., Buhlin, K., Dufrêne, Y.F. et al. (2018). Biofilm formation – what we can learn from recent developments. *Journal of Internal Medicine* 284 (4): 332–345. https://doi.org/10.1111/joim.1278229856510.

Björkström, N.K., Strunz, B., and Ljunggren, H.-G. (2022). Natural killer cells in antiviral immunity. *Nature Reviews Immunology* 22 (2): 112–123.

Bozza, S., Gaziano, R., Spreca, A. et al. (2002). Dendritic cells transport conidia and hyphae of *Aspergillus fumigatus* from the airways to the draining lymph nodes and initiate disparate Th responses to the fungus. *Journal of Immunology* 168: 1362–1371.

Branzk, N., Lubojemska, A., Hardison, S.E. et al. (2014). Neutrophils sense microbe size and selectively release neutrophil extracellular traps in response to large pathogens. *Nature Immunology* 15: 1017–1025.

Charlson, E.S., Bittinger, K., Haas, A.R. et al. (2011). Topographical continuity of bacterial populationsin the healthy human respiratory tract. *American Journal of Respiratory and Critical Care Medicine* 184: 957–963.

Chirca, I. (2019). The hospital environment and its microbial burden: challenges and solutions. *Future Microbiology* 14 (12): 1007–1010. https://doi.org/10.2217/fmb-2019-0140.

Choi, S.-Y., Kim, T.H., Hong, M.-W. et al. (2020). Transcriptomic alterations induced by aflatoxin B1 and ochratoxin A in LMH cell line. *Poultry Science* 99 (11): 5265–5274.

Crameri, R. and Blaser, K. (2002). Allergy and immunity to fungal infections and colonization. *The European Respiratory Journal* 19: 151–157.

Dallagi, H., Jha, P.K., Faille, C. et al. (2023). Removal of biocontamination in the food industry using physical methods; an overview. *Food Control* 148: 109645. https://doi.org/10.1016/j.foodcont.2023.109645.

Deacon, J.W. (2005). *Fungal biology*. Wiley.

Diglisic, G., Rossi, C.A., Doti, A., and Walshe, D.K. (1999). Seroprevalence 53 study of Hantavirus infection in the community based population. *Maryland Medical Journal* 48: 303–306.

Dixon, B.R. (2021). Giardia duodenalis in humans and animals—transmission and disease. *Research in Veterinary Science* 135: 283–289.

Doherty, T.A., Khorram, N., Chang, J.E. et al. (2012). STAT6 regulates natural helper cell proliferation during lung inflammation initiated by Alternaria. *American Journal of Physiology, Lung Cellular and Molecular Physiology* 303: L577–L588.

Doherty, T.A., Khorram, N., Lund, S. et al. (2013). Lung type 2 innate lymphoid cells express cysteinyl leukotriene receptor 1, which regulates TH2 cytokine production. *The Journal of Allergy and Clinical Immunology* 132: 205–213.

Dubey, R.C. and Maheshwari, D.K. (2014). *Textbook of Microbiology*, 5e. New Delhi: S. Chand & Co., Limited.

Efstratiou, A., Ongerth, J.E., and Karanis, P. (2017). Waterborne transmission of protozoan parasites: review of worldwide outbreaks-an update 2011–2016. *Water Research* 114: 14–22.

van Erp, E.A., van Kampen, M.R., van Kasteren, P.B., and de Wit, J. (2019). Viral infection of human natural killer cells. *Viruses* 11 (3): 243.

Fan, G., Xie, J., Yoshino, H. et al. (2017). Environmental conditions in homes with healthy and unhealthy schoolchildren in Beijing, China. *Building and Environment* 112: 270–284.

Feng, Z. and Cao, S.-J. (2019). A newly developed electrostatic enhanced pleated air filters towards the improvement of energy and filtration efficiency. *Sustainable Cities and Society* 49: 101569.

Flint, S.J., Racaniello, V.R., Rall, G.F. et al. (2020). *Principles of Virology, Pathogenesis and Control*, vol. 2. Wiley.

Fujimura, K.E., Johnson, C.C., Ownby, D.R. et al. (2010). Man's best friend? The effect of pet ownershipon house dust microbial communities. *The Journal of Allergy and Clinical Immunology* 126: 410–412.

Fujimura, K.E., Demoor, T., Rauch, M. et al. (2014). House dust exposure mediates gut microbiome *Lactobacillus* enrichment and airway immune defense against allergens and virus infection. *Proceedings of the National Academy of Sciences of the United States of America* 111: 805–810.

Funchal, G.A., Jaeger, N., Czepielewski, R.S. et al. (2015). Respiratory syncytial virus fusion protein promotes TLR-4-dependent neutrophil extracellular trap formation by human neutrophils. *PLoS One* 10 (4): e0124082.

Garcia-Perez, B.E., Mondragon-Flores, R., and Luna-Herrera, J. (2003). Internalization of *Mycobacterium tuberculosis* by macropinocytosis in nonphagocytic cells. *Microbial Pathogenesis* 35: 49–55.

Godish, D.R. and Godish, T.J. (2007). Relationship between sampling durationand concentration of culturable airborne mould and bacteria onselected culture media. *Journal of Applied Microbiology* 102: 1479–1484.

Guo, A.K., Hou, Y.Y., Hirata, H. et al. (2014). Loss of p 53 enhances NF-κB-dependent lamellipodia formation. *Journal of Cellular Physiology* 229: 696–704.

Hakansson, G., Lutay, N., Andersson, M. et al. (2013). Epithelial G protein-coupled receptor kinases regulate the initial inflammatory response during mycobacterial infection. *Immunobiology* 218: 984–994.

Halim, T.Y., Hwang, Y.Y., Scanlon, S.T. et al. (2016). Group 2 innate lymphoid cells license dendritic cells to potentiate memory TH2 cell responses. *Nature Immunology* 17: 57–64.

Hardison, S.E. and Brown, G.D. (2012). C-type lectin receptors orchestrate antifungal immunity. *Nature Immunology* 13: 817–822.

Heil F, Hemmi H, Hochrein H, etal. (2004) Species-specific recognition of single-stranded RNA via toll-like receptor 7 and 8. *Science* 303(5663): 1526–1529

Hristova, M., Habibovic, A., Veith, C. et al. (2016). Airway epithelial dual oxidase 1 mediates allergen-induced IL-33 secretion and activation of type 2 immune responses. *The Journal of Allergy and Clinical Immunology* 137: 1545–1556. e11.

Huang, W., Cao, Z., Zhang, J. et al. (2019). Aflatoxin B1 promotes autophagy associated with oxidative stress-related PI3K/AKT/mTOR signaling pathway in mice testis. *Environmental Pollution* 255 (Part 2): 113317. https://doi.org/10.1016/j.envpol.2019.113317.

Iborra, S., Izquierdo, H.M., Martínez-López, M. et al. (2012). The DC receptor DNGR-1 mediates cross-priming of CTLs during vaccinia virus infection in mice. *The Journal of Clinical Investigation* 122 (5): 1628–1643.

Iijima, K., Kobayashi, T., Hara, K. et al. (2014). IL-33 and thymic stromal lymphopoietin mediate immune pathology in response to chronic airborne allergen exposure. *Journal of Immunology* 193: 1549–1559.

Inoue, Y., Matsuwaki, Y., Shin, S.H. et al. (2005). Nonpathogenic, environmental fungi induce activation and degranulation of human eosinophils. *Journal of Immunology* 175: 5439–5447.

Janeway, C.A. (2013). Pillars article: approaching the asymptote? Evolution and revolution in immunology. Cold spring harb symp quant biol. 1989.54:1–13. *Journal of Immunology* 191 (9): 4475–4487.

Kang Frank, Y.J.J.F. (1989). Biological aerosols: A review of airborne contamination and its measurement in dairy processing plants. *Journal of Food Protection* 52 (7): 512–524.

Karp, C.L. (2010). Guilt by intimate association: what makes an allergen an allergen? *The Journal of Allergy and Clinical Immunology* 125: 955–960. quiz 61–2.20381850.

Kausar, M.A., Vijayan, V.K., Bansal, S.K. et al. (2007). Mosquitoes as sources of inhalant allergens: clinicoimmunologicand biochemical studies. *The Journal of Allergy and Clinical Immunology* 120: 1219–1221.

Kausar, M.A., Arif, J.M., Alanazi, S.M.M. et al. (2016). Assessment of microbial load in indoor environment of university and hospitals of Hail. *KSA Biochemical and Cellular Archives* 16 (1): 177–183.

Knights, D., Kuczynski, J., Charlson, E.S. et al. (2011). Bayesian community-wide culture-independent microbial source tracking. *Nature Methods* 8: 761–763.

Kouzaki, H., O'Grady, S.M., Lawrence, C.B., and Kita, H. (2009). Proteases induce production of thymic stromal lymphopoietin by airway epithelial cells through protease-activated receptor-2. *Journal of Immunology* 183: 1427–1434.

Kouzaki, H., Iijima, K., Kobayashi, T. et al. (2011). The danger signal, extracellular ATP, is a sensor for an airborne allergen and triggers IL-33 release and innate Th2-type responses. *Journal of Immunology* 186: 4375–4387.

Kouzaki, H., Tojima, I., Kita, H., and Shimizu, T. (2013). Transcription of interleukin-25 and extracellular release of the protein is regulated by allergen proteases in airway epithelial cells. *American Journal of Respiratory Cell and Molecular Biology* 49: 741–750.

Kraakman, N.J.R., González-Martín, J., Pérez, C. et al. (2021). Recent advances in biological systems for improving indoor air quality. *Reviews in Environmental Science and Bio/Technology* 20: 363–387.

Kumar, P., Kausar, M.A., Singh, A.B., and Singh, R. (2021). Biological contaminants in the indoor air environment and their impacts on human health. *Air Quality, Atmosphere & Health* 14 (11): 1723–1736. https://doi.org/10.1007/s11869-021-00978-z.

Lad, N. and Palekar, S. (2016). Preparation and evaluation of Herbal Dhoop for cleansing the air. *International Journal of Herbal Medicince* 4 (6): 98–103.

Li, J., Esteban-Fernández de Ávila, B., Gao, W. et al. (2017). Micro/nanorobots for biomedicine: delivery, surgery, sensing, and detoxification. *Science robotics* 2: eaam6431.

Lin, L., Cao, Q., Zhang, C. et al. (2022). Aflatoxin B1 causes oxidative stress and apoptosis in sheep testes associated with disrupting rumen microbiota, *Ecotoxicology and Environmental Safety* 232, 113225. https://doi.org/10.1016/j.ecoenv.2022.113225.

Yasin, M., Mazdak, R., and Mino, I. (2018). Aflatoxin B1 impairs spermatogenesis: an experimental study for crosslink between oxidative stress and mitochondria-dependent apoptosis. *Environmental Toxicology* 33 (11): 1204–1213.

Mandal, J. and Brandl, H. (2011). Bioaerosols in indoor environment – A review with special reference to residential and occupational locations. *The Open Environmental and Biological Monitoring Journal* 4: 83–96.

Marzuki, I., Ali, M., Syarif, H.E. et al. (2021). Investigation of biodegradable bacteria as bio indicators of the presence of PAHs contaminants in marine waters in the marine tourism area of Makassar City. *IOP Conference Series: Earth and Environmental Science* 750 (1): 012006. IOP Publishing.

Matsuwaki, Y., Wada, K., White, T.A. et al. (2009). Recognition of fungal protease activities induces cellular activation and eosinophil-derived neurotoxin release in human eosinophils. *Journal of Immunology* 183: 6708–6716.

Maung, T.Z., Bishop, J.E., Holt, E. et al. (2022). Indoor air pollution and the health of vulnerable groups: a systematic review focused on particulate matter (PM), volatile organic compounds (VOCs) and their effects on children and people with pre-existing lung disease. *International Journal of Environmental Research and Public Health* 19 (14): 8752.

Mehwish, H.M., Rajoka, M.S.R., Xiong, Y. et al. (2021). Green synthesis of a silver nanoparticle using *Moringa oleifera* seed and its applications for antimicrobial and sun-light mediated photocatalytic water detoxification. *Journal of Environmental and Chemical Engineering* 9 (4): 105290. https://doi.org/10.1016/j.jece.2021.105290.

Modhiran, N., Watterson, D., Muller, D.A. et al. (2015). Dengue virus NS1 protein activates cells via toll-like receptor 4 and disrupts endothelial cell monolayer integrity. *Science Translational Medicine* 7 (304): 304ra142.

Moldoveanu, A.M. (2015). Biological contamination of air in indoor spaces. *Current Air Quality Issues* 489–514: https://doi.org/10.5772/59727.

Moldoveanu, A.M. (2015). Biological contamination of air in indoor spaces. In: *Current Air Quality Issues*, 489–514. IntechOpen.

Monteiro, A., Almeida, B., Paciência, I. et al. (2021). Bacterial contamination in health care centers: differences between urban and rural settings. *Atmosphere* 12 (4): 450.

Muniz, V.S., Silva, J.C., Braga, Y.A.V. et al. (2018). Eosinophils release extracellular DNA traps in response to *Aspergillus fumigatus*. *The Journal of Allergy and Clinical Immunology* 141: 571–585. e7.

Murray, P.R., Rosenthal, K.S., and Pfaller, M.A. (2015). *Medical Microbiology*. Elsevier Health Sciences.

Napoli, C. et al. (2012). Air sampling procedures to evaluate microbial, contamination: a comparison between active and passive methodsin operating theatres. *BMC Public Health* 12: 594.

Nath, A., Baruah, N., Nonglait, M.L., and Deka, P. (2023). Biological contaminants in indoor environments of educational institutions. *Aerobiologia* 39 (1): 1–20.

Naz, F. and Arish, M. (2022). GPCRs as an emerging host-directed therapeutic target against mycobacterial infection: from notion to reality. *British Journal of Pharmacology* 179 (21): 4899–4909.

Pachori, R.R., Kulkarni, N.S., Sadar, P.S., and Mahajan, N.N. (2013). Effect of agnihotra fumes on aeromicroflora. *The Bioscan* 8 (1): 127–129.

Parida, S. and Sharma, D. (2019). The power of small changes: comprehensive analyses of microbial dysbiosis in breast cancer. *Biochimica et Biophysica Acta (BBA) – Reviews on Cancer* 1871 (2): 392–405.

Patel, P., Nandi, A., Verma, S.K. et al. (2023). Zebrafish-based platform for emerging bio-contaminants and virus inactivation research. *Science of The Total Environment* 872: 162197.

Portnoy, J.M., Williams, P.B., and Barnes, C.S. (2016). Innate immune responses to fungal allergens. *Current Allergy and Asthma Reports* 16: 62.

Prussin, A.J. and Marr, L.C. (2015). Sources of airborne microorganisms in the built environment. *Microbiome* 3 (1): 1–10.

Rautiainena, P., Hyttinenb, M., Ruokolainenb, J. et al. (2019). Indoor air-related symptoms and volatile organic compounds in materials and air in the hospital environment. *International Journal of Environmental Health Research* 29 (5): 479–488. https://doi.org/10.1080/09603123.2018.1550194.

Reed, C.E. and Kita, H. (2004). The role of protease activation of inflammation in allergic respiratory diseases. *The Journal of Allergy and Clinical Immunology* 114: 997–1008. quiz 9.15536399.

Rivas-Castillo, A., Garcia-Barrera, A., Garrido-Hernández, A. et al. (2022). Ciliated peritrichous protozoa in a tezontle-packed sequencing batch reactor as potential indicators of water quality. *Polish Journal of Microbiology* 71: 539–551.

Rodrigues, J., Caruthers, C., Azmeh, R. et al. (2016). The spectrum of allergic fungal diseases of the upper and lower airways. *Expert Review of Clinical Immunology* 12: 531–550.

Rogawansamy, S., Gaskin, S., Taylor, M., and Pisaniello, D. (2015). An evaluation of antifungal agents for the treatment of fungal contamination in indoor air environments. *International Journal of Environmental Research and Public Health* 12 (6): 6319–6332.

Romani, L. (2011). Immunity to fungal infections. *Nature Reviews Immunology* 11: 275–288.

Rostami, N., Alidadi, H., Zarrinfar, H. et al. (2021). Interventional effect of nanosilver paint on fungal load of indoor air in a hospital ward. *Canadian Journal of Infectious Diseases and Medical Microbiology* 20 (2021): 8658600. https://doi.org/10.1155/2021/8658600.

Sauve, F., Nampoothiri, S., Fernandois, D. et al. (2021). SARS-CoV-2 infects human GnRH neurons and tanycytes, disrupting hypothalamic-pituitary hormonal axes. PREPRINT (Version 1) available at Research Square. https://doi.org/10.21203/rs.3.rs-1050969/v1 (accessed 11 November 2021).

Schleimer, R.P. and Berdnikovs, S. (2017). Etiology of epithelial barrier dysfunction in patients with type 2 inflammatory diseases. *The Journal of Allergy and Clinical Immunology* 139: 1752–1761.

Seltzer, J.M. (1994). Biological contaminants. *The Journal of Allergy and Clinical Immunology* 94 (2 Pt 2): 318–326.

Seo, S., Ji, Y.G., Yoo, Y. et al. (2015). Submicron fungal fragments as another indoor biocontaminant inelementary schools. *Environmental Science. Processes & Impacts* 17 (6): 1164–1172.

Sheik, G.B., Rheam, A.I.A., Shehri, Z.S., and Otaibi, O.B.M. (2015). Assessment of bacteria and fungi in air from College of Applied Medical Sciences (Male) at AD-Dawadmi Saudi Arabia. *International Research Journal of Biological Sciences* 4 (9): 48–53.

Silva, N.C., Madureira, A.R., Pintado, M., and Moreira, P.R. (2022). Biocontamination and diversity of epilithic bacteria and fungi colonising outdoor stone and mortar sculptures. *Applied Microbiology and Biotechnology* 106 (9–10): 3811–3828.

Sousa, J. (2022). Association between home characteristics and occupant's behaviours and concentrations of bacteria, fungi and endotoxins. *Journal of Building Engineering* 45: 103409.

Srikanth, P., Sudharsanam, S., and Steinberg, R. (2008). Bio-aerosols in indoor environment: composition, health effects and analysis. *Indian Journal of Medical Microbiology* 26 (4): 302–312.

Stamatelopoulou, A., Asimakopoulos, D.N., and Maggos, T. (2019). Effects of PM, TVOCs and comfort parameters on indoor air quality of residences with young children. *Building and Environment* 150: 233–244.

Stamatelopoulou, A., Pyrri, I., Asimakopoulos, D.N., and Maggos, T. (2020). Indoor air quality and dustborne biocontaminants in bedrooms of toddlers in Athens, Greece. *Building and Environment* 173: 106756.

Stetzenbach, L.D., Buttner, M.P., and Cruz, P. (2004). Detection and enumeration of airborne biocontaminants. *Current Opinion in Biotechnology* 15 (3): 170–174.

Stetzenbach, L.D., Buttner, M.P., and Cruz, P. (2004). Detection and enumeration of airborne bio contaminants. *Current Opinion in Biotechnology* 15 (3): 170–174. ISSN 0958-1669. https://doi.org/10.1016/j.copbio.2004.04.009.

Sun, Y., Huang, K., Long, M. et al. (2022). An update on immunotoxicity and mechanisms of action of six environmental mycotoxins. *Food and Chemical Toxicology* 163: 112895.

Sutton GHC (2004) Enumeration of total airborne bacteria, yeast and mold contaminants and identification of *Escherichia coli* O157: H7, *Listeria* spp., *Salmonella* spp., and *Staphylococcus* spp. in a beef and pork slaughter facility. University of Florida.

Taghavi, M., Khosravi, A., Mortaz, E. et al. (2017). Role of pathogen-associated molecular patterns (PAMPS) in immune responses to fungal infections. *European Journal of Pharmacology* 808: 8–13.

Takeuchi Oand Akira, S. (2010). Pattern recognition receptors and inflammation. *Cell* 140 (6): 805–820.

Tao, Y., Xie, S., Xu, F. et al. (2018). Ochratoxin A: toxicity, oxidative stress and metabolism. *Food and Chemical Toxicology* 112: 320–331.

Tischer, C.G. and Heinrich, J. (2013). Exposure assessment of residential mould, fungi and microbial components in relation to children's health: achievements and challenges. *International Journal of Hygiene and Environmental Health* 216 (2): 109-114.A.

Tortora, G.J., Funke, B.R., and Case, C.L. (2015). *Microbiology: An Introduction*. Benjamin-Cummings: Books a la Carte Edition.

Trzeciak-Ryczek, A., Tokarz-Deptula, B., and Deptula, W. (2015). Antifungal immunity in selected fungal infections. *Postępy Higieny i Medycyny Doświadczalnej (Online)* 69: 469–474.

Underhill, D.M. and Pearlman, E. (2015). Immune interactions with pathogenic and commensal fungi: a two-way street. *Immunity* 43: 845–858.

Verma, A.W.M. and Deepe, G. (2015). Klein B adaptive immunity to fungi. *ColdSpring Harbor Perspectives in Medicine* 5: a019612.

Vlad, D.C., Popescu, R., Filimon, M.N. et al. (2013). Assessment of microbiological indoor air quality in public buildings: a case study (Timisoara, Romania). *African Journal of Microbiology Research* 17 (9): 1957–1963.

Vogtmann, E. and Goedert, J.J. (2016). Epidemiologic studies of the human microbiome and cancer. *British Journal of Cancer* 114: 237.

Warnes, S.L., Little, Z.R., and Keevil, C.W. (2015). Human coronavirus 229E remains infectious on common touch surface materials. *MBio* 6 (6): 10–1128.

Williams, P.B., Barnes, C.S., Portnoy, J.M., and Environmental Allergens W (2016). Innate and adaptive immune response to fungal products and allergens. *The Journal of Allergy and Clinical Immunology. In Practice* 4: 386–395.

Williams, P.B., Barnes, C.S., Portnoy, J.M., and Environmental Allergens W (2016). Innate and adaptive immune response to fungal products and allergens. *The Journal of Allergy and Clinical Immunology In Practice* 4: 386–395.

Wills-Karp, M., Nathan, A., Page, K., and Karp, C.L. (2010). New insights into innate immune mechanisms underlying allergenicity. *Mucosal Immunology* 3: 104–110.

Wilson, B.A. and Ho, M. (2010). Recent insights into *Pasteurella multocida* toxin and other G-protein-modulating bacterial toxins. *Future Microbiology* 5 (8): 1185–1201.

Wilson, B.A. and Ho, M. (2010). Recent insights into *Pasteurella multocida* toxin and other G-protein-modulating bacterial toxins. *Future Microbiology* 5 (8): 1185–1201.

Wu, F.F.S., Siebers, R., Chang, C.F. et al. (2009). Indoor allergens and microbial bio-contaminants in homes of asthmatic children in central Taiwan. *Journal of Asthma* 46 (7): 745–749.

Yassin, M.F. and Almouqatea, S. (2010). Assessment of airborne bacteria and fungi in an indoor and outdoor environment. *International Journal of Environmental Science and Technology* 7 (3): 535–544.

Yoon, J., Ponikau, J.U., Lawrence, C.B., and Kita, H. (2008). Innate antifungal immunity of human eosinophils mediated by a beta 2 integrin, CD11b. *Journal of Immunology* 181: 2907–2915.

Yu ITS, Li Y, Wong TW, Tam W, Chan AT, Lee JHW,52. etal. Evidence of airborne transmission of the severe acute respiratory syndrome virus. *The New England Journal of Medicine* 2004; 350:1731–1739

Yu, L., Shang, S., Tao, R. et al. (2017). High doses of recombinant mannan-binding lectin inhibit the binding of influenza A (H1N1)pdm 09 virus with cells expressing DC-SIGN. *Acta Pathologica, Microbiologica et Immunologica Scandinavica* 125 (7): 655–664.

Zeynep, C.S., Betul, O.O., Pamir, C. et al. (2022). Epithelial barrier hypothesis: effect of the external exposome on the microbiome and epithelial barriers in allergic disease. *Allergy* 77 (5): 1418–1449.

Zhang, Z., Biagini Myers, J.M., Brandt, E.B. et al. (2017). Beta-Glucan exacerbates allergic asthma independent of fungal sensitization and promotes steroid-resistant TH2/TH17 responses. *The Journal of Allergy and Clinical Immunology* 139: 54–65. e8.

14

Indoor Airborne Insect Allergens

Environmental Evaluation and Health Implications

Mohd Adnan Kausar[1], *Sadaf Anwar*[1], *Mohammad Zeeshan Najm*[2], *Sadaf*[3], *Simran Kaur*[2], *Samir Qiblawi*[4], *and Vinita Katiyar*[5]

[1] *Department of Biochemistry, College of Medicine, University of Ha'il, Ha'il, Saudi Arabia*
[2] *School of Biosciences, Apeejay Stya University, Gurugram, Haryana, India*
[3] *Department of Biotechnology, Jamia Millia Islamia, Delhi, New Delhi, India*
[4] *Medical and Laboratory Sciences Department, Palestine Technical University - Kadoorie, Tulkarm, Palestine.*
[5] *Regional Services Division, Indira Gandhi National Open University (IGNOU), Maidan Garhi, Delhi, New Delhi, India*

14.1 Introduction

Allergy is one of the most significant immunological illnesses because it affects approximately 30% of the population worldwide and causes inflammation in various body organ systems such as the skin, respiratory tract, cardiovascular system, and the gastrointestinal tract from exposure to an allergen (Shin et al. 2023). Allergen is any biological or chemical substance that can mediate hypersensitivity reactions associated with any clinical (IgE) reaction in humans.

A diverse variety of biological pollutants, including bacteria, fungi, viruses, algae, insects, and their byproducts including endotoxins, mycotoxins, volatile organic compounds, etc., can be found in the indoor air environment. Biological pollutants have been divided into groups based on their potential to cause allergic, infectious, poisonous, or inflammatory reactions in people (Kumar et al. 2021). Biological contamination in the interior environment and its possible causes for the transmission of different ailments are currently not well known. Indoor airborne insect allergens pose a serious public health problem and understanding the sources, risks, and prevention technologies associated with indoor insect allergens is crucial for promoting healthy indoor environments and reducing the burden of allergic diseases. This book chapter discusses the various indoor airborne insect allergens, the immune response instigated, and the health and

Airborne Biocontaminants and Their Impact on Human Health, First Edition.
Edited by Rajeev Singh and Anamika Singh.

environmental implications. It also gives detailed information on the diverse biological contaminants that can be found in various settings, including homes, hospitals, workplaces, schools, etc.

14.2 Classification of Allergens

Allergens are identified as substances that stimulate allergy in the human body, i.e. they are responsible for initiating an immune response in a hypersensitive person. Allergens are grouped into various categories based on their origin, mode of transmission, and effect on the body (Allergen Nomenclature Database 2023). Some common classifications of allergens are discussed in the subsequent sections.

14.2.1 Allergens Based on Environment

These allergens that are not present in the human body, rather they are present in the external environment and are further divided into two groups, i.e. indoor, and outdoor allergens (Pomés et al. 2016).

Indoor allergens are classified as substances that may cause allergic reactions in the body of sensitive individuals when the allergen is either inhaled or touched. Many allergens, such as animal dander, pollen, dust, and mold, are typically found in indoor environments (Kim et al. 2013). Some common indoor allergens identified are dust mites, mold, cockroaches, pet dander, pollen, etc. Dust mites are insects, which are small in size and are generally found in dust. They are well recognized for consuming dead skin cells on the skin and for frequently causing allergic reactions (Do et al. 2016). Another group of insects that are frequently encountered in homes is the cockroach. Their feces and saliva are known to cause allergic reactions in sensitive people (Jacob et al. 2002). Mold is a fungus that thrives in moist, humid environments like bathrooms and kitchens. When inhaled, the mold spores are a common allergen that causes the allergic reaction (Erwin et al. 2023). Pets may be another factor contributing to an increase in allergy reactions in sensitive people since pet saliva and hair may contain allergens that can provoke immunological reactions (Burge and Rogers 2000).

Outdoor allergens are classified as substances that are generally found in the external environment including air, soil, and plants (Singh and Hays 2016). These allergens interact with the body and trigger allergic reactions when inhaled, touched, or ingested by sensitive individuals. Pollen, insect stings, mold, animal dander, air pollution, and others have all been identified as common outdoor allergies (Adams et al. 2022). The most prevalent outdoor allergy is pollen with its sources being trees, grasses, and weeds. This allergen enters the body and elevates the immune response, thereby causing seasonal allergic rhinitis (hay fever).

Insect venom is another category of outdoor allergen where the sting of insects such as bees, wasps, and hornets, can cause allergic reactions (Pomés et al. 2016). Mold in the outdoor environment can be found in soil, damp areas, and decaying leaves and plants. The inhalation of the spores produced by the mold is responsible for triggering the allergic response (Erwin et al. 2023). Animals can also induce an allergic response when exposed to their skin flakes, saliva, and urine. Apart from this, smog can also irritate the respiratory system and contribute to triggering allergic reactions in some people.

14.2.2 Allergens Based on Route of Exposure

Allergens can be grouped based on the route of exposure, i.e. how an allergen enters the body. Various routes of allergen exposure are identified such as inhalant, ingestant, injectant, and contactant (Shamji and Durham 2017). Inhalation allergens are described as allergens that enter the human body through the respiratory system. These allergens are said to be airborne and can be inhaled through the nose or mouth. The common examples of inhalation allergens are mold spores, pollens, dust mites, pet dander, etc (Vighi et al. 2008). Ingestion allergen also known as food allergen is another category of allergens where the allergen is ingested and cause allergic reactions in the digestive system. Seafood, tree nuts, milk, eggs, soy, and peanuts are a few examples of common food allergens (David et al. 2017). Injection allergen is the type wherein the allergen is injected into the body, and this results in severe allergic reactions, including anaphylaxis (Abramson et al. 2010). Insect venom, particular drugs, and several immunizations are examples of injection allergens (Cui et al. 2020). Contact allergens are the allergens that encounter the skin and can result in skin allergies or eczema. The common examples of contact allergens are latex, fragrances, certain metals, and chemicals (Berard et al. 2003; Kimber et al. 2014; Kahn et al. 2016; Blumenthal et al. 2019).

It is important to identify the allergens that trigger an allergic reaction and take steps to avoid or minimize exposure to them. This can be achieved through various measures such as avoiding certain foods, wearing protective clothing or masks, and using air filters or purifiers (Figure 14.1).

14.3 Airborne Insect Allergens

Insect allergens are substances produced by insects, which upon exposure can elevate immunological response in the body of the sensitive individuals. A variety of insects produce allergens such as bees, wasps, hornets, yellow jackets, fire ants, and cockroaches (Lee et al. 2016; de Gier and Verhoeckx 2018). The immune system

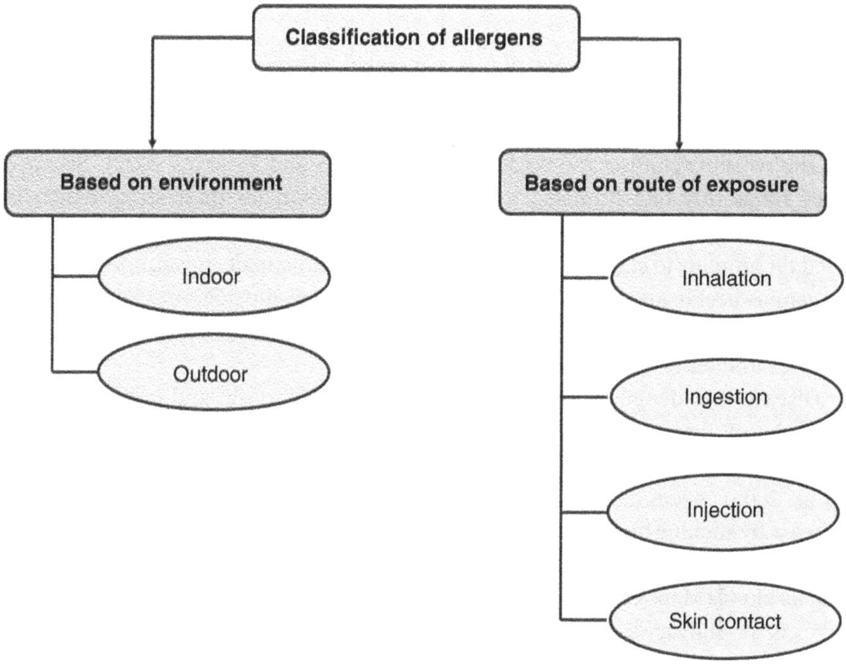

Figure 14.1 Classification of allergen based on environment and route of exposure.

interprets insect allergens as antigens and creates immunoglobulin E (IgE), an antibody that is produced in response to exposure to the allergens (Platts-Mills 2001). The interaction between the antibody and the allergen elevates the release of certain molecules such as histamine and is responsible for displaying the symptoms associated with the allergic response such as hives, swelling, difficulty breathing, and in severe cases, anaphylaxis. Insect allergies are frequently brought on by insect stings (Cui et al. 2020). Nonetheless, even nonvenomous insects like cockroaches can release allergens that might trigger allergic reactions in those with sensitivity. Exposure to insect allergens can also exacerbate asthma symptoms in people with asthma. Indoor airborne insect allergens are defined as the substances produced by the insects, which through the airway enter the body of the susceptible individual and trigger the immune response (Kim et al. 2013). These allergens can become airborne when disturbed, such as by sweeping or vacuuming, and are frequently present in the feces, saliva, or body parts of insects. Cockroaches, dust mites, and some beetle and moth species are among the common indoor insects that cause allergies. When exposed to certain allergens, some individuals may experience symptoms like sneezing, runny noses, itchy eyes, and even asthma attacks.

14.3.1 Mosquitoes

A mosquito is a flying insect that is a member of the order Diptera. They are unique among insects as they only have one pair of functioning wings. Over 3000 species of mosquitoes have been identified till date and are found in every region of the world excluding the coldest areas (Hawkes and Hopkins 2022). Because they frequently bite humans, mosquitoes are known for their sting, which is caused by the saliva being injected while sucking the blood. Female mosquitoes feed on humans to suck blood and produce eggs, while male mosquitoes typically consume nectar and plant fluids. Body heat, carbon dioxide, and certain compounds found in perspiration and other secretions are the factors that draw mosquitoes to their hosts (Tauxe et al. 2013).

Diseases like dengue fever, malaria, yellow fever, filariasis, and several types of encephalitis are carried by female mosquitoes (Billingsley et al. 2006; Javed et al. 2013). To produce fertile eggs, the female mosquito consumes blood (Attardo et al. 2005). *Aedes aegypti*, the female mosquito, carries out the progression of disease by sucking blood of an infected person and probing for blood from another. The mosquito injects saliva into the victim's skin through a small duct while sucking its blood (Mans et al. 2008). The key factor for allergic reactions after mosquito bites is the salivary secretions that widen the blood vessels to aid in feeding. The delayed-type local cutaneous responses and the responses mediated by IgE are a part of allergic reactions (Xu et al. 1998; Peng and Simons 2004). Rarely, mosquito bites can cause a severe allergic reaction known as anaphylaxis. Anaphylaxis symptoms can include breathing difficulties, facial or throat swelling, and a rapid or weak pulse (Vander Does et al. 2022).

A large number of mosquito species including *Aedes albifasciatus, Aedes vexans, Aedes aegypti, Aedes albopictus, Anopheles sinensis, Armigeres subalbatus, Culiseta inornate, Culex pipiens, Cx. Fatigans,* and *Cx. tarsalis*, have been studied for their ability to elicit allergic reactions (Liu et al. 2017; Cantillo and Puerta 2021). Sarcoplasmic calcium-binding protein (Aed a 5) (Cantillo et al. 2017), Porin3 (Aed a 6) (Cantillo and Puerta 2021), and Tropomyosin (Aed a 10) (Nene et al. 2007) are the allergens produced by the *Aedes aegypti*, which uses airway as the route of exposure so as to trigger the allergic response (Cantillo et al. 2017).

14.3.2 Cockroaches

Cockroaches are categorized as a distinctive type of insect that belongs to the Blattodea order under the phylum Arthropoda. Cockroaches may serve as an intermediate or secondary host and facilitate the transmission of some parasites and pathogenic organisms to humans (Holt 1989; Baumholtz et al. 1997; Graczyk et al. 2005; Tatfeng et al. 2005).

Because they can contaminate food, spread disease, and cause allergic reactions in some people, cockroaches are regarded as pests. They frequently inhabit bathrooms, kitchens, and other locations near water supplies as they travel from building to building in warm weather (Holt 1989; Katial 2003). There are over 4000 different cockroach species, and about 16 of them are harmful to human health (Baumholtz et al. 1997). However, the German cockroach, American cockroach, and Oriental cockroach are the three species that infest homes the most frequently (Baumholtz et al. 1997; Katial 2003). Cockroach allergens are proteins that can be detected in cockroach saliva, feces, and skin fragments that have been shed. Some people, especially those who have asthma, may experience allergic reactions as a result of these allergens becoming airborne. Shortness of breath, coughing, and wheezing are among the respiratory symptoms that can result from exposure to cockroach allergens. Other times, coming into contact with cockroach allergens might also result in skin reactions like hives or eczema.

Oriental cockroach is known as a common pest throughout South Asia, Europe, and the United States. It prefers to eat organic matter that has begun to deteriorate. Because it moves via sewage pipes and prefers crawl areas and damp basements, it is also referred to as a black beetle or water bug (Baumholtz et al. 1997; Katial 2003).

The *Periplaneta americana*, or American cockroach, lives in sewers and basements and maintains its habitat around buildings. These can frequently be found at bakeries, groceries, and restaurants. They can be found throughout the world in temperate, tropical, and subtropical areas (Baumholtz et al. 1997; Katial 2003).

Periplaneta fuliginosa, commonly known as the dusky brown (smoky brown) cockroach, is common in harbors and airports and is also known as the traveller cockroach because it infests containers used for shipping, semi-truck trailers, onboard aircraft, and cargo ships. It is typically seen outside near suburban homes and other buildings, but it can also be seen inside buildings like factories, restaurants, hospitals, ships, and homes (Appel and Smith 2002; Jeong et al. 2004).

Chitinase (Bla g 12) (Pomés et al. 2021), Arginine kinase (Bla g9) (Chuang et al. 2010), and Tropomyosin (Bla g7) (Jeong et al. 2003) are some of the allergens produced by the Blattella germanica whereas Peroxiredoxin-6 (Prx6) (Per a20) (Wang et al. 2023), Porin 3 (Per a 19) (Ansotegui et al. 2020), and Alpha-tubulin (Per a 17) (Wangorsch et al. 2021) are some of the allergens produced by *Periplaneta americana*, which uses airway as the route of exposure so as to trigger the allergic response.

14.3.3 House Dust Mite

About 40 000 mite species exist in the world (Nathanson 1969). Mites are tiny (0.1–1 mm) animals belonging to the Arachnida group (Zhang et al. 2019) and have four stages of development: egg, larva, nymph, and adult. Depending on the surroundings, the life expectancy of an adult mite ranges from a few weeks to many months. Most homes have different types of dust mites, and the dominant species vary across and within households within a given geographic area.

It has been found and characterized that house dust mites produce more than 20 families of allergenic proteins, and they are regarded as the main source of powerful allergens. (Thomas et al. 2002). Dust mites thrive best in environments with a relative humidity of 80–90% and a temperature range of 23–30 °C. Water is ingested by dust mites daily or taken in through the air. They also constantly lose water, therefore maintaining a balance between water uptake and loss is necessary for survival. The efficient ways to reduce the amount of house dust mites in specific areas can be reached via frequent washing, keeping relative humidity below 50%, and vacuum cleaning (Tovey and Marks 1999; Eggleston 2001).

The feces and body parts of house dust mites include proteins known as house dust mite allergens. When dust is disturbed, these allergens can become airborne and trigger allergic reactions in some people, especially in those who have asthma and allergies. The majority of homes across the world have house dust mites, one of the most prevalent indoor allergens. House dust mite allergen exposure can result in skin reactions like itching and hives as well as respiratory symptoms including wheezing, coughing, and shortness of breath (Courtioux and Boudot 2022).

Cysteine protease (Der m 1) (Warner et al. 1998) and NCP2 (Der m 2) (Hu et al. 2022) are the allergens produced by the *Dermatophagoides microceras* (Warner et al. 1998); Peritrophin-like protein domain (PF01607) (Der p 23) (Weghofer et al. 2013), Inorganic pyrophosphatase (Der p 32) (Liu et al. 2018), and Thioredoxin-like protein (Der p 40) (Cai et al. 2023) are some of the allergens produced by *Dermatophagoides pteronyssinus* (Weghofer et al. 2013; Liu et al. n.d.); Peritrophin-like protein (Der f 23) (He et al. 2019), Ferritin (Der f 30) (An et al. 2013) are some of the allergens produced by *Dermatophagoides farinae*, which uses airway as the route of exposure so as to trigger the allergic response (An et al. 2013).

14.3.4 Storage Mite

Storage mites are a distinct kind of mite that are frequently discovered in hay and other animal feed, as well as in stored grains, nuts, and other commodities. They have been demonstrated to reproduce up to 30 times annually when kept at ideal conditions (temperature: 25 °C; relative humidity: 90%) (Sánchez-Ramos et al. 2007). Storage mites come in a wide variety of species, but some of the more prevalent ones include *Tyrophagus putrescentiae, Glycyphagus domesticus*, and *Lepidoglyphus destructor*. Storage mites are thought to be a form of household allergen that might make some people sensitive to them. The proteins in storage mite allergens can be detected in the body parts, secretions, and feces of storage mites (van Hage-Hamsten and Johansson 1992). When mite-infested materials are disturbed, these allergens can become airborne and cause allergy reactions in some people. In addition to skin reactions like itching and hives, exposure to storage mite allergens can result in respiratory symptoms like asthma, coughing, and shortness of breath. Asthma and allergy sufferers are especially vulnerable to the allergens in storage mites, and exposure to them might make their symptoms worse (Carnés et al. 2017).

Gly d 2 (Gafvelin et al. 2001) is the allergen produced by the *Glycyphagus domesticus* (Gafvelin et al. 2001); Mite group 5 allergen (Lep d 5) (Eriksson et al. 2001), and Fatty acid-binding protein (Lep d 13) (Lin et al. 1994) are some of the allergens produced by *Lepidoglyphus destructor* (Eriksson et al. 2001); Cofilin (Tyr p 31) (Zhou et al. 2024), Inorganic pyrophosphatase (Tyr p 32) (Wan et al. 2024) are some of the allergens produced by *Tyrophagus putrescentiae*, which uses airway as the route of exposure so as to trigger the allergic response (Table 14.1).

14.4 Allergens and Immune Response

Under normal conditions, lung inflammation is very minimal despite its exposure to various environmental antigens and particles ingested. As air is constantly inspired and surface area of the respiratory epithelium is comparatively large, this prevents the inflammation of the lungs upon exposure (Upham et al. 2008). The main function of the immune system is to distinguish between harmless antigens and potential pathogens to reduce tissue damage while protecting the body from serious infections (Institute for Quality and Efficiency in Health Care (IQWiG) 2020). Inhalation of soluble proteins usually leads to an immune state known as inhalation tolerance rather than an important immune reaction, even if the tissue is not damaged. On the other hand, when exposed to pathogenic bacteria, the immune system in the lungs becomes very active, which in turn activates the innate immune system through TLRs and other pattern recognition receptors (Li and Wu 2021).

The lungs are protected from microbial infections by special defense systems where the mucous membranes and cough reflexes eliminate air particles while the nose and upper airways filter and clean the inspiration air. The innate immune system depends on a small number of receptors that can react quickly but stereotypically to infections (Marshall et al. 2018). Alveolar macrophages and neutrophils express pattern-receptors and are the main internal immune cells of healthy lung cells, containing collected cells (substitute proteins), defensins, lactoferrines, mannose-related lectins, complements, and enzymes to provide nonspecific protection against infection (Allard et al. 2018). A well-tuned adaptive immune response to foreign antigens depends on dendritic cells (DCs), lymphocytes, and B cells. The adaptive immune system develops long-lasting immunological memory of foreign antigens despite a relatively delayed first defense against antigens and offers some degree of defense against further antigen exposure.

Table 14.1 Representation of various airborne insect allergens.

Source	Species	Allergen	Biochemical name	Route of exposure	References
Mosquito	*Aedes aegypti*	Aed a 5 Aed a 6 Aed a 10	Sarcoplasmic Calcium binding protein Porin3 Tropomyosin	Airway	Xu et al. (1998), Cantillo and Puerta (2021), Cantillo et al. (2017) and Nene et al. (2007)
Cockroach	*Blattella* germanica *Periplaneta Americana*	Bla g 12 Bla g 9 Bla g 7 Per a 20 Per a 19 Per a 17	Chitinase Arginine kinase Tropomyosin Peroxiredoxin-6 Porin 3 Alpha-tubulin	Airway	Pomés et al. (2021), Chuang et al. (2010), Jeong et al. (2003), Wang et al. (2023), Ansotegui et al. (2020) and Wangorsch et al. (2021)
House Dust Mites	*Dermatophagoides microceras*; *Dermatophagoides pteronyssinus*; *Dermatophagoides farina*	Der m 1 Der m 2 Der p 23 Der p 32 Der p 40 Der f 23	Cysteine protease NCP2 Peritrophin-like protein domain Inorganic pyrophosphatase Thioredoxin like protein Peritrophin-like protein	Airway	Warner et al. (1998), Hu et al. (2022), Weghofer et al. (2013), Liu et al. (2018), Cai et al. (2023) and He et al. (2019)
Storage mites	*Glycyphagus domesticus* *Lepidoglyphus destructor* *Tyrophagus putrescentiae*	Gly d 2 Lep d 5 Lep d 13 Tyr p 31 Tyr p 32	— Mite group 5 allergen Fatty acid-binding protein Cofilin Inorganic pyrophosphatase	Airway	Gafvelin et al. (2001), Eriksson et al. (2001), Lin et al. (1994), Zhou et al. (2024) and Wan et al. (2024)
Silverfish	*Lepisma saccharina*	Lep s 1	Tropomyosin	Airway	Barletta et al. (2005)

Inhaled antigens usually settle or impact on the epithelium layer. Mucociliary layer and tight connections between epithelial cells provide the only means for immune-competent cells to access antigens. The antigen-presenting cells (APCs) are responsible for presenting antigens that penetrate the epithelium. If the host has not been exposed to an antigen before, naive T cells are activated in local lymph nodes in response to migrating DCs presenting it. Contrary to this, airway submucosa and lung interstitium can also be involved in recalling or remembering previously encountered antigens. T cell anergy, deletion, and inhibition by particular regulatory T cells are all implicated in the development and maintenance of inhalation tolerance (Chen and Kolls 2013).

The encounter between the antigen and human body results in the production of IgE antibodies, which mounts to an immunological response. IgE antibodies upon production can induce prevalent allergic reactions including hives, hay fever, and asthma due to their specificity to particular allergens (Allergies and the Immune System 2023). This immune response is facilitated by the production of interleukin-4 (IL-4) and interleukin-10 (IL-10). Upon exposure, the interaction of the allergen with TLP and PRP receptors present on the surface of the DCs and T cells results in the production of interleukins. DCs produce IL-4 and IL-10 whereas T cells are responsible for the production of IL-10. The production and release of IL-4 and IL-10 from the DCs results in elevated number of B cells. The release of B cells due to IL-4 is followed by the production of plasma cells, which leads to the activation of IgE-mediated immune response. On the other hand, the B cells released due to IL-10 directly lead to the IgA- and IgG-mediated activation of immune response. Furthermore, allergic asthma may develop as a consequence of the interaction between immunological and neuronal cells in the lungs (Upham et al. 2008) (Figure 14.2).

14.5 Allergens and Disease

Allergens can cause a disease or allergic reaction because they trigger an overreaction of the immune system. In a person without allergies, the immune system responds to harmful invaders such as bacteria or viruses, while ignoring harmless substances like pollen or food proteins. However, in individuals with allergies, the immune system mistakenly identifies an otherwise harmless substance (the allergen) as a dangerous invader and mounts an immune response against it (Shamji and Durham 2017).

This immune response includes the production of a type of antibody called IgE, which binds to mast cells and basophils in the body. When the allergen is encountered again, it binds to the IgE on the surface of these cells, causing them to release histamine and other chemicals into the bloodstream (Fernandez 2022). This results in the characteristic symptoms of an allergic reaction, such as itching, hives, swelling, sneezing, and coughing.

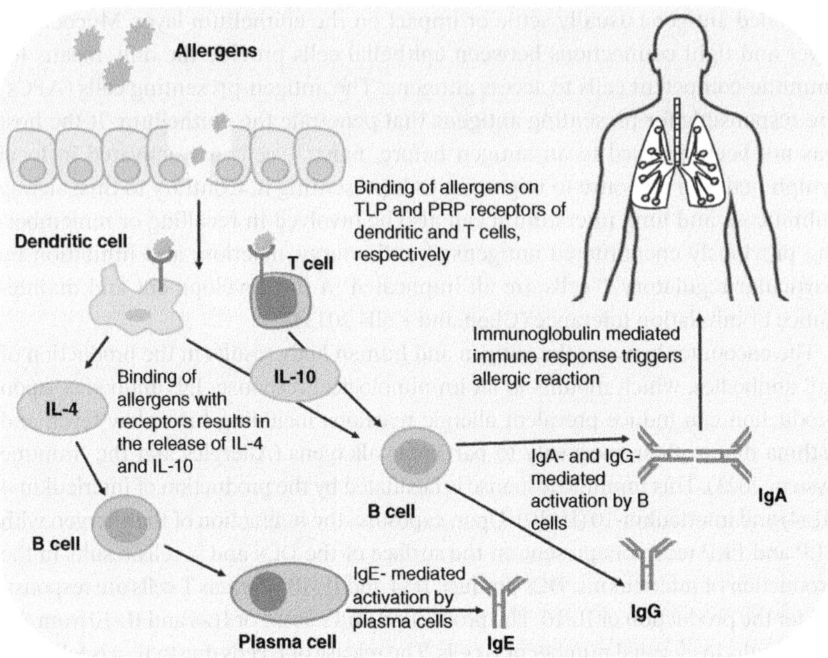

Figure 14.2 Representation of allergen and the elevated immune response.

Repeated exposure to allergens can lead to chronic inflammation and damage to tissues in the body, which can contribute to the development of allergic diseases such as asthma, allergic rhinitis, and atopic dermatitis (Arshad 2009). The severity of an allergic reaction can vary widely, from mild symptoms to life-threatening anaphylaxis, depending on the individual and the allergen involved.

14.5.1 Mosquito Allergen: Immune Response and Disease

The mosquito saliva is identified as the major component that triggers the allergic reaction against the mosquito bite. These allergic reactions are found to be desensitized in healthy adults. However, in the case of immune-compromised patients, severe reactions are observed. Among the various species of mosquitoes, *Aedes aegypti* (yellow fever mosquito) is known to transmit various diseases such as chikungunya, yellow fever, dengue fever, and zipa fever (Cantillo et al. 2017). Although this species has its origin in Africa, it is now the occupant of regions including tropical, subtropical, and temperate regions.

The allergen Aed a 10 (Nene et al. 2007) is present in the saliva of the *Aedes aegypti* mosquito, which acts as a transmitter for various viral diseases in human beings. The adult female mosquitoes belonging to the *Aedes aegypti* species are known to feed on blood so as to produce fertile eggs. They bite one person and then move to another in search of blood. The infected mosquitoes can spread the disease in a healthy person's body as its saliva can enter the bloodstream of the victim, thus initiating the immune response (Vander Does et al. 2022).

Aed a 10 (tropomyosin) is a recognized allergen present in the saliva of mosquitoes. Asthma and allergic rhinitis sufferers were found to bind IgE to the allergen Aed a 10 (tropomyosin) present in mosquito saliva. Other IgE-binding proteins found in *Aedes aegypti* have not yet been described. The saliva of this insect includes at least eight allergens (Cantillo and Puerta 2021), three of which have air as their route of exposure. Although the detailed immune response reaction is not well understood, some hypotheses suggest that the histamine found in the saliva of the mosquito is responsible for initiating the itch response (Hassan and Haji 2014). Type 1 hypersensitivity is a direct reaction to allergens, mediated by IgE, common in classical allergic and allergic conditions such as atopic skin disease (AD), rhinoconjunctivitis, and asthma. The high production of Th2 cells supports the IgE class transition, which increases the IgE production in response to allergens and antigens. Th2 cells produce IL-4, IL-5, and IL-13. IgE binds to low-affinity receptors (FcRII/CD23) and high-affinity receptors (FcRI) of mast cells and basophils, including B cells, activated macrophages, and eosinophils. The cross-linking of receptor-associated IgEs between antigens results in MCs and basophils producing pharmaceutically powerful compounds such as histamine, leukotrienes, and peptides, and attracting neutrophils and eosinophils. Type 4 hypersensitivity, i.e. delayed-type hypersensitivity (DTH), is mediated by T cells and monocytes/macrophages and manifests in two to three days but is independent of antibodies. MHCII is an important histocompatibility complex on the surface of cells presenting antigens such as DCs and macrophages, which works as complex and is used by CD4+ T1cells to detect antigens. These latter release IL-12, which promotes the growth of CD4+ Th1 cells. The Th1 cells release IL-2 and IFN-γ, which trigger the release of additional Th1 cytokines, mediating the immune response (Janeway et al. 2001).

All mosquito-borne virus infections involve the inoculation of viruses into mosquito bite sites, which causes widespread swelling, itching, and redness. The developed reaction against the mosquito saliva is referred to as mosquito allergies (Kulthanan et al. 2010). The various diseases mediated by *Aedes aegypti* mosquito are Zika virus, eastern equine encephalitis (EEE), dengue fever, West Nile virus (WNV) and chikungunya. Zika virus is a viral infection wherein the patient suffers from fever, joint pain, and body rash. When this infection is contracted by a pregnant woman, it can result in birth defects like microcephaly. Dengue fever is another viral infection caused by the dengue virus, which is followed by high fever, body rash, muscle and joint pain, and

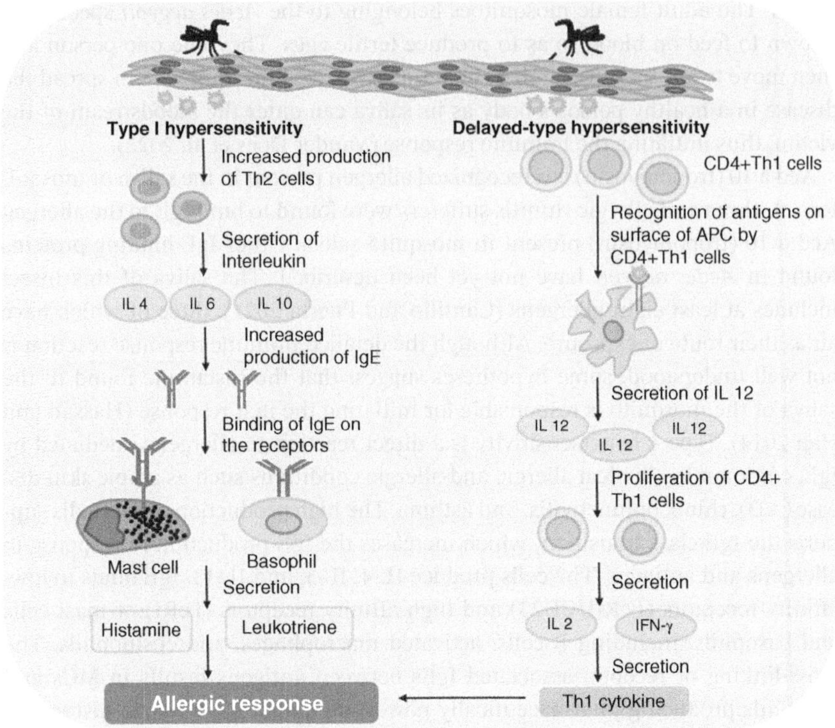

Figure 14.3 Representation of mosquito allergen and elevated immune response.

headache (Vander Does et al. 2022). Under severe conditions, dengue fever can develop into dengue hemorrhagic fever, which is a life-threatening condition. Chikungunya is a viral infection where the patient shows symptoms of fever, body rash, headache, and muscle and joint pain. The pain in joints can persist for a few months to even years (Figure 14.3).

14.5.2 House Dust Mite Allergen: Immune Response and Disease

Allergies are caused by small insects called mites, and house dust mites such as Dermatophagoides pteronyssinus and Dermatophagoides farinae are identified as the main perpetrators (Baxi and Phipatanakul 2010). Both species exhibit cross-reaction and have different allergic diversity (Chruszcz et al. 2009). Furthermore, *Euroglyphus maynei* and *Blomia tropicalis* are widespread mites in humid regions. Millions of mites are stored on household furniture, beds, and clothes, but they are invisible to the naked eye (Sharma et al. 2011). Recent discoveries indicate their presence in foods such as cooking flour that leads to anaphylactic reactions after consumption (Suesirisawad et al. 2015). Almost all parts of the body of mites, including the intestine, feces, skin, and eggs are allergens, causing allergies in 85% of the population suffering from asthma (Gregory and Lloyd 2011). More than 20 household dust mite allergens have been identified and divided into many groups. Most of them are proteins, either serine proteases of groups 2, 3, 6, and 9, or cysteine proteases of groups 1 (the Papain-like family) (trypsin, chymotrypsin, and collagenase). Studies indicate that the cysteine proteases p 1 (*D. pteronyssinus*) and f 1 (*D. farinae*) control the proteolytic activity of other types of allergens by activation of the zymogens (Herman et al. 2014).

IgE antibodies usually detect harmless environmental antigens (usually proteins) such as pollen or mite allergens and mediate type I hypersensitivity reactions. In people who are susceptible, an imbalanced immune response characterized by excessive CD4 T (Th)2 cells causes the development of a higher amount of allergen-specific IgE antibodies (Wawrzyniak et al. 2017). These Th2 cells are polarized, differentiated T helper cells that generate specific interleukin-4, IL-5, and IL-13 cytokines. The switch from the antibody class in B cells from IgM or IgG antibodies to IgE antibodies is caused by the cytokines Th2 IL-4 and IL-13 (Froidure et al. 2016).

Th2 cell activation and allergic inflammation are influenced by DCs. DCs are found in human tissues exposed to external environment, including skin and lung epithelial cells (Steinman 2012). As sentinels, they control the tissues for the incoming antigens. In addition to expressing the large number of pathogenic recognition receptors (PRRs) present in the internal immune system, DCs can also recognize allergens, break them into smaller peptides, and then transmit them through the major histocompatibility complexes of Class I and Class II of Class MHC to T-cell receptors. The cytokine pattern produced by DC plays an important role in the development of naïve activated T cells into functional T cell subgroups (at least Th1, Th2, Th17, and Treg) (Van Beek et al. 2012). For a long time, it has been understood that the activation of basophils by cross-linking FcRI or other cell surface receptors has produced significant IL-4 (Lambrecht and Hammad 2010). Because of the activation of the brain and immune cells, the adaptive immune response becomes imbalanced. Innate immune cells, including basophils, mast cells, and eosinophils, are also activated by cells and cell-derived chemokines to maintain Th2 priming. ILCs such as nuclei and innate2-type helper cells can be induced by IL-25 and IL-33 by the release of epithelial cells to produce IL-5 and IL-13 (Spits and Cupedo 2012). Collectively, it is clear that the onset of Th2-based allergic reactions is the result of complex interactions of many molecular pathways, including intrinsic and adaptive immunity.

By releasing characteristic cytokine interferons (IFNs), Th1 cells regulate Th2 cells by blocking their proliferative capability. Most of the IgE antibodies released stably attach to high-affinity receptors of mast cells or basophils. Re-exposure to allergens leads to a cross-connection of receptor-related inflammatory molecules, thereby activating the effector cells. The activation triggers the immediate release of pre-produced inflammatory mediators such as histamine from the internal cell granules and causes the occurrence of typical allergic symptoms (hay fever, asthma, etc.) (Wawrzyniak et al. 2017). Dust mites are the main source of recurrent allergies (HDM). *Dermatophagoides pteronyssinus*, *D. farinae*, and *Blomia tropicalis* are the three most important insects. In recent studies, it has been estimated that between 65 and 130 million people around the world have HDM sensitization (i.e. they have identified IgE specific to a specific allergen in their serum) and this affects 15–20% of the population in industrialized countries (Jacquet 2021). House dust mite allergies can emerge as a variety of conditions, including allergic rhinoconjunctivitis, atopic dermatitis, and, most importantly, allergic asthma.

The most prevalent chronic disease among children, asthma, is a chronic inflammatory disease of the conducting airways that affects 300 million individuals globally. Reversible airway obstruction, airway hyper-responsiveness (AHR), eosinophil, and CD4+ T helper (Th) type 2 cell infiltration into the airway submucosa, mucus hypersecretion, and airway remodelling are identified hallmarks for asthma disease. The main factors contributing to asthma are genetic predispositions and environmental effects, such as birth patterns, childhood infections, and exposure to inhaled allergens, including pollutants and environmental allergens such as pollen, animal dust, and mosquitoes (von Mutius 2009). Acari is a taxonomical subdivision of mites, which includes more than 50 000 species. In addition to asthma, rhinitis, allergic conjunctivitis, and atopic dermatitis are common allergy diseases caused by HDM (Gavino et al. 2008). Despite regional variations, around 50–85% of asthma patients have HDM allergies (Gregory and Lloyd 2011). When asthma patients are exposed to dust mites in the home, their immune system can over-react, causing inflammation and restriction of the airways. Symptoms of asthma such as wheezing, shortness of breath, cough, and chest tightness may result from it (Figure 14.4).

14.5.3 Cockroach Allergen: Immune Response and Disease

Cockroaches are considered to be one of the most important allergens (Tungtrongchitr et al. 2004). The two species, German cockroach and American cockroach (*Blattella germanica* and *Periplaneta americana*, respectively), were discovered as the cause of most allergic reactions (Patterson and Slater 2002). Several species of cockroaches are also associated with allergies, including the brown band cockroach (*Supella longipalpa*), the Harlequin Roach (*Neostylopyga rhombifolia*), the Australian cockroach (*Periplaneta australasiae*), and the brown cockroach (*Periplaneta brunnea*) (Tungtrongchitr et al. 2004). Even the species commonly kept as pets, the Madagascar hissing cockroach, (*Gromphadorhina portentosa*) has

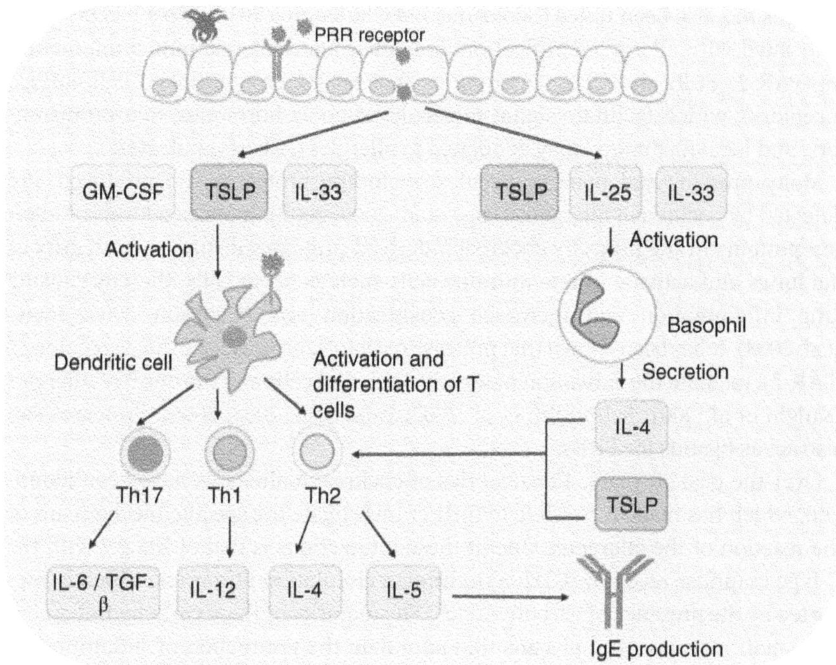

Figure 14.4 Representation of house dust mite allergen and elevated immune response.

signs of allergic risk (Morgan et al. 2007). Almost all cockroaches are identified to be potential allergens. Bla g 1, Bla g 2, Bla g 4, Bla g 5, Bla g 6, and A 1, A 3, and A 7 are the allergens associated with *P. americana* and B. *germanica*, respectively, and have received the most attention (Arruda and Chapman 2001). The cockroaches' saliva, gut, body extract, feces, and skin casts contain the allergens, which are grouped under the categories of calycin, glutathione-*S*-transferase, troponin, tropomyosin, aspartic protease, and arylophorin (Arruda and Chapman 2001). Body extracts from these insects contain a variety of proteins, but Bla g1, Bla g 2, and Per a 1 were considered to be the most important allergens and cause hypersensitivity (Arruda et al. 2014). Bla g 1 and Per a 1 are members of group 1 and are known to have several tandem repeats of amino acids, share structural morphology (52–72%), cross-reaction, and binding to different lipids (Mueller et al. 2013; Arruda et al. 2014). Other notable allergens are Per a 3 (arylphorin), Per a 7 (tropomyosin), Bla g 2 (inactive aspartic proteinase), Bla g 4 (calycin), Bla g 5 (glutathione-S-transferase), Bla g 6 (insect troponin C), and Bla g 7 (tropomyosin), which are able to provoke IgE responses.

Allergy inflammation caused by cockroach allergens can be IgE or non-IgE, the former being more common. The activation of T cell receptors of different lineages has also been noted (Sookrung and Chaicumpa 2010). Host receptors are associated with allergy identification and subsequent allergic induction, including PAR-2, TLR, and CLR. Disruptions in PAR (G protein-linked cell surface receptors), which facilitate signal molecules such as hormones, neurotransmitters, and local mediators, may be related to allergies (Alberts et al. 2002).

Many proteins and macromolecules, including proteases, are produced and detected by cockroach allergens (Page et al. 2005, 2007). Increased levels of allergen proteins in the lungs by cockroach-derived proteases disturb the integrity of the lungs and activate innate immune cells such as nerve cells, thereby causing lung inflammation and increased sensitization risk (Cookson 2004; Jeong et al. 2008). It has been shown that protease receptors, protease-activated receptors-2 (PAR-2), mediate the activation of airway epithelial cells and the onset of allergies (Knight et al. 2001; Schmidlin et al. 2002; Page et al. 2003, 2006). Proteases can also act as ligands for PRRs.

Over the past 20 years, 12 categories of cockroach allergens have been identified, which has made it possible to further investigate the specific mechanisms of the reaction of the allergens. One of these interactions is that of Bla g 2 with the C-type mannose receptor (CD206) in human circulation fibrocytes, which is supported by the presence of carbohydrates. The majority of these carbohydrates contain small mannose-end glycans that stimulate the production of inflammatory cytokines (TNF-, IL-6) and signalling molecules, such as NF-kB (NF-kB) (Tsai et al. 2013).

TSLP, IL25, IL33, and TGFb1 are molecules (cytokines and chemokines) that are produced when epithelial cells are activated by cockroaches. These cytokines and chemicals attract inflammatory cells to the airways damaged by allergens to repair and suppress inflammation. In contrast, cockroaches can interfere with the integrity of the airway epithelium by activating proteinase receptor-2 (PAR-2) pathways, which increases the penetration of allergens and activates the innate immune cells (including mesenchymal stem cells (MSCs), DCs as well as through Toll-like receptors (TLRs), C-type lectin receptors (CLRs) and ary hydrocarbon receptor (AhR). The adaptive immune system becomes unbalanced and favors more Th2 and Th17 phenotypes due to these activated innate immune cells. In addition, the receptors expressed by immune cells of type 2 (ILC2) in epithelial cells TSLP, IL25, and IL33 interact, resulting in the release of IL5 and IL13, and allergic inflammation (Lambrecht and Hammad 2015).

In particular, cockroach levels are linked to allergy sensitivity and asthma (Crain et al. 2002; Amr et al. 2003). At least half of the inner-city houses contain clinically significant amounts of cockroach allergens. A maximum of 30% of middle-class suburban homes also have detectable cockroaches, but these levels are often significantly lower than those in suburban homes (Matsui et al. 2003; Cohn et al. 2006). 60–80% of asthmatic adolescents living in urban areas are

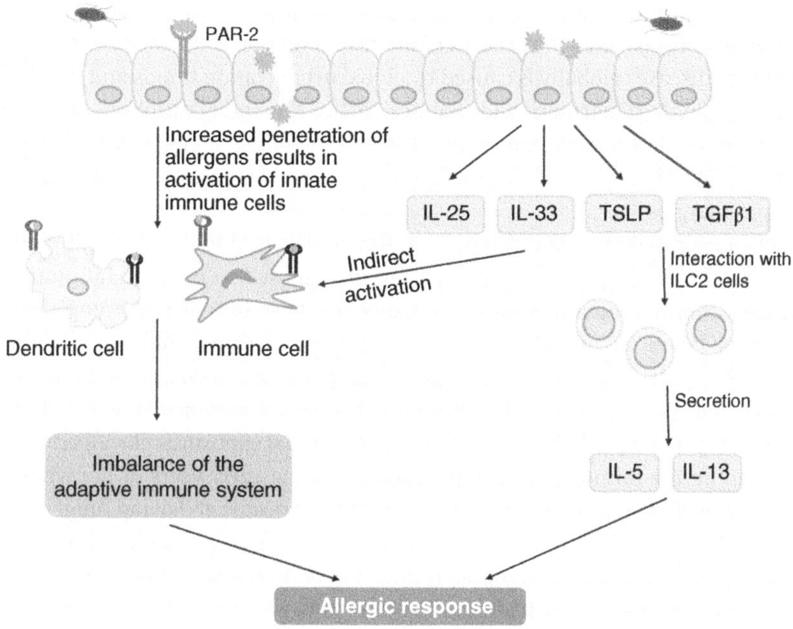

Figure 14.5 Representation of cockroach allergen and elevated immune response.

allergic to cockroaches (Gruchalla et al. 2005). 21% of the suburban population were informed (Matsui et al. 2003). The wheeze of young children is associated with sensitivity to cockroach allergens (Donohue et al. 2008). The asthma outcomes of children with asthma in urban areas, including increased use of asthma-related medical services, were also directly linked to the levels of cockroaches (Donohue et al. 2008). There is an established correlation between cockroach allergen levels and asthma outcomes in asthma-infected children, including an increase in asthma-related healthcare use (Wang et al. 2009) (Figure 14.5).

14.6 Environment and Allergens: Ways to Control

The symptoms of allergic diseases such as atopic eczema, food allergies, asthma, and allergic rhinitis can be significantly affected by environmental exposure (Kanchongkittiphon et al. 2015; Burbank et al. 2017). More than 20% of the world's population is currently affected by these diseases (Kay 2001; Bantz et al. 2014), but in recent years, the frequency of these diseases has increased too rapidly, making genetic changes to the only explanation (Platts-Mills 2015; Burbank et al. 2017). Changes in lifestyles (dietary habits, indoor leisure time, physical activities) and exposure to the environment (air pollution, smoking, mold, infections) alter the process of atopic allergies and the increase in the incidence of diseases globally is caused by global urbanization trends and rapid population growth (Guarnieri and Balmes 2014; Kanchongkittiphon et al. 2015; Burbank et al. 2017; Morgan et al. 2018). Many of these changes are associated with perinatal and lasting risk factors for the onset and escalation of asthma and atopic allergies.

The terms "allergy" and "atopy" are frequently used to describe IgE-mediated diseases in which people with atopy are predisposed to produce IgE antibodies against common environmental allergens and have one or more atopic allergic diseases (such as atopic eczema, food allergy, asthma, and allergic rhinitis (Kay 2001), although some nonatopic allergic diseases, such as nonatopic asthma and contact dermatitis, develop through IgE-independent mechanisms. The interaction of genetic predispositions and environmental exposures plays a critical role in shaping the immune system, particularly in the early years of life when neonates transit from limited utero environmental exposure to having fungi and bacteria colonize their skin, lungs, and intestinal tract to form their microbiome.

Insect allergens are substances produced by certain insects that can cause allergic reactions in some individuals. The allergens produced by the insects, which have airways as their route of exposure are referred to as indoor airborne insect allergens. The saliva, urine, feces, and body parts of insects including cockroaches, dust mites, and stinging insects like bees and wasps are some of the sources of these allergens. In particular, indoors, where they can concentrate and affect the air quality, the presence of insect allergens can have a substantial negative impact on the environment. The presence and spread of airborne insect allergens can be significantly influenced by the environment. For instance, allergens from cockroaches and mosquitoes are frequently discovered in homes and buildings, especially in metropolitan areas where these pests are more prevalent. Some people can develop allergies when exposed to mosquito saliva, dead cockroach body parts, and dust from their droppings. Asthma and other respiratory disorders can be made worse by dust mite allergens, which can also lead to allergic reactions.

Controlling insect infestations and eliminating sources of allergens from the environment, including excrement, saliva, and shed skin particles, are crucial for reducing the environmental impact of insect allergies.

When applied topically, *N*-diethyl meta toluamide (DEET) is effective in reducing the frequency of insect bites. Although they are generally safe, exposure to concentrated formulations, excessive inhalation or consumption can cause neurotoxicity and systemic toxicity, as well as urinary syndrome, dermatitis and bleeding eruptions (Swale and Bloomquist n.d.). Antihistamines can reduce early and immediate symptoms of mosquito bites in adults and children (Föex and

Lee 2006). Although it may be difficult to control the allergens of mosquitoes, there are several methods that can reduce exposure, including staying indoors during the mosquito season. Avoiding outdoor activities during these periods can reduce the likelihood of exposure to insect allergens, as mosquitoes are more active in the morning and night (Seda and Horrall 2022).

Insect repellents contain DEET and picaridin and can help prevent mosquito bites and reduce exposure to mosquito allergens, but are essential to read and follow the instructions on the package carefully (CDC 2023). To avoid bites of mosquitoes and reduce exposure to mosquito allergens, wear long shirts, long trousers, and socks. Mosquitoes can be kept away from the house by closing windows and doors, using screens or net sand maintaining tight seals. The air purifier with the HEPA filter can reduce indoor mosquito allergens. Eliminating water sources outside the house, such as bird baths, cylinders, and flowerpots, will reduce the number of mosquitoes and, as a result, reduce the exposure to insect allergens. Since mosquitoes reproduce in standing water, these sources should be removed.

Blocking access points; reducing supplies of food, water, and shelter; eradicating contamination sources with traps and insecticides; and eradicating reservoirs with HEPA vacuuming and mattress covers are all ways to manage the cockroach allergen 75. The professional-led intervention, which reduces cockroach exposure by 80–90%, is the most successful intervention, though.

Apart from the mosquitoes and cockroaches, exposure to HDM can be reduced in a number of ways. For pillow and mattress covers, fine woven textiles with pore sizes of less than 6–10 mm are ideal (Baxi and Phipatanakul 2010). Every other item on the bed should be regularly washed with a hot cycle and followed by drying, which will essentially eradicate all mites. It is crucial to use detergent and bleach, and frequent washing is also advisable (Arlian et al. 2003; Choi et al. 2008). The prevailing consensus is that a relative humidity threshold of 45–50% is needed to effectively reduce HDM allergens. High-efficiency particulate air (HEPA) filters are required on vacuum cleaners in order to assist control of HDM allergens. These cleaners, however, are unable to get rid of living mites from a carpet. Patients are adviced to wear masks when cleaning and to leave the room for 20 minutes afterward since vacuum cleaners can disturb the dust in the room (Gore et al. 2006; Wilson and Platts-Mills 2018). Several studies have demonstrated that significant sources of HDM allergen include carpet and upholstered furniture. As a result, there should not be many carpets, and if any, they should be removed, cleaned, and dried outside in the sun. Tannic acid and benzyl benzoate have only temporarily and modestly reduced the concentration of mites, and it is still unknown whether they could be used as a preventative measure (Kroidl et al. 1998; Kalayci et al. 2022). If humidity cannot be managed, the general rule for all other potential mite-growing places is to limit the use of upholstered material (Wilson and Platts-Mills 2018).

14.7 Conclusion

Indoor insect allergens are a significant problem because they have an impact on both the environment and human health. Cockroaches, dust mites, and moths are just some of the insects that live in human homes and produce these allergens. High concentrations of these allergens can occur in indoor areas like homes, workplaces, hospitals, and schools. Indoor insect allergens can have a serious impact on people's health, particularly young children, elderly people, and people with respiratory problems. These allergens can cause skin irritations and inflammations, as well as respiratory symptoms such as wheezing, coughing, and breathing.

In addition to health benefits, indoor insect allergies can have an impact on the environment. Insecticides and other pesticide management methods can transmit harmful chemicals to the environment, threatening people's health and the ecosystem as a whole. For example, cockroach allergies are associated with the growth of antibiotic-resistant bacteria in the indoor environment, which increases the risk of disease. In order to maintain a satisfactory level of indoor air quality and protect public health, it is essential to reduce exposure to insect allergens in indoor air. Various steps can be adopted for eradicating these allergens including appropriate sanitation and cleaning, insect control, and ventilation. To protect human health and the environment, measures must be taken to reduce exposure to insect allergens in the air. Regular cleaning, prudent food storage, and nontoxic pest management techniques can help reduce the presence of insects and the allergens they carry.

References

Abramson, M.J., Puy, R.M., and Weiner, J.M. (2010). Injection allergen immunotherapy for asthma. *Cochrane Database of Systematic Reviews* 8: CD001186.

Adams, K.E., Tracy, J.M., and Golden, D.B.K. (2022). Anaphylaxis to stinging insect venom. *Immunology and Allergy Clinics of North America* 42 (1): 161–173.

Alberts, B., Johnson, A., Lewis, J. et al. (2002). *Molecular Biology of the Cell*, 4e. New York: Garland Science Signaling through G-Protein-Linked Cell-Surface Receptors. Available from: https://www.ncbi.nlm.nih.gov/books/NBK26912/.

Allard, B., Panariti, A., and Martin, J.G. (2018). Alveolar macrophages in the resolution of inflammation, tissue repair, and tolerance to infection. *Frontiers in Immunology* 9: 1777.

Allergen Nomenclature Database (2023). http://www.allergen.org/ (accessed 24 Febraury 2023).

Allergies and the Immune System (2023). Johns Hopkins medicine. https://www.hopkinsmedicine.org/health/conditions-and-diseases/allergies-and-the-immune-system (accessed 28 Febraury 2023).

Amr, S., Bollinger, M.E., Myers, M. et al. (2003). Environmental allergens and asthma in urban elementary schools. *Annals of Allergy, Asthma & Immunology* 90 (1): 34–40.

An, S., Chen, L., Long, C. et al. (2013). Dermatophagoides farinae allergens diversity identification by proteomics. *Molecular & Cellular Proteomics: MCP* 12 (7): 1818–1828. https://doi.org/10.1074/mcp.M112.027136.

An, S., Chen, L., Long, C. et al. (2013). Dermatophagoides farinae allergens diversity identification by proteomics. *Molecular & Cellular Proteomics.* 12 (7): 1818–1828.

Ansotegui, I.J., Melioli, G., Canonica, G.W. et al. (2020). IgE allergy diagnostics and other relevant tests in allergy, a World Allergy Organization position paper. *The World Allergy Organization Journal*, 13 (2), 100080. https://doi.org/10.1016/j.waojou.2019.100080.

Appel, A.G. and Smith, L.M. (2002). Biology and management of the smokybrown cockroach. *Annual Review of Entomology.* 47: 33–55.

Arlian, L.G., Vyszenski-Moher, D.A.L., and Morgan, M.S. (2003). Mite and mite allergen removal during machine washing of laundry. *Journal of Allergy and Clinical Immunology* 111 (6): 1269–1273.

Arruda, L.K. and Chapman, M.D. (2001). The role of cockroach allergens in asthma. *Current Opinion in Pulmonary Medicine* 7 (1): 14–19.

Arruda, L.K., Barbosa, M.C.R., Santos, A.B.R. et al. (2014). Recombinant allergens for diagnosis of cockroach allergy. *Current Allergy and Asthma Reports* 14 (4): 428.

Arshad, S.H. (2009). Allergen avoidance and prevention of allergy. *Allergy Frontiers of Therapeutic Prevention.* 5: 3–18.

Attardo, G.M., Hansen, I.A., and Raikhel, A.S. (2005). Nutritional regulation of vitellogenesis in mosquitoes: implications for anautogeny. *Insect Biochemistry and Molecular Biology* 35 (7): 661–675.

Bantz, S.K., Zhu, Z., and Zheng, T. (2014). The atopic march: progression from atopic dermatitis to allergic rhinitis and asthma. *Journal of Clinical and Cellular Immunology* 5 (2): 202.

Barletta, B., Butteroni, C., Puggioni, E.M. et al. (2005). Immunological characterization of a recombinant tropomyosin from a new indoor source, *Lepisma saccharina*. *Clinical and Experimental Allergy: Journal of the British Society for Allergy and Clinical Immunology* 35 (4): 483–489. https://doi.org/10.1111/j.1365-2222.2005.02214.x.

Baumholtz, M.A., Parish, L.C., Witkowski, J.A., and Nutting, W.B. (1997). The medical importance of cockroaches. *International Journal of Dermatology* 36 (2): 90–96.

Baxi, S.N. and Phipatanakul, W. (2010). The role of allergen exposure and avoidance in asthma. *Adolescent Medicine: State of the Art Reviews* 21 (1): 57.

Berard, F., Marty, J.P., and Nicolas, J.F. (2003). Allergen penetration through the skin. *European Journal of Dermatology* 13 (4): 324–330.

Billingsley, P.F., Baird, J., Mitchell, J.A., and Drakeley, C. (2006). Immune interactions between mosquitoes and their hosts. *Parasite Immunology* 28 (4): 143–153.

Blumenthal, K.G., Peter, J.G., Trubiano, J.A., and Phillips, E.J. (2019). Antibiotic allergy. *Lancet (London, England)* 393 (10167): 183–198.

Burbank, A.J., Sood, A.K., Kesic, M.J. et al. (2017). Environmental determinants of allergy and asthma in early life. *Journal of Allergy and Clinical Immunology* 140 (1): 1–12.

Burge, H.A. and Rogers, C.A. (2000). Outdoor allergens. *Environmental Health Perspectives.* 108 (Suppl 4): 653–659.

Cai, Z.L., Liu, S., Li, W.Y. et al. (2023). Identification of an immunodominant IgE epitope of Der f 40, a novel allergen of *Dermatophagoides farinae*. *The World Allergy Organization Journal* 16 (8): 100804. https://doi.org/10.1016/j.waojou.2023.100804.

Cantillo, J.F. and Puerta, L. (2021). Mosquitoes: important sources of allergens in the tropics. *Front Allergy* 2: 690406. https://doi.org/10.3389/falgy.2021.690406.

Cantillo, J.F., Puerta, L., Puchalska, P. et al. (2017). Allergenome characterization of the mosquito Aedes aegypti. *Allergy* 72 (10): 1499–1509. https://doi.org/10.1111/all.13150.

Carnés, J., Iraola, V., Cho, S.H., and Esch, R.E. (2017). Mite allergen extracts and clinical practice. *Annals of Allergy, Asthma & Immunology* 118 (3): 249–256.

CDC (2023). Prevent mosquito bites|Zika virus. https://www.cdc.gov/zika/prevention/prevent-mosquito-bites.html (accessed 1 March 2022).

Chen, K. and Kolls, J.K. (2013). T cell-mediated host immune defenses in the lung. *Annual Review of Immunology* 31: 605.

Choi, S.Y., Lee, I.Y., Sohn, J.H. et al. (2008). Optimal conditions for the removal of house dust mite, dog dander, and pollen allergens using mechanical laundry. *Annals of Allergy, Asthma & Immunology* 100 (6): 583–588.

Chruszcz, M., Chapman, M.D., Vailes, L.D. et al. (2009). Crystal structures of mite allergens Der f 1 and Der p 1 reveal differences in surface-exposed residues that may influence antibody binding. *Journal of Molecular Biology* 386 (2): 520–530.

Chuang, J.G., Su, S.N., Chiang, B.L. et al. (2010). Proteome mining for novel IgE-binding proteins from the German cockroach (*Blattella germanica*) and allergen profiling of patients. *Proteomics* 10 (21): 3854–3867. https://doi.org/10.1002/pmic.201000348.

Cohn, R.D., Arbes, S.J., Jaramillo, R. et al. (2006). National prevalence and exposure risk for cockroach allergen in U.S. households. *Environmental Health Perspectives* 114 (4): 522–526.

Cookson, W. (2004). The immunogenetics of asthma and eczema: a new focus on the epithelium. *Nature Reviews Immunology.* 4 (12): 978–988.

Courtioux, B. and Boudot, C. (2022). Dust mite allergy. *Actualités Pharmaceutiques* 58 (588): 43–45.

Crain, E.F., Walter, M., O'Connor, G.T. et al. (2002). Home and allergic characteristics of children with asthma in seven U.S. urban communities and design of an environmental intervention: the Inner-City Asthma Study. *Environmental Health Perspectives* 110 (9): 939–945.

Cui, L., Xu, Y.-Y., Wang, X.-J., and Guan, K. (2020). Stinging insect allergens. *Current Protein & Peptide Science* 21 (2): 142–152.

David, N.A., Penumarti, A., Burks, A.W., and Slater, J.E. (2017). Food allergen extracts to diagnose food-induced allergic diseases: how they are made. *Annals of Allergy, Asthma & Immunology* 119 (2): 101–107.

Do, D.C., Zhao, Y., and Gao, P. (2016). Cockroach allergen exposure and risk of asthma. *Allergy* 71 (4): 463–474.

Donohue, K.M., Al-alem, U., Perzanowski, M.S. et al. (2008). Anti-cockroach and anti-mouse IgE are associated with early wheeze and atopy in an inner-city birth cohort. *Journal of Allergy and Clinical Immunology* 122 (5): 914–920.

Eggleston, P.A. (2001). Methods and effectiveness of indoor environmental control. *Annals of Allergy, Asthma & Immunology* 87 (6 Suppl 3): 44–47.

Eriksson, T.L., Rasool, O., Huecas, S. et al. (2001). Cloning of three new allergens from the dust mite Lepidoglyphus destructor using phage surface display technology. *European Journal of Biochemistry* 268 (2): 287–294. https://doi.org/10.1046/j.1432-1327.2001.01879.x.

Erwin, E.A., Woodfolk, J.A., Custis, N., and Platts-Mills, T.A.E. (2023). Animal danders. *Immunology and Allergy Clinics of North America.* 23 (3): 469–481.

Fernandez, J. (2022). Overview of allergic reactions – immune disorders – MSD manual consumer version. https://www.msdmanuals.com/en-in/home/immune-disorders/allergic-reactions-and-other-hypersensitivity-disorders/overview-of-allergic-reactions (accessed 28 Febraury 2023).

Föex, B.A. and Lee, C. (2006). Oral antihistamines for insect bites. *Emergency Medicine Journal* 23 (9): 721.

Froidure, A., Mouthuy, J., Durham, S.R. et al. (2016). Asthma phenotypes and IgE responses. *European Respiratory Journal* 47 (1): 304–319.

Gafvelin, G., Johansson, E., Lundin, A. et al. (2001). Cross-reactivity studies of a new group 2 allergen from the dust mite *Glycyphagus domesticus*, Gly d 2, and group 2 allergens from *Dermatophagoides pteronyssinus*, *Lepidoglyphus destructor*, and *Tyrophagus putrescentiae* with recombinant allergens. *The Journal of Allergy and Clinical Immunology* 107 (3): 511–518. https://doi.org/10.1067/mai.2001.112264.

Gavino, A.C.P., Needham, G.R., and High, W.A. (2008). Atopic dermatitis, patch testing, and house dust mites: a brief review. *Dermatitis* 19 (3): 121–128.

de Gier, S. and Verhoeckx, K. (2018). Insect (food) allergy and allergens. *Molecular Immunology* 100: 82–106.

Gore, R.B., Durrell, B., Bishop, S. et al. (2006). High-efficiency vacuum cleaners increase personal mite allergen exposure, but only slightly. *Allergy* 61 (1): 119–123. https://doi.org/10.1111/j.1398-9995.2005.00946.x.

Graczyk, T.K., Knight, R., and Tamang, L. (2005). Mechanical transmission of human protozoan parasites by insects. *Clinical Microbiology Reviews* 18 (1): 128–132.

Gregory, L.G. and Lloyd, C.M. (2011). Orchestrating house dust mite-associated allergy in the lung. *Trends in Immunology* 32 (9): 402–411.

Gruchalla, R.S., Pongracic, J., Plaut, M. et al. (2005). Inner City Asthma Study: relationships among sensitivity, allergen exposure, and asthma morbidity. *Journal of Allergy and Clinical Immunology* 115 (3): 478–485.

Guarnieri, M. and Balmes, J.R. (2014). Outdoor air pollution and asthma. *Lancet (London, England)* 383 (9928): 1581–1592.

van Hage-Hamsten, M. and Johansson, S.G.O. (1992). Storage mites. *Experimental & Applied Acarology* 16 (1–2): 117–128.

Hassan, I. and Haji, M.L. (2014). Understanding itch: an update on mediators and mechanisms of pruritus. *Indian Journal of Dermatology, Venereology and Leprology* 80 (2): 106.

Hawkes, F.M. and Hopkins, R.J. (2022). Chapter 2. The mosquito: an introduction. In: *Mosquitopia: The Place of Pests in a Healthy World* (eds Hall, M. and Tamïr, D.). New York: Routledge. https://www.ncbi.nlm.nih.gov/books/NBK585164/; doi: 10.4324/9781003056034-3.

He, Y., Dou, C., Su, Y. et al. (2019). Identification of Der f 23 as a new major allergen of *Dermatophagoides farinae*. *Molecular Medicine Reports* 20 (2): 1270–1278. https://doi.org/10.3892/mmr.2019.10305.

Herman, J., Thelen, N., Smargiasso, N. et al. (2014). Der p 1 is the primary activator of Der p 3, Der p 6 and Der p 9 the proteolytic allergens produced by the house dust mite *Dermatophagoides pteronyssinus*. *Biochimica et Biophysica Acta* 1840 (3): 1117–1124.

Holt, R.H.F. (1989). Hymenolepis diminuta utilizes the envelope surrounding *Moniliformis moniliformis* in order to survive in the cockroach host. *The Journal of Parasitology* 75 (1): 160–162.

Hu, R.H., Wu, C.T., Wu, T.S. et al. (2022). Systematic characterization of the group 2 house dust mite allergen in *Dermatophagoides microceras*. *Frontiers in Cellular and Infection Microbiology* 11: 793559. https://doi.org/10.3389/fcimb.2021.793559.

Institute for Quality and Efficiency in Health Care (IQWiG) (2020). How does the immune system work? https://www.ncbi.nlm.nih.gov/books/NBK279364/ (accessed 28 Febraury 2023).

Jacob, B., Ritz, B., Gehring, U. et al. (2002). Indoor exposure to molds and allergic sensitization. *Environmental Health Perspectives* 110 (7): 647.

Jacquet, A. (2021). Characterization of innate immune responses to house dust mite allergens: pitfalls and limitations. *Front Allergy* 2: 7.

Janeway, C.A. Jr., Travers, P., Walport, M. et al. (2001). *Immunobiology: The Immune System in Health and Disease.* 5th ed. New York: Garland Science. The production of IgE. Available from: https://www.ncbi.nlm.nih.gov/books/NBK27117/.

Javed, S., Khan, F., Ramirez-Fort, M., and Tyring, S.K. (2013). Bites and mites: prevention and protection of vector-borne disease. *Current Opinion in Pediatrics* 25 (4): 488–491. https://doi.org/10.1097/MOP.0b013e328362c4ab.

Jeong, K.Y., Lee, J., Lee, I.Y. et al. (2003). Allergenicity of recombinant Bla g 7, German cockroach tropomyosin. *Allergy* 58 (10): 1059–1063. https://doi.org/10.1034/j.1398-9995.2003.00167.x.

Jeong, K.Y., Hwang, H., Lee, J. et al. (2004). Allergenic characterization of tropomyosin from the dusky brown cockroach, *Periplaneta fuliginosa*. *Clinical and Diagnostic Laboratory Immunology*. 11 (4): 680–685.

Jeong, S.K., Kim, H.J., Youm, J.K. et al. (2008). Mite and cockroach allergens activate protease-activated receptor 2 and delay epidermal permeability barrier recovery. *The Journal of Investigative Dermatology* 128 (8): 1930–1939.

Kahn, S.L., Podjasek, J.O., Dimitropoulos, V.A., and Brown, C.W. (2016). Natural rubber latex allergy. *Disease-a-Month* 62 (1): 5–17.

Kalayci, O., Miligkos, M., Pozo Beltrán, C.F. et al. (2022). The role of environmental allergen control in the management of asthma. *World Allergy Organization Journal* 15 (3): 100634.

Kanchongkittiphon, W., Mendell, M.J., Gaffin, J.M. et al. (2015). Indoor environmental exposures and exacerbation of asthma: an update to the 2000 review by the Institute of Medicine. *Environmental Health Perspectives* 123 (1): 6–20.

Katial, R.K. (2003). Cockroach allergy. *Immunology and Allergy Clinics of North America*. 23 (3): 483–499.

Kay, A.B. (2001). Allergy and allergic diseases. First of two parts. *The New England Journal of Medicine* 344 (1): 30–37.

Kim, J., Lee, S., Young, W.S. et al. (2013). The indoor level of house dust mite allergen is associated with severity of atopic dermatitis in children. Journal of Korean Medical Science 28 (1): 74.

Kimber, I., Dearman, R.J., Basketter, D.A., and Boverhof, D.R. (2014). Chemical respiratory allergy: reverse engineering an adverse outcome pathway. *Toxicology* 318 (1): 32–39.

Knight, D.A., Lim, S., Scaffidi, A.K. et al. (2001). Protease-activated receptors in human airways: upregulation of PAR-2 in respiratory epithelium from patients with asthma. *The Journal of Allergy and Clinical Immunology* 108 (5): 797–803.

Kroidl, R.F., Göbel, D., Balzer, D. et al. (1998). Clinical effects of benzyl benzoate in the prevention of house-dust-mite allergy. *Allergy* 53 (4): 435–440. https://doi.org/10.1111/j.1398-9995.1998.tb03919.x.

Kulthanan, K., Wongkamchai, S., and Triwongwaranat, D. (2010). Mosquito allergy: clinical features and natural course. *The Journal of Dermatology* 37 (12): 1025–1031.

Kumar, P., Kausar, M.A., Singh, A.B., and Singh, R. (2021). Biological contaminants in the indoor air environment and their impacts on human health. *Air Quality,*

Atmosphere, & Health 14 (11): 1723–1736. https://doi.org/10.1007/s11869-021-00978-z.

Lambrecht, B.N. and Hammad, H. (2010). The role of dendritic and epithelial cells as master regulators of allergic airway inflammation. *Lancet* 376 (9743): 835–843.

Lambrecht, B.N. and Hammad, H. (2015). The immunology of asthma. *Nature Immunology* 16 (1): 45–56.

Lee, H., Halverson, S., and Mackey, R. (2016). Insect allergy. *Primary Care* 43 (3): 417–431.

Li, D. and Wu, M. (2021). Pattern recognition receptors in health and diseases. *Signal Transduction and Targeted Therapy* 6 (1): 1–24.

Lin, K.L., Hsieh, K.H., Thomas, W.R. et al. (1994). Characterization of Der p V allergen, cDNA analysis, and IgE-mediated reactivity to the recombinant protein. *The Journal of Allergy and Clinical Immunology* 94 (6 Pt 1): 989–996. https://doi.org/10.1016/0091-6749(94)90117-1.

Liu, Z., Zhou, T., Lai, Z. et al. (2017). Competence of *Aedes aegypti*, *Ae. albopictus*, and *Culex quinquefasciatus* mosquitoes as zika virus vectors, China. *Emerging Infectious Diseases.* 23(7):1085.

Liu, X.Y., Yang, K.Y., Wang, M.Q. et al. (2018). High-quality assembly of *Dermatophagoides pteronyssinus* genome and transcriptome reveals a wide range of novel allergens. *The Journal of Allergy and Clinical Immunology* 141 (6): 2268–2271.e8. https://doi.org/10.1016/j.jaci.2017.11.038.

Liu, X.Y., Yang, K.Y., Wang, M.Q. et al. High-quality assembly of *Dermatophagoides pteronyssinus* genome and transcriptome reveals a wide range of novel allergens. *The Journal of Allergy and Clinical Immunology* 141 (6): 2268–2271. e8.

Mans, B.J., Ribeiro, J.M.C., and Andersen, J.F. (2008). Structure, function, and evolution of biogenic amine-binding proteins in soft ticks. *The Journal of Biological Chemistry.* 283 (27): 18721.

Marshall, J.S., Warrington, R., Watson, W., and Kim, H.L. (2018). An introduction to immunology and immunopathology. *Allergy, Asthma & Clinical Immunology* 14 (2): 1–10.

Matsui, E.C., Wood, R.A., Rand, C. et al. (2003). Cockroach allergen exposure and sensitization in suburban middle-class children with asthma. *Journal of Allergy and Clinical Immunology* 112 (1): 87–92.

Morgan, M.S., Arlian, L.G., Bernstein, J.A., and Yoder, J.A. (2007). Allergenicity of the Madagascar hissing cockroach. *Annals of Allergy, Asthma & Immunology* 98 (3): 258–261.

Morgan, B.W., Siddharthan, T., Grigsby, M.R. et al. (2018). Asthma and allergic disorders in Uganda: a population-based study across urban and rural settings. *Journal of Allergy and Clinical Immunology and Practices* 6 (5): 1580–1587. e2.

Mueller, G.A., Pedersen, L.C., Lih, F.B. et al. (2013). The novel structure of the cockroach allergen Bla g 1 has implications for allergenicity and exposure assessment. *The Journal of Allergy and Clinical Immunology* 132 (6): 1420–1426. e9.

von Mutius, E. (2009). Gene-environment interactions in asthma. The Journal of Allergy and Clinical Immunology 123 (1): 3–11.

Nathanson, M.E. (1969). Arthropods and allergy. *The Journal of Asthma Research.* 7 (2): 55–63.

Nene, V., Wortman, J.R., Lawson, D. et al. (2007). Genome sequence of *Aedes aegypti*, a major arbovirus vector. *Science* 316 (5832): 1718–1723. https://doi.org/10.1126/science.1138878.

Page, K., Strunk, V.S., and Hershenson, M.B. (2003). Cockroach proteases increase IL-8 expression in human bronchial epithelial cells via activation of protease-activated receptor (PAR)-2 and extracellular-signal-regulated kinase. *The Journal of Allergy and Clinical Immunology* 112 (6): 1112–1118.

Page, K., Hughes, V.S., Odoms, K.K. et al. (2005). German cockroach proteases regulate interleukin-8 expression via nuclear factor for interleukin-6 in human bronchial epithelial cells. *American Journal of Respiratory Cell and Molecular Biology* 32 (3): 225–231.

Page, K., Hughes, V.S., Bennett, G.W., and Wong, H.R. (2006). German cockroach proteases regulate matrix metalloproteinase-9 in human bronchial epithelial cells. *Allergy* 61 (8): 988–995.

Page, K., Lierl, K.M., Herman, N., and Wills-Karp, M. (2007). Differences in susceptibility to German cockroach frass and its associated proteases in induced allergic inflammation in mice. *Respiratory Research* 8 (1): 91.

Patterson, M.L. and Slater, J.E. (2002). Characterization and comparison of commercially available German and American cockroach allergen extracts. *Clinical and Experimental Allergy* 32 (5): 721–727.

Peng, Z. and Simons, F.E.R. (2004). Mosquito allergy: immune mechanisms and recombinant salivary allergens. *International Archives of Allergy and Immunology* 133 (2): 198–209.

Platts-Mills T.A. (2001). The role of immunoglobulin E in allergy and asthma. *American Journal of Respiratory and Critical Care Medicine* 164 (8 Pt 2), S1–S5. https://doi.org/10.1164/ajrccm.164.supplement_1.2103024.

Platts-Mills, T.A.E. (2015). The allergy epidemics: 1870–2010. *Journal of Allergy and Clinical Immunology* 136 (1): 3–13.

Pomés, A., Chapman, M.D., and Wünschmann, S. (2016). Indoor allergens and allergic respiratory disease. *Current Allergy and Asthma Reports* 16 (6): 43.

Pomés, A., Wünschmann, S., and Chapman, M.D. (2016). Allergens. *Encyclopedia Immunobiology* 5: 281–289.

Pomés, A., Schulten, V., Glesner, J. et al. (2021). IgE and T cell reactivity to a comprehensive panel of cockroach allergens in relation to disease. *Frontiers in immunology* 11: 621700. https://doi.org/10.3389/fimmu.2020.621700.

Sánchez-Ramos, I., Álvarez-Alfageme, F., and Castañera, P. (2007). Effects of relative humidity on development, fecundity and survival of three storage mites. *Experimental & Applied Acarology* 41 (1–2): 87–100.

Schmidlin, F., Amadesi, S., Dabbagh, K. et al. (2002). Protease-activated receptor 2 mediates eosinophil infiltration and hyperreactivity in allergic inflammation of the airway. *Journal of Immunology* 169 (9): 5315–5321.

Seda, J. and Horrall, S. (2022). Mosquito bites. *Vet Allergy* 10: 267–270.

Shamji, M.H. and Durham, S.R. (2017). Mechanisms of allergen immunotherapy for inhaled allergens and predictive biomarkers. *The Journal of Allergy and Clinical Immunology* 140 (6): 1485–1498.

Sharma, D., Dutta, B.K., and Singh, A.B. (2011). Dust mites population in indoor houses of suspected allergic patients of South assam. *India. ISRN Allergy* 2011: 1–7.

Shin, Y.H., Hwang, J., Kwon, R. et al. (2023). Global, regional, and national burden of allergic disorders and their risk factors in 204 countries and territories, from 1990 to 2019: a systematic analysis for the Global Burden of Disease Study 2019. *Allergy* 78 (8): 2232–2254. https://doi.org/10.1111/all.15807.

Singh, M. and Hays, A. (2016). Indoor and outdoor allergies. *Primary Care* 43 (3): 451–463.

Sookrung, N. and Chaicumpa, W. (2010). A revisit to cockroach allergens. *Asian Pacific Journal of Allergy and Immunology* 28 (2–3): 95–106.

Spits, H. and Cupedo, T. (2012). Innate lymphoid cells: emerging insights in development, lineage relationships, and function. *Annual Review of Immunology* 30: 647–675.

Steinman, R.M. (2012). Decisions about dendritic cells: past, present, and future. *Annual Review of Immunology* 30: 1–22.

Suesirisawad, S., Malainua, N., Tungtrongchitr, A. et al. (2015). Dust mite infestation in cooking flour: experimental observations and practical recommendations. *Asian Pacific Journal of Allergy and Immunology* 33 (2): 123–128.

Swale, D.R. and Bloomquist, J.R. Is DEET a dangerous neurotoxicant? *Pest Management Science* 75 (8): 2068–2070.

Tatfeng, Y.M., Usuanlele, M.U., Orukpe, A. et al. (2005). Mechanical transmission of pathogenic organisms: the role of cockroaches. *Journal of Vector Borne Diseases* 42: 129–134.

Tauxe, G.M., MacWilliam, D., Boyle, S.M. et al. (2013). Targeting a dual detector of skin and CO_2 to modify mosquito host seeking. *Cell* 155 (6): 1365–1379. https://doi.org/10.1016/j.cell.2013.11.013.

Thomas, W.R., Smith, W.A., Hales, B.J. et al. (2002). Characterization and immunobiology of house dust mite allergens. *International Archives of Allergy and Immunology* 129 (1): 1–18.

Tovey, E. and Marks, G. (1999). Methods and effectiveness of environmental control. *The Journal of Allergy and Clinical Immunology* 103 (2 Pt 1): 179–191.

Tsai, Y.M., Hsu, S.C., Zhang, J. et al. (2013). Functional interaction of cockroach allergens and mannose receptor (CD206) in human circulating fibrocytes. *PLoS One.* 8 (5): e64105.

Tungtrongchitr, A., Sookrung, N., Munkong, N. et al. (2004). The levels of cockroach allergen in relation to cockroach species and allergic diseases in Thai patients. *ProQuest* 22: 115–121.

Upham, J.W., Stick, S.M., and Moodley, Y. (2008). Lung cell biology. In: *Pediatric Respiratory Medicine* (ed. L.M. Taussig and L.I. Landau), 35–43. Philadelphia, USA: Mosby/Elsevier https://doi.org/10.1016/B978-032304048-8.50008-6.

Van Beek, A.A., Knol, E.F., De Vos, P. et al. (2012). Recent developments in basophil research: do basophils initiate and perpetuate type 2 T-helper cell responses? *International Archives of Allergy and Immunology.* 160 (1): 7–17.

Vander Does, A., Labib, A., and Yosipovitch, G. (2022). Update on mosquito bite reaction: itch and hypersensitivity, pathophysiology, prevention, and treatment. *Frontiers in Immunology* 13: 1024559.

Vighi, G., Marcucci, F., Sensi, L. et al. (2008). Allergy and the gastrointestinal system. *Clinical and Experimental Immunology* 153 (Suppl 1): 3.

Wan, A.T., Xiong, Q., Xiao, X. et al. (2024). Multi-omic analysis of *Tyrophagus putrescentiae* reveals insights into the allergen complexity of storage mites. *Clinical and Experimental Allergy: Journal of the British Society for Allergy and Clinical Immunology*, 54 (1), 77–81. https://doi.org/10.1111/cea.14418.

Wang, J., Visness, C.M., Calatroni, A. et al. (2009). Effect of environmental allergen sensitization on asthma morbidity in inner-city asthmatic children. *Clinical & Experimental Allergy* 39 (9): 1381–1389.

Wang, L., Xiong, Q., Saelim, N. et al. (2023). Genome assembly and annotation of *Periplaneta americana* reveal a comprehensive cockroach allergen profile. *Allergy* 78 (4): 1088–1103. https://doi.org/10.1111/all.15531.

Wangorsch, A., Jamin, A., Eichhorn, S. et al. (2021). Component-resolved diagnosis of american cockroach (*Periplaneta americana*) allergy in patients from different geographical areas. *Frontiers in Allergy* 2: 691627. https://doi.org/10.3389/falgy.2021.691627.

Warner, A., Boström, S., Munir, A.K. et al. (1998). Environmental assessment of Dermatophagoides mite-allergen levels in Sweden should include Der m 1. *Allergy* 53 (7): 698–704. https://doi.org/10.1111/j.1398-9995.1998.tb03956.x.

Wawrzyniak, M. and ÓMahony L, Akdis M. (2017). Role of regulatory cells in oral tolerance. *Allergy, Asthma & Immunology Research* 9 (2): 107–115.

Weghofer, M., Grote, M., Resch, Y. et al. (2013). Identification of Der p 23, a peritrophin-like protein, as a new major *Dermatophagoides pteronyssinus* allergen associated with the peritrophic matrix of mite fecal pellets. *Journal of Immunology (Baltimore, Md.: 1950)* 190 (7): 3059–3067. 10.4049/jimmunol.1202288.

Weghofer, M., Grote, M., Resch, Y. et al. (2013). Identification of Der p 23, a peritrophin-like protein, as a new major *Dermatophagoides pteronyssinus* allergen associated with the peritrophic matrix of mite fecal pellets. *Journal of Immunology* 190 (7): 3059–3067.

Wilson, J.M. and Platts-Mills, T.A.E. (2018). Home environmental interventions for house dust mite. *The Journal of Allergy and Clinical Immunology: In Practice* 6 (1): 1–7.

Xu, W., Simons, F.E.R., and Peng, Z. (1998). Expression and rapid purification of an *Aedes aegypti* salivary allergen by a baculovirus system. *International Archives of Allergy and Immunology* 115 (3): 245–251.

Zhang, Y.X., Chen, X., Wang, J.P. et al. (2019). Genomic insights into mite phylogeny, fitness, development, and reproduction. *BMC Genomics* 20 (1): 954. https://doi.org/10.1186/s12864-019-6281-1.

Zhou, D., Ren, Y., Zhou, Y. et al. (2024). Expression, purification, and activity of novel allergen Tyr p 31 from Tyrophagus putrescentiae. *International Journal of Biological Macromolecules*, 258 (Pt 1), 128856. https://doi.org/10.1016/j.ijbiomac.2023.128856.

15

Diagnosis and Treatment of Health Effects Caused by Airborne Contaminants

Sujata Adhana[1], *Anil Kumar Mavi*[2], *Umesh Kumar*[3], *Garima Rathi*[4], *and Devendra Kumar*[5]

[1] *Bhaskaracharya College of Applied Sciences, University of Delhi, Delhi, New Delhi, India*
[2] *Department of Botany, Sri Aurobindo College, University of Delhi, Delhi, New Delhi, India*
[3] *Department of Biosciences, Institute of Management Studies Ghaziabad (University Courses Campus), Ghaziabad, Uttar Pradesh, India*
[4] *Delhi Public School Ghaziabad, Ghaziabad, Uttar Pradesh, India*
[5] *Department of Environmental Science, Ramanujan College, University of Delhi, Delhi, New Delhi, India*

15.1 Introduction

The health of people is significantly threatened by airborne pollutants, which have a variety of negative effects on many organ systems (Ghio and Soukup 2019; Bernstein et al. 2021). Understanding and managing airborne pollutants as their health implications become more acute as urbanization, industrialization, and pollution increase is important (WHO 2018). This chapter seeks to enlighten readers with a thorough and in-depth analysis of the diagnosis and management of health impacts caused by airborne pollutants.

According to the CDC (2021), airborne pollutants might be either solid or liquid particles, gases, fumes, allergies, or biological agents. They originate from a variety of areas, including natural sources, indoor pollutants, building sites, industrial emissions, and vehicle exhaust (EPA 2021). These contaminants can enter the respiratory system through inhalation and cause harm to the human body.

According to Holgate et al. (2015), the health impacts of airborne pollutants are varied and can impact several organ systems. The respiratory system is especially susceptible to problems like irritation, inflammation, and lung damage from pollutants (CDC 2021). Additionally, exposure to some pollutants can increase the incidence of heart attacks and strokes as well as other cardiovascular disorders

Airborne Biocontaminants and Their Impact on Human Health, First Edition.
Edited by Rajeev Singh and Anamika Singh.

(Bernstein et al. 2021). The most prevalent symptoms are skin and eye irritations, but some pollutants can also impact the central nervous system (CNS) and result in neurological issues (Ghio and Soukup 2019).

Accurate diagnosis is essential for effectively managing the negative effects that airborne contaminants have on human health. The clinical history of patients, symptoms, and physical examination are used by healthcare professionals to estimate the likelihood of exposure and associated health issues (Holgate et al. 2015). Confirming diagnoses and determining the severity of the health impacts are aided by diagnostic procedures and tests such as pulmonary function tests, allergy tests, and environmental evaluations (CDC 2021).

A multifaceted strategy is employed in the treatment of airborne contaminant-related health consequences. Through source management, enhanced ventilation systems, and the application of PPE, exposure is mostly reduced. Corticosteroids, bronchodilators, and antihistamines, among other pharmaceutical treatments, are examples of medical interventions that can control underlying problems and reduce symptoms. Making lifestyle changes, such as giving up smoking and engaging in regular exercise, can help lessen the harmful effects of airborne pollutants on health.

We will thoroughly examine many types of airborne contaminants, including asbestos, mold, radon, volatile organic compounds (VOCs), and allergies, in this chapter. We will also look at their particular health implications, diagnostic techniques, and therapeutic approaches. We will also talk about the connections between these toxins and occupational health, as well as considerations for the elderly, children, and vulnerable populations. The chapter will quickly explore IAQ assessment's significance and go over the relevant legal requirements and standards. Additionally, we'll look into new study data that aims to advance our comprehension of and management of airborne pollutants. This chapter aims to foster the development of a healthier and safer environment for all people by providing healthcare practitioners with the knowledge and resources necessary to diagnose, treat, and prevent adverse health impacts connected to airborne pollutants (Figure 15.1).

15.2 Airborne Contaminants

Contaminants in the air are compounds or particles that are in the atmosphere. These toxins may be breathed or may enter the body and affect it. These pollutants may take the shape of biological agents, solid particles, liquid droplets, vapors, odors, or allergies. They can arise from a number of processes, such as those that are part of the natural world, human activity, industrial emissions, combustion processes, building processes, and indoor pollutants.

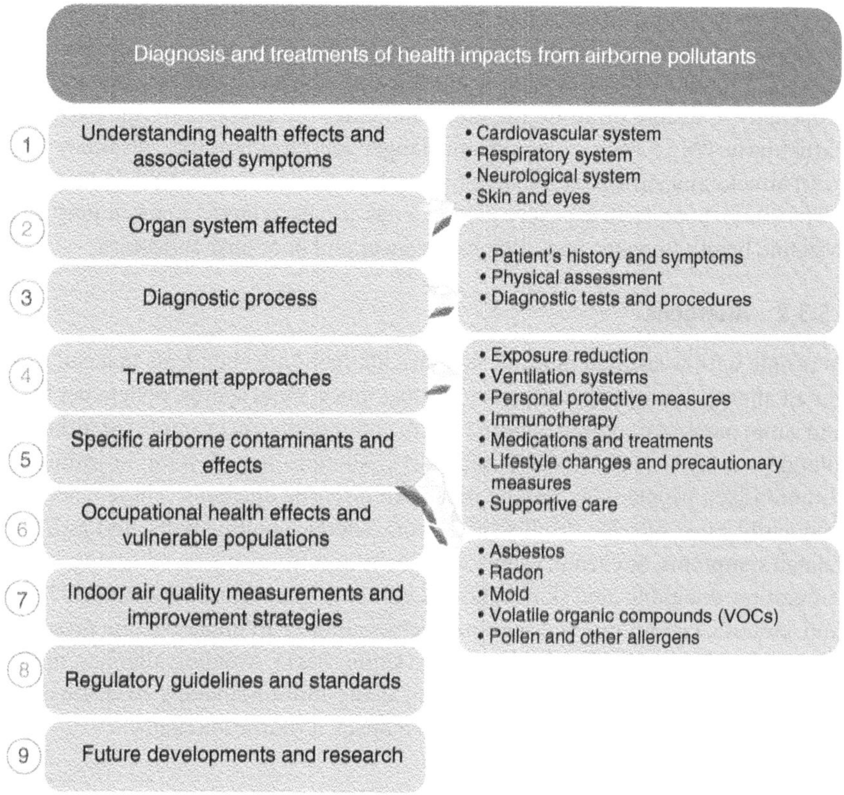

Figure 15.1 Flow diagram for chapter overview.

15.3 Common Airborne Contaminants

15.3.1 Particulate Matter

It is made up of substances that are found in the environment, both solid and liquid. These particles come in a variety of shapes, materials, and origins. They are frequently divided into two categories based on their aerodynamic diameter: PM10 refers to particles with a diameter less than 10 μm, and PM2.5 refers to particles with a diameter less than 2.5 μm (EPA 2021; WHO 2021).

Particulate matter can originate from a number of places, including combustion processes (such as motor vehicle emissions, power plant emissions, wood burning emissions, industrial emissions, and building sites), as well as natural sources such as dust and pollen. Particulate matter can have a variety of compositions, with soot, metals, chemical compounds, and biological particles among its constituents (EPA 2021; WHO 2021).

Particulate matter exposure can cause a variety of respiratory issues, including coughing, wheezing, and difficulty breathing. Additionally, it may exacerbate preexisting respiratory diseases such as chronic bronchitis and asthma. Due to their propensity to deeply infiltrate the lungs and enter the circulation, fine particles, particularly PM2.5, pose a significant danger for cardiovascular disorders like heart attacks and strokes as well as a high risk of cardiovascular death (EPA 2021; WHO 2021). Long-term exposure to PM2.5 has also been linked to respiratory and systemic health impacts, including lung cancer and poor birth outcomes.

15.3.2 Allergens

In people with weakened immune systems, allergens are chemicals that can trigger an allergic immunological reaction. Dust mites, mold spores, pollen, pet hair, and some particular allergens from insects and rodents are examples of airborne allergies (AAAAI 2021; Mayo Clinic 2021). Allergies may develop in immunocompromised people who breathe in certain airborne allergens. These reactions frequently cause allergic rhinitis (hay fever), allergic conjunctivitis, and asthma-related symptoms. Sneezing, watery eyes, itchy and runny noses, wheezing, nasal congestion, coughing, and skin rashes are common symptoms. Geographical location, climatic conditions, and seasonal factors all have an impact on the presence of airborne allergens (AAAAI 2021; Mayo Clinic 2021). Avoiding allergies, maintaining a clean indoor environment, and using immunotherapy to desensitize people to certain allergens are all ways to manage airborne allergens.

15.3.3 Fumes and Gases

A wide variety of chemicals and contaminants that are airborne can be found in gases and fumes. These pollutants are frequently produced by industrial processes, vehicular emissions, and combustion processes. VOCs, hazardous gases generated from industrial operations, ozone (O_3), carbon monoxide (CO), sulfur dioxide (SO_2), nitrogen dioxide (NO_2), and ozone (O_2) are common gases and fumes (EPA 2021; WHO 2021).

Gas and fume inhalation can negatively impact the cardiovascular and respiratory systems as well as other organ systems. For instance, the colorless and odorless gas known as carbon monoxide was created when fossil fuels combined. It attaches to the hemoglobin in RBCs, decreasing their ability to carry oxygen and causing symptoms like headaches, nausea, and dizziness as well as, in more extreme situations, unconsciousness and death. The respiratory system can become irritated by nitrogen dioxide and sulfur dioxide, aggravating asthma and raising the risk of respiratory infections. Ozone is a reactive gas that can irritate and inflame the airways, especially in people who already have asthma (EPA 2021; WHO 2021).

VOCs are chemicals with a high degree of vaporization and gas formation potential. They are released from a range of things, including household goods, industrial processes, paints, solvents, and building materials. High concentrations of some VOCs can cause symptoms like headaches, dizziness, and, in some extreme situations, damage to the kidneys, liver, and CNS, as well as irritation of the nose, throat, and eyes. Some VOCs are also suspected of having long-term negative impacts on health, such as being carcinogenic or causing allergy and respiratory sensitivity.

15.3.4 Biological Agents

Microorganisms such as bacteria, viruses, and fungi are examples of biological agents. These microbes can produce bioaerosols when they are airborne, which people can breathe in.

Biological agents that are airborne can come from a variety of places, such as wet structures, water-damaged areas, and polluted HVAC systems. Legionella bacteria, which causes Legionnaires' disease, and Aspergillus fungi, which can cause respiratory infections in susceptible people, are two examples of airborne biological agents. Respiratory illnesses can be brought on by influenza viruses that spread by respiratory droplets and become airborne (WHO 2021; CDC 2022).

Respiratory infections, allergies, and toxic responses are just a few of the health impacts that can result from exposure to airborne biological agents. The type of microbe, concentration in the air, and personal sensitivity are only a few of the variables that affect how severe the health impacts will be.

15.4 Specific Airborne Contaminants and Their Impact on Health

15.4.1 Asbestos

Due to its strength and heat resistance, asbestos, a naturally occurring mineral fiber, has become more important in building materials. However, breathing in asbestos fibers can have serious negative health impacts, such as:

Mesothelioma: Mesothelioma is an uncommon and aggressive type of cancer that develops when an individual is exposed to asbestos and affects the lining of the heart, abdomen, or lungs.

Lung Cancer: Long-term asbestos exposure greatly raises the chance of getting lung cancer, especially in people who smoke.

Asbestosis: Long-term exposure to asbestos fibers can result in lung tissue scarring, which is what is referred to as asbestosis. According to the WHO (2014) and the American Cancer Society (2022), growing breathlessness, persistent cough, and chest discomfort are all signs of asbestosis.

15.4.2 Radon

Radon is a radioactive gas that naturally occurs in soil and rock. It can enter buildings and build up to dangerous levels. Lung cancer risk can be greatly increased by high radon exposure (EPA 2022).

15.4.3 Mold

A form of fungus that prefers moist, humid settings is mold. Inhaling or coming into close contact with mold spores can have a number of negative health impacts, including:

Allergic Reactions: People who are allergic to mold spores can have allergic reactions to them. Sneezing, coughing, runny nose, itchy eyes, and skin rashes are a few symptoms that may appear.

Exacerbation of Asthma: When exposed to mold spores, people with asthma may notice a worsening of their symptoms, including coughing, wheezing, and shortness of breath.

Infections of the Respiratory System: According to the WHO and the CDC, many molds can create toxins that can lead to respiratory infections, especially in people who already have compromised immune systems.

15.4.4 Volatile Organic Compounds (VOCs)

VOCs are chemicals that can be released as gases from a variety of sources, such as paints, solvents, household cleansers, and construction materials. VOC exposure can have a variety of negative health impacts, such as:

VOCs can irritate the mucous membranes, which can result in symptoms including watery eyes, nasal congestion, sore throats, and headaches.

Respiratory symptoms, such as coughing, wheezing, and shortness of breath, may be brought on by prolonged exposure to high levels of VOCs.

Specific VOCs may trigger allergic reactions in some people, which can worsen asthma symptoms or cause skin rashes, hives, or other allergic reactions (WHO 2010; EPA 2021).

15.4.5 Allergens Such as Pollen

Pollen can cause allergic reactions in immunocompromised people, just as other allergens such as dust mites, pet dander, and certain insect and rodent allergens. According to Pawankar et al. (2013), allergic rhinitis, allergic conjunctivitis, and asthma symptoms are just a few of the health impacts of airborne allergens. Table 15.1 contains a list of common and specialized airborne contaminants.

Table 15.1 Common and specific airborne contaminants and health effects.

Airborne contaminants	Health effects
Particulate matter (PM)	Respiratory and cardiovascular problems like coughing, breathlessness, difficulty in breathing, constriction in the chest, asthma, COPD, bronchitis, heart stroke, heart attacks, cardiovascular death, allergic reactions, etc.
Allergens	Allergic rhinitis, asthma exacerbation, allergic conjunctivitis, skin allergies, allergic bronchitis, allergic dermatitis, etc.
Gases and fumes	
Carbon monoxide (CO)	Headache, dizziness, nausea, and, in severe cases, unconsciousness and death
Sulfur dioxide (SO_2), nitrogen dioxide (NO_2)	Exacerbating respiratory conditions like asthma and high risk of respiratory infections
Ozone	Respiratory irritation and inflammation, especially in individuals with asthma
Volatile organic compounds (VOCs)	Eyes, nose, throat irritation, headache, dizziness, and liver and kidney damage.
Biological agents – Bacteria, viruses, and fungi	Respiratory infections, allergies, and toxic reactions
Asbestos	Allergic reactions, lung cancer, mesothelioma, asbestosis
Radon	Lung cancer
Mold	Allergic reactions, asthma exacerbation, respiratory infections
Pollen and other allergens	Allergic rhinitis, asthma exacerbation, asthma symptoms

15.5 Health Effects Caused by Airborne Contaminants

Those exposed to airborne pollutants may have a variety of different health effects. Based on the precise type and concentration of pollutants, as well as the length and frequency of exposure, these health effects can vary. Here is all the information you need to know about the harmful impact of airborne pollutants on health:

Cardiovascular effects: Airborne pollutants can harm the cardiovascular system and cause serious health problems. People who are exposed to these toxins may experience a variety of cardiovascular consequences, including accelerated atherosclerosis, oxidative stress, endothelial dysfunction, increased blood pressure, and inflammation in the body.

Respiratory effects: The respiratory system is significantly impacted by a number of airborne pollutants. Respiratory symptoms like coughing, dyspnea, difficulty breathing, chest constriction, and an increased risk of lung infections can all result from inhaling particles, gases, and fumes. Asthma, bronchitis, and chronic obstructive pulmonary disease (COPD) can all be caused by prolonged exposure to certain contaminants, such as particulate matter or occupational dust (WHO 2018; EPA 2021).

Nervous system effects: Neurotoxic effects are possible with some airborne pollutants, especially with some VOCs. Long-term exposure to certain VOCs with high concentrations, like benzene or toluene, can cause neurological symptoms including headaches, fatigue, and dizziness as well as cognitive decline and possibly neurobehavioral problems. According to Ghio and Soukup (2019), the precise processes causing these consequences are complicated and may entail either direct neurotoxicity or indirect impacts on the CNS.

Allergies to the skin and eyes and their immunological effects: Sensitized people may experience allergic reactions when exposed to airborne allergens such as pollens, mold spores, and animal dander. These reactions can result in allergic rhinitis (hay fever), allergic conjunctivitis, and allergic skin allergies. In certain situations, exposure to specific pollutants can aggravate pre-existing allergy disorders or lead to the emergence of new allergies. In addition, exposure to some biological substances can cause immunological reactions that result in hypersensitivity reactions or respiratory infections (Holgate et al. 2015; CDC 2021).

15.6 Identifying Health Effects

15.6.1 Patient History and Symptoms

The diagnosis of airborne contaminant-related health consequences requires a thorough patient history. The healthcare professional should ask the patient about their job, way of life, interests, and exposure to the environment. The length, frequency, and intensity of exposure to pollutants should be thoroughly explored. The onset, duration, and course of symptoms must also be described in great detail. The type of pollutant and the damaged organ system will determine the range of symptoms that patients may experience. Establishing a possible connection between symptoms and exposure to airborne pollutants might be aided by a thorough and accurate patient history (NIOSH 2018).

15.6.2 Physical Evaluation

A thorough physical examination is performed to evaluate the patient's health and look for particular symptoms of exposure to airborne pollutants. Depending on the reported symptoms and potential exposure, the examination may concentrate on the cardiovascular system, respiratory system, skin, eyes, and neurological function.

The healthcare professional may auscultate the lungs for unusual breath sounds, listen for indications of respiratory distress, and assess the respiratory system as a whole. During a cardiovascular system examination, blood pressure may be taken, heart sounds evaluated, and vascular anomalies may be looked for. Any rashes, irritations, or allergic responses may be discovered by a skin examination. The eyes can be checked for conjunctivitis, redness, or watering. Assessment of cognitive function, physical coordination, and sensory perception may be part of a neurological evaluation (Ball et al. 2019).

The physical examination adds to the diagnosis and treatment planning process by identifying any specific symptoms that could indicate the presence of airborne pollutants.

15.6.3 Diagnostic Tests and Procedures

Diagnostic tests and procedures are employed to determine whether airborne pollutants are present and to assess their potential health effects on the patient. The probable pollutants and the particular health impacts under investigation will determine which tests are used.

Diagnostic procedures and testing include the following:

Tests of pulmonary function: These examinations gauge lung capacity, airflow, and gas exchange to assess respiratory function. These evaluations offer important knowledge about how the respiratory system is operating. They can aid in the diagnosis of lung conditions that are obstructive or restrictive and are brought on by exposure to airborne pollutants.

Testing for allergies: Allergy tests, such as blood or skin prick tests, can identify particular allergens to which the patient may be hypersensitive. These tests assist in identifying allergy reactions brought on by allergens in the air.

Imaging tests: Such as chest X-rays, CT scans, or MRIs, may be carried out to examine the heart, lungs, or other afflicted organs for structural damage or abnormalities.

Analysis of biomarkers: Biomarkers, such as particular proteins or chemicals in blood or urine, can show that a person has been exposed to a particular pollutant or can point to the presence of associated health impacts.

Microbiological testing: If biological agents are suspected, samples from the patient or the environment may be taken for lab examination in order to pinpoint specific microorganisms and weigh the health risks associated with them.

Environmental sampling: Surface or air samples may be taken in situations when there is a suspicion of environmental contamination, and they may then be examined to determine whether airborne pollutants are present (Stephen T. and Barry S. (Eds.)).

15.7 Health Consequences and Related Signs and Symptoms

15.7.1 Cardiovascular System

Airborne pollutants can also impact the cardiovascular system, which can result in cardiovascular illnesses and their associated symptoms. The following are possible cardiovascular system health effects:

Heart attack risk is rising: Exposure to some gases, such as NO_2 and CO, can increase the risk of developing cardiovascular events, such as heart attacks, by decreasing the flow of oxygen to the heart and causing inflammation and blood clotting.

Hypertension: Long-term exposure to air pollution, particularly fine particulate matter (PM2.5), has been linked to a higher risk of hypertension (high blood pressure), which is a key factor in the development of cardiovascular illnesses.

Alterations in heart rhythm: It has been established that air pollution, particularly that from combustion sources, is associated with arrhythmias and irregular heartbeat rhythms.

Reduced heart function: According to Brook et al. (2010) and Zhang et al. (2012), exposure to air pollution over an extended period of time can harm the cardiovascular system, heart function, and cardiac output.

15.7.2 Respiratory System

The respiratory system may be impacted in a variety of ways by exposure to airborne pollutants. The following are typical respiratory system health effects:

Coughing: Chronic or persistent coughing may be a sign of exposure to irritants, gases, or particulates in the air.

Wheezing: Wheezing, which is characterized by a high-pitched whistling sound made during inhaling, may be brought on by inflammation or restriction of the airways. This symptom is frequently seen in diseases like asthma or after being exposed to allergens.

Breathing problems: Respiratory irritation or injury from pollutants can make it difficult or impossible to breathe.

Exacerbation of asthma: People who already have asthma may see a worsening of their symptoms when exposed to allergens or irritants in the air. Coughing, wheezing, chest pain, and shortness of breath are a few of these symptoms.

A chronic lung condition known as COPD: Chronic obstructive pulmonary disease (COPD) is a chronic inflammatory lung disorder defined by a restriction in airflow. Gases and particulate matter, particularly if they are breathed or inhaled frequently, can contribute to COPD.

Respiratory infections can result from airborne biological agents, such as bacteria and viruses (Kelly and Fussell 2011; WHO 2013; Peden et al. 2016). These infections can produce symptoms like fever, cough, colds, sore throats, and breathing difficulties.

15.7.3 Neurological System

The neurological system may be impacted by some airborne pollutants, leading to neurological symptoms and disorders. These are:

Headaches: Headaches and migraines may be brought on by exposure to specific gases, such as CO and VOCs.
Dizziness and lightheadedness: Inhaling some gases, such as CO and some solvents, might make you feel unbalanced and dizzy.
Cognitive impairment: Prolonged exposure to specific airborne toxins, such as VOCs and some metals, has been linked to cognitive impairment, which includes issues with attention, memory, and cognitive ability.

When assessing individuals who may have been exposed to airborne pollutants, it is crucial to recognize these health impacts and the accompanying symptoms. For the management of these disorders and the reduction of potential long-term health consequences, proper diagnosis and treatment are crucial (Kelly and Fussell 2011; Zhang et al. 2012).

15.7.4 Skin and Eyes

These consist of skin irritation, redness, itching, and rashes that can be brought on by contact with certain gases, fumes, and Particulate matter.

Allergic contact dermatitis: When exposed to specific airborne allergens, such as some chemicals or plant-based compounds, some people may get allergic contact dermatitis.
Eye redness and irritation: According to OSHA (2011), irritants in the air, such as gases, fumes, and allergens, can cause eye redness, stinging, and tearing as well as conjunctivitis (inflammation of the conjunctiva).

15.8 Treatment Approaches

15.8.1 Exposure Control

Reducing or eliminating exposure to airborne pollutants is one of the typical ways used to treat health problems brought on by those contaminants. This can be done using a variety of tactics, including:

Source control: Finding and eliminating the source of contamination by measures like better ventilation, engineering controls, or the use of the right PPE.

Improving the quality of the air inside a building by taking steps like frequent cleaning, good ventilation, and the use of air purifiers or filtration systems.

Implementing workplace safety rules, such as sufficient ventilation, the use of PPE, and adherence to occupational exposure limits (OELs), are examples of occupational safety measures.

Keeping away from allergies for those who have allergies, and preventing exposure to particular allergens can be accomplished by taking steps like using bedding that is allergen-proof, cleaning often, and limiting contact with pets or other sources of allergens.

15.8.2 Immunotherapy

Those who experience severe allergic reactions to airborne allergens may consider immunotherapy, more especially allergen-specific immunotherapy. To desensitize the immune system and gradually lessen the intensity of allergic symptoms, this treatment involves administering allergens under close observation on a regular basis in gradually increasing dosages. Depending on the exact allergens and the patient's preferences, immunotherapy may be administered as subcutaneous injections or sublingual pills.

15.8.3 Medications and Therapies

To alleviate symptoms, manage illnesses, or lessen the impact of exposure to airborne pollutants, doctors may prescribe medications and therapies. These may consist of:

Bronchodilators: Bronchodilators, which relax and enlarge the airways, can help relieve respiratory condition symptoms like wheezing and breathlessness.

Anti-inflammatory drugs: To control disorders like asthma or allergic responses, corticosteroids or other anti-inflammatory antibiotics may be administered.

Antihistamines: These drugs can ease allergy symptoms brought on by exposure to airborne allergens, such as sneezing, itching, and runny nose.

Therapies for the respiratory system: Nebulizers or inhalers, which deliver medication directly to the lungs, may be prescribed to treat respiratory problems.

15.8.4 Lifestyle Changes and Preventive Measures

Making specific lifestyle adjustments, managing conditions, and taking preventive measures can support medical interventions and lessen the effects of airborne pollutants. These may consist of:

Smoking: It should be avoided because it can exacerbate respiratory symptoms and significantly raise the risk of respiratory disorders. For the sake of your general respiratory health, you must stop smoking.

Exercise on a regular basis: Physical activity on a regular basis can promote cardiovascular health, improve lung function, and increase overall well-being.

Healthy diet: A well-balanced diet that contains a variety of nutrient-dense fruits, vegetables, and whole grains that are high in antioxidants and important nutrients that support cardiovascular and respiratory health.

Staying properly hydrated promotes overall health and aids in maintaining the respiratory system's integrity.

Keeping a clean-living space: Routine cleaning, dusting, and organizing your home can help limit your exposure to airborne toxins.

Personal safety precautions: In occupational or high-risk areas, wearing the proper personal protection equipment, such as masks or respirators, can decrease exposure to airborne toxins.

15.8.5 Supportive Care

Supportive care is just as important to addressing the health impacts brought on by airborne pollutants as specialized therapies. Adequate rest and sleep to help healing and overall well-being are examples of supportive care interventions.

Keeping your body properly hydrated will keep your heart and lungs healthy.

Use of saline nasal sprays or humidifiers (Kelly and Fussell 2011; WHO 2013; Peden et al. 2016).

15.9 Effects on Occupational Health

Workers may be exposed to several airborne contaminants at work, which could have a serious impact on their health. This includes some important information about the effects of airborne pollutants on occupational health (NIOSH 2024; OSHA 2011).

15.9.1 Airborne Contaminants Associated With the Workplace

Depending on the sector and particular job responsibilities, workplaces may contain a variety of airborne contaminants. Dust particles, chemical fumes, gases, and biological agents are a few examples of typical work-related airborne pollutants. A variety of harmful health impacts, including respiratory conditions, skin rashes, allergies, and even various malignancies, can result from exposure to these toxins at work.

15.9.2 Occupational Exposure Limits

Regulatory bodies and organizations set OELs for numerous airborne pollutants in order to safeguard employees' health. OELs identify the highest concentrations of a specific air contaminant that can be tolerated for a specific amount of time without causing harm to human health. OELs take into account elements like toxicity, length of exposure, and the possibility of cumulative effects.

To maintain compliance with OELs, monitoring and measuring the amounts of airborne contaminants in the workplace is crucial. Programs for routine air sampling and monitoring aid in identifying potential exposure hazards and making it easier to put effective control measures in place.

15.9.3 Control and Prevention Measures

Various prevention and control measures should be put into place to reduce occupational exposure to airborne pollutants. Engineering controls, administrative controls, PPE, instruction and training, routine inspection and maintenance, and medical surveillance are some of these measures. The protection of locations where airborne pollutants are prevalent depends on the enactment of comprehensive preventative and control measures as well as occupational health and safety laws.

15.10 Special Implications for Vulnerable Groups

Certain demographic groups, such as children, the elderly, and people with pre-existing diseases, may be especially vulnerable to the harmful effects of airborne pollutants on their health. In order to create specialized strategies to protect the well-being of these populations, it is essential to recognize their vulnerabilities. Children, the elderly, and those with pre-existing diseases are three vulnerable populations that need special attention. The following section lists these considerations (American Academy of Pediatrics 2018; US Environmental Protection Agency 2020; World Health Organization 2021).

15.10.1 Children

Children are particularly vulnerable to the effects of airborne pollutants because their organ systems are still growing, they have faster breathing rates, and they may be exposed for longer periods of time. Vaccinations, regular health check-ups, education and awareness, minimizing exposure, and minimization of risk are some of the crucial components in protecting children from airborne contaminants.

15.10.2 Senior Citizens

Elderly people may have immune systems that are impaired, which make them more vulnerable to the negative health impacts of airborne pollutants. The effects of exposure can also be worsened by age-related illnesses such as cardiovascular disease and long-term respiratory disorders. IAQ, routine health checks, and medication management are all protective factors for the elderly.

15.10.3 People With Pre-Existing Conditions

People who already have a medical condition, such as asthma, allergies, heart disease, or respiratory issues, may be more vulnerable to the effects of airborne pollutants. They ought to consist of collaboration with healthcare professionals, prevention techniques, and management of underlying diseases.

These at-risk groups should receive special consideration to safeguard their health from the harmful impacts of airborne pollutants. For these people, the right support systems and education can help reduce the dangers and foster healthier settings.

15.11 Evaluation and Improvement of Indoor Air Quality

A comfortable and healthy interior environment is maintained by good IAQ. There are many factors that can contribute to poor IAQ, including airborne pollutants. In order to reduce the health hazards associated with indoor airborne contaminants, this section discusses major components of monitoring and improving IAQ (US Environmental Protection Agency 2020; World Health Organization 2021; National Institute for Occupational Safety and Health 2022).

In order to pinpoint probable sources and assess the general air quality, IAQ monitoring entails the systematic assessment and measurement of indoor air pollutants. Monitoring can be done in a number of ways, including sample air: obtaining samples of the air and analyzing them in labs to determine the amounts of particular contaminants, like particulate matter, VOCs, or mold spores.

Real-time observation monitoring and measuring IAQ parameters, such as temperature, humidity, carbon dioxide (CO_2) levels, and particulate matter concentrations, continually.

Index of IAQ: Evaluating IAQ with the aid of established metrics that offer a global score based on pollutant concentrations and related health hazards.

Regular IAQ monitoring aids in spotting possible problems, evaluating the efficiency of preventative measures, and ensuring adherence to advised guidelines and standards.

15.11.1 Ventilation Systems

By removing and diluting airborne pollutants, proper ventilation is essential for maintaining excellent IAQ. Ventilation systems should take into account:

Exchanging air outside: Supplying enough fresh outdoor air to help reduce indoor pollution. There are two basic ways to get proper ventilation: mechanical ventilation systems and natural ventilation. Opening windows and doors to let fresh air enter inside spaces is known as natural ventilation. On the contrary, mechanical ventilation systems actively manage and move air around structures to guarantee proper airflow and exchange, such as HVAC (Heating, Ventilation, and Air Conditioning) systems. Both strategies help keep the atmosphere properly ventilated.

Making sure ventilation rates adhere to the suggested rules and requirements for the particular occupancy and building type.

Filtration: The process of removing allergens and Particulate matter from incoming air by using filters in ventilation systems.

Maintaining ventilation systems on a regular basis will ensure optimal operation and stop the accumulation of impurities like dust or mildew.

Improved air circulation, the removal and control of airborne pollutants, and the creation of a healthier interior environment can all be achieved through ventilation system optimization.

15.11.2 Techniques for Filtration and Purification

IAQ can be improved by actively eliminating airborne contaminants with the help of additional methods. Methods of filtration and purification include the following:

Air filters: To remove particles and allergens from the air, high-efficiency air filters are used in HVAC systems or portable air purifiers.

Utilizing ultraviolet (UV) light to eliminate or render inactive biological pollutants, such as bacteria and viruses, in ductwork or air-handling equipment is known as UV germicidal irradiation.

Electrostatic precipitators: These devices remove airborne particles by charging and collecting them using electrostatic fields.

Utilizing activated carbon to absorb and eliminate certain gases, smells, and VOCs from the air.

Depending on the specific contaminants of concern and the objectives of IAQ improvement, the proper filtering and purification procedures should be chosen.

Minimizing airborne contaminants, fostering a healthier interior environment, and lowering the related health risks can all be achieved with regular IAQ assessments, effective ventilation systems, and the application of appropriate filtering and purification techniques.

15.12 Regulatory Standards and Guidelines

Regulations have been established to address IAQ and the management of airborne contaminants in order to guarantee people's safety and well-being. These recommendations and specifications act as benchmarks for identifying and reducing indoor air pollution. The significance of regulatory standards and guidelines is discussed in this section, along with several important cases (WHO 2010; EPA 2021; OSHA 2021).

15.12.1 Significance of Regulatory Standards and Guidelines

A foundation for preserving acceptable IAQ levels and reducing the health risks brought on by airborne pollutants is provided by regulatory guidelines and standards. They are often established based on scientific research, risk analyses, and industry best practices by government agencies, health organizations, and expert committees. Following are the main advantages of regulatory standards and guidelines:

Health protection: By setting limits for permissible pollutant concentrations, they seek to protect people's health and well-being.

Compliance and enforcement: Regulatory standards and guidelines serve as a foundation for enforcement activities to guarantee that IAQ requirements are met in a variety of contexts, including the workplace, public spaces, and residential premises.

They encourage consistency in IAQ management techniques across various jurisdictions and industries, assisting in the development of a shared understanding and strategy for ensuring healthy indoor environments.

15.12.2 Illustrations of Regulatory Standards and Guidelines

IAQ and airborne contaminants-specific regulatory guidelines and standards have been created by a number of organizations and authorities. Examples that stand out include the following:

Agency for Environmental Protection (EPA): The IAQ Tools for Schools Program, which provides tools for evaluating and improving IAQ in educational settings, is one of the many IAQ-related resources offered by the EPA in the United States.

OSHA is the Occupational Safety and Health Administration: OSHA establishes OELs for different airborne pollutants and gives companies instructions on how to safeguard employees from dangerous substances at work.

World Health Organization (WHO): The WHO provides advice on IAQ, including the IAQ Guidelines for Selected Pollutants, which outlines health-based standards for a number of contaminants that are frequently detected indoors.

The American Society of Heating, Refrigerating, and Air Conditioning Engineers (ASHRAE) is a professional organization that sets standards. For example, ASHRAE Standard 62.1 specifies ventilation requirements for a satisfactory IAQ in commercial buildings.

Building codes and regulations: There are specialized building codes and regulations for many nations and areas, which include rules for IAQ, ventilation systems, and the management of airborne contaminants in new construction and restoration projects.

In order to promote healthier indoor environments and lower the associated health risks, compliance with regulatory rules and standards ensures that the proper actions are taken to assess, control, and manage airborne contaminants.

For building owners, employers, facility managers, and individuals to design and maintain interior settings that prioritize occupant health and well-being, adherence to these rules and regulations is necessary.

15.13 Research and Future Management

In order to expand knowledge and enhance prevention, diagnostic, and treatment options, continual research is crucial. The field of diagnosing and treating health impacts brought on by airborne pollutants is a dynamic one. This section examines prospective research directions and areas that show promise for resolving issues related to airborne pollutants (Annesi-Maesano et al. 2012).

Emerging technologies and advancements: Technological advancements have the potential to fundamentally change how airborne pollutants are diagnosed and treated. There are new developments and emerging technologies in the following industries:

Sensor technologies: The development of more sensitive, inexpensive sensors that can quickly and reliably identify airborne contaminants. This technology allows for continuous monitoring and the early detection of potentially dangerous situations.

Artificial intelligence (AI) and big data: Utilizing machine learning and AI algorithms to analyze huge datasets, identify trends, and predict health impacts in response to exposure to specific chemicals.

By adding nanoparticles to air filters, personal protection equipment, and other technologies, airborne pollutants can be captured and neutralized more effectively.

Advancements in telemedicine platforms and remote monitoring technology make it easier to diagnose and treat patients remotely, especially in underdeveloped areas.

Research recommendations: Although a lot of progress has been made in understanding how airborne pollutants affect human health, there are still a number of questions that require more research. These territories consist of:

Impacts on long-term health: Long-term cohort studies are being conducted to evaluate the possible carcinogenicity and other systemic health hazards as well as the chronic health impacts of prolonged exposure to particular pollutants.

Multiple pollutants interacting: Considering that people are frequently exposed to a variety of particles, gases, and biological agents at the same time, it is important to examine the synergistic effects of exposure to various airborne pollutants.

Vulnerable populations: Analyzing how susceptible vulnerable groups – like kids, the elderly, and people with established medical conditions – are to various toxins and creating specialized intervention plans.

Investigating the intricate microbial communities found in indoor environments and how they interact with airborne pollutants in order to better understand how these communities may affect the emergence of immunological and respiratory diseases.

Studying efficient ways to inform the public about the risks posed by airborne pollutants, create awareness, and encourage behavior adjustments to reduce exposure.

Insights gained from ongoing research in these areas will help improve diagnosis, treatment, and prevention by shedding light on the mechanisms of action, health impacts, and best management practices for airborne pollutants.

To sum up, future directions and research in the area of diagnosing and treating health effects brought on by airborne contaminants seek to take advantage of new technologies, close knowledge gaps, and create efficient interventions to safeguard public health and ensure healthier indoor environments (Baek et al. 2018; Gao et al. 2020).

15.14 Conclusion

It is critical to identify and treat health effects brought on by airborne contaminants in order to protect both the environment and public health. The definition of airborne contaminants and their sources, a summary of health effects, common airborne contaminants like particulate matter, gases and fumes, allergens, and biological agents, as well as the process of diagnosing health effects, have all been covered in this chapter thorough overview of this important subject.

For an accurate diagnosis and successful treatment, it is essential to have a thorough awareness of the health impacts linked with airborne pollutants. Among the organ systems that are frequently harmed by exposure to airborne pollutants are the respiratory system, cardiovascular system, skin and eye, and nervous system. A detailed diagnostic process encompassing patient history, physical examination, diagnostic tests, and treatments is conducted after recognizing the connected symptoms.

Treatment strategies include reducing exposure through source management, ventilation systems, and individual protective equipment. While lifestyle modifications and preventive actions are essential in reducing exposure risks, medications and therapies are used to treat symptoms and control underlying disorders. For people harmed by airborne pollutants, supportive care and immunotherapy also help improve outcomes.

The health impacts of particular airborne pollutants such as asbestos, mold, radon, VOCs, pollen, and other allergens, as well as the accompanying diagnostic and therapeutic techniques, are also covered in this chapter. Effects on occupational health as well as special considerations for vulnerable groups like children, the elderly, and people with pre-existing diseases are also covered.

For the maintenance of a healthy interior environment, IAQ evaluation and improvement tactics, such as IAQ monitoring, ventilation systems, and filtration and purification techniques, are crucial. Regulations and standards guarantee the application of safety precautions to shield people from airborne pollutants.

Future technologies and developments have the potential to lead to better preventative, treatment, and diagnostic methods. Long-term health impacts, the combination of various pollutants, vulnerable populations, the indoor microbiome, and effective risk communication all require further research.

In conclusion, a thorough and multidisciplinary approach is necessary for the identification and treatment of health problems brought on by airborne pollutants. Healthcare professionals, policymakers, and individuals may all work together to reduce exposure dangers, promote healthy indoor settings, and guarantee the population's well-being by staying informed, taking the right actions, and performing additional research.

References

AAAAI. (2021). Tips to remember: indoor allergens. https://www.aaaai.org/conditions-and-treatments/library/allergy-library/indoor-allergens-tips.

American Academy of Pediatrics (2018), 141(1), 0–762.

American Cancer Society. (2022). Asbestos and cancer risk. https://www.cancer.org/cancer/cancer-causes/asbestos.html.

Annesi-Maesano, I., Bonay, M., & Charles, M. A. (2012). Future challenges in occupational asthma research. *European Respiratory Review*, 21(124), 147–153.

Baek H, Cho M, Kim S, Hwang H, Song M, Yoo S. (2018) Analysis of length of hospital stay using electronic health records: a statistical and data mining approach. *PLoS One* 2018 13(4):e0195901. doi: https://doi.org/10.1371/journal.pone.0195901.

Ball, J.W., Dains, J.E., Flynn, J.A. et al. (2019). *Seidel's Guide to Physical Examination*, 9e. Elsevier.

Bernstein, J.A., Alexis, N., Bacchus, H. et al. (2021). The health effects of nonindustrial indoor air pollution. *Journal of Allergy and Clinical Immunology* 147 (5): 1592–1607.

Brook, R.D., Rajagopalan, S., Pope, C.A. 3rd et al. (2010). Particulate matter air pollution and cardiovascular disease: an update to the scientific statement from the American Heart Association. *Circulation* 121 (21): 2331–2378. https://doi.org/10.1161/CIR.0b013e3181dbece1.

CDC (Centers for Disease Control and Prevention). (2021). Indoor air quality: contaminants. https://www.cdc.gov/niosh/topics/indoorenv/contaminants.html.

CDC. (2022). Mold: basic facts. https://www.cdc.gov/mold/faqs.htm.

EPA. (2021). Volatile organic compounds (VOCs). https://www.epa.gov/indoor-air-quality-iaq/volatile-organic-compounds-impact-indoor-air-quality.

EPA. (2022). A citizen's guide to radon: the guide to protecting yourself and your family from radon. https://www.epa.gov/radon/citizens-guide-radon-guide-protecting-yourself-and-your-family-radon.

Gao, J., Zheng, P., Jia, Y. et al. (2020). Mental health problems and social media exposure during COVID-19 outbreak. *Plos One* 15 (4): e0231924. https://doi.org/10.1371/journal.pone.0231924.

Ghio, A.J. and Soukup, J.M. (2019). Cardiovascular effects of inhaled ambient air particulate matter: a critical review. *Journal of Toxicology and Environmental Health, Part A* 82 (1): 1–26.

Holgate, S.T., Samet, J.M., Koren, H.S. et al. (2015). Health effects of air pollution. *Journal of Allergy and Clinical Immunology* 136 (3): 771–781.

Kelly, F.J. and Fussell, J.C. (2011). Air pollution and public health: emerging hazards and improved understanding of risk. *Environment International* 43 (1): 117–121. https://doi.org/10.1016/j.envint.2011.09.012.

Mayo Clinic. (2021). Allergies. https://www.mayoclinic.org/diseases-conditions/allergies/symptoms-causes/syc-20351497.

National Institute for Occupational Safety and Health. (2022). Indoor environmental quality. https://www.cdc.gov/niosh/topics/indoorenv/.

National Institute for Occupational Safety and Health. (2024). Occupational exposure limits. https://www.cdc.gov/niosh/topics/oels/default.html.

NIOSH (National Institute for Occupational Safety and Health) (2018). Chapter A2: physical examination. In: *NIOSH Manual of Analytical Methods (NMAM)*, 5e. https://www.cdc.gov/niosh/nmam/pdfs/A2.pdf.

Occupational Safety and Health Administration (OSHA) (2011). Indoor air quality. https://www.osha.gov/indoor-air-quality.

Occupational Safety and Health Administration. (2021). Indoor air quality (IAQ) in commercial and institutional buildings. https://www.osha.gov/SLTC/indoorairquality/.

Pawankar, R., Canonica, G.W., and Holgate, S.T. (2013). Allergic diseases and asthma: a major global health concern. *World Allergy Organization Journal* 6 (1): 1–2. https://doi.org/10.1186/1939-4551-6-1.

Peden, D.B., Reed, C.E., and Browning, S. R. (Eds.). (2016). *The Clinician's Guide to Asthma*, 1e. CRC Press.

U.S. Environmental Protection Agency. (2020). Indoor air quality (IAQ). https://www.epa.gov/indoor-air-quality-iaq.

WHO (World Health Organization). (2018). Ambient air pollution: a global assessment of exposure and burden of disease. https://www.who.int/publications-detail/9789241511353.

WHO. (2014). Asbestos: elimination of asbestos-related diseases. https://www.who.int/news-room/fact-sheets/detail/asbestos-elimination-of-asbestos-related-diseases.

World Health Organization (WHO). (2013). Health effects of particulate matter: policy implications for countries in Eastern Europe, Caucasus and Central Asia. https://www.euro.who.int/__data/assets/pdf_file/0005/189051/Health-effects-of-particulate-matter-final-Eng.pdf.

World Health Organization (2010). *WHO Guidelines for Indoor Air Quality: Selected Pollutants*. World Health Organization.

World Health Organization. (2021). Indoor air pollution and health. https://www.who.int/news-room/fact-sheets/detail/indoor-air-pollution-and-health.

Zhang, J.J., McCreanor, J.E., Cui, X., and Zhang, Q. (2012). Cardiovascular and respiratory responses to short-term exposure to traffic-related air pollution with and without moderate physical activity. *Journal of Exposure Science & Environmental Epidemiology* 22 (5): 544–552. https://doi.org/10.1038/jes.2012.42.

16

Strategies for Reducing Exposure to Airborne Biological Contaminants in Indoor Environments

Maneet Kumar Chakrawarti, Madhuri Singh, and Kasturi Mukhopadhyay

Antimicrobial Research Laboratory, School of Environmental Sciences, Jawaharlal Nehru University, New Delhi, India

16.1 Introduction

In modern interconnected world, airborne biological contaminants pose a significant threat to public health (Humbal et al. 2018; Moelling and Broecker 2020). These airborne biological particles, ranging from viruses and bacteria to allergens and fungal spores, can be transmitted through the air we breathe (Lacey and West 2006; Mandal and Brandl 2011). The airborne bio-contaminants are of variable sizes ranging from order of 0.003 μm for viruses (Taylor 1988), from 0.25 to 20 μm for bacteria (Thompson 1981), from 17 to 58 μm for plant pollens (Stanley and Linskins 1974), and from 1 to 30 μm for fungi (Gregory 1973), and are respirable in nature. Indoor Air Quality (IAQ) is a crucial aspect of human health and well-being as people spend a significant portion of their time indoors (Mannan and Al-Ghamdi 2021). The main sources of indoor biological contaminants may include humans; pets; plants; plumbing systems; heating, ventilation, and air-conditioning (HVAC) systems; mold; resuspension of settled dust; and outdoor air (Prussin and Marr 2015). The indoor aerobiological contaminants can be affected by temperature, relative humidity, air exchange rate, and occupant density (Kembel et al. 2012; Meadow et al. 2014; Fujiyoshi et al. 2017).

The quality of the air we breathe within our homes and buildings can have a profound impact on our health (Burger 1990; Mandal and Brandl 2011; Kim et al. 2018; Kausar et al. 2021). Exposure to airborne biological contaminants can lead to a spectrum of health risks particularly related to respiratory system (Douwes et al. 2003; Lee et al. 2006; Peccia et al. 2011). Respiratory infections,

Airborne Biocontaminants and Their Impact on Human Health, First Edition.
Edited by Rajeev Singh and Anamika Singh.

such as the common cold, flu, and severe respiratory syndromes like COVID-19, building-related illnesses, and even cancer can result from inhalation of contaminated air (Eccles 2005; Kim et al. 2018; Flerlage et al. 2021; Nath et al. 2023). Allergens, such as pollen and fungal spores, trigger allergic reactions, while long-term exposure to indoor air pollutants can exacerbate chronic respiratory conditions like asthma (Lacey 1991; Bergmann et al. 2017; Menzel et al. 2017). The exposure to aerobiological contaminants is primarily initiated by inhalation and further leads to its deposition into the different regions of the respiratory tract depending on particle size, lung morphology, and breathing characteristics (Nazaroff 2016). Vulnerable populations, including children, the elderly, and those with compromised immune systems, are at greater risk (Nazaroff 2016).

Effective strategies to minimize exposure to aerobiological contaminants are paramount to safeguarding individual and community well-being (Hedworth et al. 2021; Izadyar and Miller 2022). This chapter explores a range of proactive measures designed to mitigate the risks associated with aerobiological contaminants, from engineering controls to personal protection and beyond. Furthermore, we also discuss air decontamination techniques to reduce airborne bio-contaminants along with key challenges in controlling aerobiological contaminants.

16.2 Engineering Controls and Real-Time Air Quality Monitoring

Engineering controls encompass air ventilation, filtration, and purification system of a built environment. These strategies along with real-time monitoring of airborne biological contaminants could significantly reduce its exposure to the occupants.

16.2.1 Ventilation Strategies

Proper ventilation is crucial in maintaining good IAQ by minimizing airborne contaminants. Proper air exchange rates, coupled with well-maintained ventilation systems, ensure the dilution and removal of contaminants. Natural ventilation, like opening windows, can contribute to better air circulation and is a simple and historical method to reduce indoor airborne bio-contaminants (Nazaroff 2016). Mechanical ventilation systems, such as HVAC systems, ensure regular air exchange, diluting and removing contaminants (Kembel et al. 2012; Dai et al. 2021).

In hospitals, mechanical ventilation systems may reduce risk of healthcare-associated infections and improve IAQ within hospitals (Stockwell et al. 2019). The main ventilation system along with local ventilation could prevent spread of airborne bio-contaminant-related infections in high-risk areas in hospitals

(Kek et al. 2023). Negative pressure rooms are also used in healthcare facilities to isolate patients with contagious airborne diseases such as SARS-CoV-2 (Bhagat and Linden 2020; Dumont-Leblond et al. 2020). In these rooms, air is drawn toward inside, preventing contaminants from escaping into surrounding areas.

16.2.2 Air Filtration and Purification

Preventing infiltration and transmission of airborne infectious particles in built environments are principal challenges for mitigation and infection control (Fernstrom and Goldblatt 2013). High-efficiency particulate air (HEPA) filters play a crucial role in trapping airborne particles, including biological contaminants (Dai et al. 2021). Additional air purification technology, such as UV-C germicidal irradiation, offers supplementary layers of protection against airborne biological contaminants (Beggs et al. 2006; Beggs and Avital 2020).

16.2.3 Real-Time Indoor Air Quality Monitoring

Traditional microbiological colony count methods for enumerating airborne contaminants could be laborious. So, we need time-efficient methodologies to monitor these airborne contaminants. Real-time monitoring of IAQ enables proactive responses to changing conditions (Taştan and Gökozan 2019). Sensor-equipped systems can detect particulate matter, volatile organic compounds (VOCs), and microbial contaminants in the indoor air (Zhang et al. 2021; Arabi et al. 2022; Nnachi et al. 2022). Zhang and coworkers developed a Python-powered Raspberry Pi platform for low-cost air quality system (LCAQS) integrated with sensors, which measure, analyze, and determine IAQ by detecting critical indoor air pollutants (Zhang et al. 2021). Biosensors could rapidly detect bacterial pathogens in the environment as compared to culture-based methods by utilizing the lock-and-key principle of enzyme-complex in the recognition of analytes (Nnachi et al. 2022). Moreover, artificial intelligence-based methods using data from IAQ sensors could effectively predict concentrations of airborne bio-contaminants in real-time and near-future ($\leq$60 minutes) basis in built environments (Lee et al. 2023). By promptly addressing elevated contaminant levels, occupants can take measures to reduce exposure.

16.3 Personal Protective Strategies

Exposure to airborne contaminants can be avoided using personal protection including using respiratory, eye, and face protective equipment, particularly in high-risk settings such as healthcare facilities.

16.3.1 Respiratory Protection

N95 and other similar respirators provide a physical barrier against inhalation of airborne particles (Allen and Barn 2020; Cheng et al. 2020). Respiratory protection is especially crucial for healthcare workers and individuals in high-risk environments. Community-wide mask-wearing may contribute to the control of COVID-19 by reducing the amount of emission of saliva and respiratory droplets from infected individuals (Cheng et al. 2020).

16.3.2 Eye and Face Protection

While respiratory protection primarily addresses inhalation exposure, eye and face protection guards against mucous membrane contact (Chu et al. 2020). Safety goggles and face shields provide a barrier against aerosols that may come into contact with the eyes, nose, or mouth (Roberge 2016; Hall et al. 2023).

16.4 Behavioral and Hygiene Practices

Behavioral and good hygiene practices are cost-effective strategies for reducing exposure to airborne biological particles in indoor environments. These practices help minimize the generation of airborne biological contaminants from the source and further restrict their transmission in indoor environments.

16.4.1 Respiratory Etiquette and Personal Hygiene

Promoting proper respiratory etiquette, including covering coughs and sneezes with a tissue or the elbow, reduces the dispersal of respiratory droplets laden with contaminants (Chavis and Ganesh 2020). These respiratory etiquettes coupled with immediate hand hygiene limit the transmission of infectious agents (Mathur 2011; Gozdzielewska et al. 2022; Nath et al. 2023).

16.4.2 Minimizing Contaminant Generation

The key sources of indoor bio-contaminants are humans; pets; plants; plumbing systems; HVAC systems; mold; resuspension of settled dust; and outdoor air (Prussin and Marr 2015). Controlling temperature, relative humidity, air exchange rate, and occupant density could minimize the generation of these contaminants in built environments (Kembel et al. 2012; Meadow et al. 2014; Fujiyoshi et al. 2017).

16.5 Public Health Strategies

The recent coronavirus pandemic has heightened the debate on public health, emphasizing the need for effective disease outbreak management and further educating individuals and communities to collectively take steps to curb the spread of airborne pathogens.

16.5.1 Disease Outbreak Preparedness

Effective response to disease outbreaks involving airborne contaminants requires well-defined plans for quarantine, isolation, and contact tracing along with source control measures that can limit the spread of infectious agents (Tang et al. 2006; Agarwal et al. 2021). A recent study reported that optimized ventilation strategies could decrease the incidence of airborne SARS-CoV-2 in hospital settings (Dumont-Leblond et al. 2020).

Antibiotic-resistant bacteria cost hundreds of thousands of annual lives worldwide (World Health Organization 2014). The antibiotic resistance genes dissemination in the environment could potentially affect public health (Li et al. 2018; Yang et al. 2018). Larsson and Flach (2022) reviewed methods for surveillance, potential drivers, and mitigation strategies for antibiotic pollution, resistance evolution, and transmission (Larsson and Flach 2022). A study conducted in a university library and an animal house facility showed a high prevalence of drug-resistant airborne staphylococci, particularly during the monsoon season (Kumari et al. 2020).

16.5.2 Communication and Education

Public awareness campaigns play a central role in informing the public about the risks associated with airborne biological contaminants (Ramírez et al. 2019). Clear communication from health authorities empowers individuals to take preventive actions, fostering a collective effort to reduce exposure (Mach et al. 2021).

16.6 Current Air Decontamination Techniques to Reduce Airborne Biological Contaminants

Air decontamination techniques to reduce airborne biological contaminants have become increasingly important, especially in healthcare settings and high-risk environments. These techniques aim to remove or inactivate airborne pathogens, allergens, and other biological contaminants, thereby enhancing IAQ and lowering the risk of infections. Here are some current air decontamination techniques and technologies:

16.6.1 High-Efficiency Particulate Air Filtration

HEPA filters are highly efficient at trapping particles as small as 0.3 μm, which includes most bacteria and viruses (Dai et al. 2021). These filters are commonly used in air purifiers and HVAC systems for controlling airborne biological contaminants in indoor environment (Azimi and Stephens 2013). A study demonstrated that a HEPA filtration system could protect immunocompromised patients from aspergillosis in a hematologic oncology patient care unit of a hospital (Hahn et al. 2002).

16.6.2 Ultraviolet Germicidal Irradiation

Ultraviolet Germicidal Irradiation (UVGI) systems use ultraviolet (UV) light to disinfect the air by inactivating microorganisms, including bacteria and viruses (Beggs et al. 2006; Beggs and Avital 2020). UV-C light at specific wavelengths (260–270 nm) is effective for inactivating microbes by damaging their nucleic acids. UVGI is often used in HVAC systems, air purifiers, and germicidal lamps. A study found that UV irradiation has been reported to be effective in inactivating SARS-CoV-2 by impeding viral replication in a highly controlled experimental model (Biasin et al. 2021). Furthermore, a recent study also reported that a UVGI system in an occupied classroom can noticeably disinfect airborne viruses, achieving up to 96% reduction, which is as effective as increasing ventilation rate by more than five times (Park et al. 2022).

The hybrid effect of ultraviolet irradiation and thermal energy has a greater inactivation efficiency on bacterial bioaerosols compared to individual treatments. For example, simultaneous exposure to UV irradiation and thermal energy for less than 1.05 seconds inactivates over 99% of *S. epidermidis* bioaerosols at 120 °C using a single UV lamp in a hybrid UV-thermal energy system (Hwang et al. 2010). A study conducted in a hospital reported that no-touch decontamination methods, including UV devices and hydrogen peroxide systems, may reduce nosocomial pathogens in patient rooms (Weber et al. 2016).

16.6.3 Photocatalytic Oxidation

Photocatalytic oxidation (PCO) systems could help maintain IAQ (Yu and Kim 2013). PCO systems use a catalyst such as Titanium dioxide (TiO_2) and UV light to generate hydroxyl radicals and other reactive oxygen species capable of destroying organic contaminants, including airborne pathogens (Foster et al. 2011; Mamaghani et al. 2017). PCO is used in air purifiers and some HVAC systems (Mamaghani et al. 2017; Wu et al. 2022).

16.6.4 Electrostatic Precipitation

Electrostatic precipitation is a promising technique to inactivate or kill airborne bio-contaminants (Li and Wen 2003). Electrostatic precipitators use an electric charge to remove particles from the air (Preston et al. 2023). These systems can capture larger particles, including some biological contaminants, by imparting an electric charge and attracting them to plates or surfaces (Redmann et al. 2022). A well-designed electrostatic disinfector can effectively inactivate airborne bio-contaminants with a disinfection efficiency of up to 100% (Feng et al. 2021b). Higher relative humidity enhances the effectiveness of an electrostatic precipitator (Li and Wen 2003).

16.6.5 Portable Air Purifiers

The recent COVID-19 pandemic necessitates the use of portable air purifiers even in the household settings. These compact air purifiers are equipped with HEPA filters, UV-C lamps, or other technologies and are used in homes, offices, and healthcare settings to reduce the concentration of airborne pathogens (Cheng et al. 1998; Chen et al. 2010; Feng et al. 2021a; Castellini Jr et al. 2022). A study reported that the use of portable air cleaners reduced average aerosol exposure by 31–66% relative to the baseline case in a conference room with thermal stratification (Castellini Jr et al. 2022). Furthermore, a systematic review concluded that portable HEPA Purifiers have potential to eliminate airborne SARS-CoV-2 from indoor environments and complement primary decontamination strategies such as ventilation (Liu et al. 2022).

16.6.6 HVAC System Upgrades

Upgrading HVAC systems combined with improved filtration, UVGI, or other air decontamination technologies can help reduce the spread of airborne biological contaminants and further improve IAQ (Goswami 2003; de Robles and Kramer 2017; Swamy 2021).

It is important to note that the effectiveness of these air decontamination techniques can vary based on factors such as the type of contaminant, the airflow patterns, and the specific technology used. Additionally, a combination of strategies, including improved ventilation, filtration, and infection control practices, is often recommended for comprehensive air decontamination in high-risk environments such as healthcare settings. The selection of the appropriate air decontamination technique should be based on the specific needs and constraints of the indoor space.

16.7 Challenges in Controlling Exposure to Airborne Biological Contaminants

Investigating airborne bio-contaminants in the built environment is a challenging task due to the selection of appropriate sampling methods, the lack of efficient source-tracking methodologies, and further correlation with human health implications (Kumar et al. 2021).

Controlling exposure to airborne biological contaminants poses several challenges, primarily due to the complex nature of these bio-contaminants and their sources. Airborne biological contaminants encompass allergens, pathogens, and toxins (Lacey and West 2006; Mandal and Brandl 2011). Each type of bio-contaminant requires a different mitigation approach for its control, which can include filtration, cleaning and disinfection, or isolation measures (Recer 2004; Pertegal et al. 2023). Indoor bio-contaminants can originate from a wide range of sources, including humans; pets; plants; plumbing systems; HVAC systems; mold; resuspension of settled dust; and outdoor air (Prussin and Marr 2015). Extensive research is needed to assess exposure to airborne biological contaminants, as well as to standardize protocols for air sampling and develop reliable diagnostic tests for airborne infections (Kumar et al. 2021; Nath et al. 2023). A recent study reported that liquid-based air samplers can introduce bias in determining bacterial diversity due to re-aerosolization and evaporative loss of collection liquid, which can lead to under- or overestimation of the prevalence of certain bacterial genera (Lemieux et al. 2019).

In addition, seasonal and environmental variations can affect levels of airborne bio-contaminants. For instance, pollen levels can spike during specific seasons, and fungal growth may be more prevalent in damp conditions (Singh and Babu 1982; Wang et al. 2016). The airborne bio-contaminants in public restrooms and office buildings showed diurnal and seasonal variations (Bonetta et al. 2010; Lee et al. 2012). The study evaluating bacterial bioaerosol within the university library showed a high prevalence of Gram-positive bacteria in all seasons and found the maximum bacterial load during the monsoon compared to the pre-monsoon and post-monsoon periods (Kumari et al. 2020). Airborne bacterial communities in a university health center showed a significant correlation with both particulate matter and temperature (Chakrawarti et al. 2020). Environmental factors (temperature and relative humidity) can affect the survival and transmission of viruses in indoor environments (Guo et al. 2021). The infectivity of coronavirus is reduced at higher temperatures, while low relative humidity in dry environments promotes viral survival and transmission (Guo et al. 2021).

Moreover, poor building design and inadequate ventilation further deteriorate indoor microbiological air quality (Kembel et al. 2012; Izadyar and Miller 2022). A previous study showed a greater prevalence of mycotoxins-producing molds in settling dust samples collected from different water-damaged buildings (Bloom et al. 2009). Continuous exposure to indoor molds may lead to sick building syndrome or building-related illness (Crook and Burton 2010). Significant seasonal and daily variation was reported in airborne microorganisms in an Italian office building equipped with an HVAC system (Bonetta et al. 2010). Kembel and coworkers quantified the airborne bacterial community composition in patient rooms under different ventilation conditions using next-generation DNA sequencing. They found relationships between indoor bacterial diversity and building attributes, primarily the source of ventilation air, airflow rates, relative humidity, and temperature (Kembel et al. 2012).

Controlling infectious indoor bio-contaminants is a major public health challenge that demands rapid and coordinated responses, including quarantine measures, contact tracing, and vaccination campaigns (Tang et al. 2006; Astuti and Ysrafil 2020; Agarwal et al. 2021). Airborne microorganisms showing resistance to antimicrobials and carrying antibiotic-resistance genes in indoor environments further augment the challenges faced by public health mitigation strategies (Ben Maamar et al. 2020; Kumari et al. 2020). Lack of awareness regarding health risks associated with airborne bio-contaminants, along with financial and technical resource limitations to implement effective control measures, further exacerbate the situation (James et al. 2018).

Addressing these challenges requires a multidisciplinary approach that involves public health authorities, environmental scientists, engineers, healthcare professionals, policymakers, and the community. In-depth research, education, and the development of innovative technologies and strategies are essential to control exposure to airborne biological contaminants effectively.

16.8 Conclusion

The battle against airborne biological contaminants in indoor environments requires a multifaceted approach. By understanding the different types of airborne bio-contaminants, their sources, and transmission, and by implementing engineering controls such as improved air ventilation and filtration systems, individuals and communities can effectively reduce exposure risks.

Additionally, practicing effective personal protection, like wearing N95 respirators, and adopting health-promoting behaviors, could minimize disease transmission. These strategies, when combined with public health measures like quarantine, isolation, contact tracing, and source control, are especially effective in limiting the spread of infectious airborne microorganisms, particularly in high-risk environments.

Furthermore, informed decision-making involving public health authorities, environmental scientists, engineers, healthcare professionals, policymakers, and the community is integral to a comprehensive defence against airborne bio-contaminants. Elaborative research and the development of innovative strategies are necessary to effectively reduce exposure to airborne biological contaminants in indoor environments.

References

Agarwal, N., Meena, C.S., Raj, B.P. et al. (2021). Indoor air quality improvement in COVID-19 pandemic. *Sustainable Cities and Society* 70: 102942.

Allen, R.W. and Barn, P. (2020). Individual-and household-level interventions to reduce air pollution exposures and health risks: a review of the recent literature. *Current Environmental Health Reports* 7: 424–440.

Arabi, M., Alghamdi, M., Kabel, K. et al. (2022). Detection of volatile organic compounds by using MEMS sensors. *Sensors* 22 (11): 4102.

Astuti, I. and Ysrafil (2020). Severe acute respiratory syndrome coronavirus 2 (SARS-CoV-2): an overview of viral structure and host response. *Diabetes & Metabolic Syndrome: Clinical Research & Reviews* 14 (4): 407–412.

Azimi, P. and Stephens, B. (2013). HVAC filtration for controlling infectious airborne disease transmission in indoor environments: predicting risk reductions and operational costs. *Building and Environment* 70: 150–160.

Beggs, C.B. and Avital, E.J. (2020). Upper-room ultraviolet air disinfection might help to reduce COVID-19 transmission in buildings: a feasibility study. *PeerJ* 8: e10196.

Beggs, C.B., Noakes, C.J., Sleigh, P.A. et al. (2006). Methodology for determining the susceptibility of airborne microorganisms to irradiation by an upper-room UVGI system. *Journal of Aerosol Science* 37 (7): 885–902.

Ben Maamar, S., Hu, J., and Hartmann, E.M. (2020). Implications of indoor microbial ecology and evolution on antibiotic resistance. *Journal of Exposure Science & Environmental Epidemiology* 30 (1): 1–15.

Bergmann, K.C., Sehlinger, T., Gildemeister, J., and Zuberbier, T. (2017). A novel experimental technology for testing efficacy of air purifiers on pollen reduction. *Allergo Journal* 26: 20–25.

Bhagat, R.K. and Linden, P.F. (2020). Displacement ventilation: a viable ventilation strategy for makeshift hospitals and public buildings to contain COVID-19 and other airborne diseases. *Royal Society Open Science* 7 (9): 200680.

Biasin, M., Bianco, A., Pareschi, G. et al. (2021). UV-C irradiation is highly effective in inactivating SARS-CoV-2 replication. *Scientific Reports* 11 (1): 6260.

Bloom, E., Nyman, E., Must, A. et al. (2009). Molds and mycotoxins in indoor environments – a survey in water-damaged buildings. *Journal of Occupational and Environmental Hygiene* 6 (11): 671–678.

Bonetta, S.A., Bonetta, S.I., Mosso, S. et al. (2010). Assessment of microbiological indoor air quality in an Italian office building equipped with an HVAC system. *Environmental Monitoring and Assessment* 161: 473–483.

Burger, H. (1990). Bioaerosols: prevalence and health effects in the indoor environment. *Journal of Allergy and Clinical Immunology* 86 (5): 687–701.

Castellini, J.E. Jr., Faulkner, C.A., Zuo, W. et al. (2022). Assessing the use of portable air cleaners for reducing exposure to airborne diseases in a conference room with thermal stratification. *Building and Environment* 207: 108441.

Chakrawarti, M.K., Singh, M., Yadav, V.P., and Mukhopadhyay, K. (2020). Temporal dynamics of air bacterial communities in a university health centre using illumina MiSeq sequencing. *Aerosol and Air Quality Research* 20 (5): 966–980.

Chavis, S. and Ganesh, N. (2020). Respiratory hygiene and cough etiquette. *Infection Control in the Dental Office: A Global Perspective* 18: 91–103.

Chen, C., Zhao, B., Cui, W. et al. (2010). The effectiveness of an air cleaner in controlling droplet/aerosol particle dispersion emitted from a patient's mouth in the indoor environment of dental clinics. *Journal of the Royal Society Interface* 7 (48): 1105–1118.

Cheng, Y.S., Lu, J.C., and Chen, T.R. (1998). Efficiency of a portable indoor air cleaner in removing pollens and fungal spores. *Aerosol Science and Technology* 29 (2): 92–101.

Cheng, V.C.C., Wong, S.C., Chuang, V.W.M. et al. (2020). The role of community-wide wearing of face mask for control of coronavirus disease 2019 (COVID-19) epidemic due to SARS-CoV-2. *Journal of Infection* 81 (1): 107–114.

Chu, D.K., Akl, E.A., Duda, S. et al. (2020). Physical distancing, face masks, and eye protection to prevent person-to-person transmission of SARS-CoV-2 and COVID-19: a systematic review and meta-analysis. *The Lancet* 395 (10242): 1973–1987.

Crook, B. and Burton, N.C. (2010). Indoor moulds, sick building syndrome and building related illness. *Fungal Biology Reviews* 24 (3–4): 106–113.

Dai, R., Liu, S., Li, Q. et al. (2021). A systematic review and meta-analysis of indoor bioaerosols in hospitals: The influence of heating, ventilation, and air conditioning. *Plos One* 16 (12): e0259996.

Douwes, J., Thorne, P., Pearce, N., and Heederik, D. (2003). Bioaerosol health effects and exposure assessment: progress and prospects. *Annals of Occupational Hygiene* 47 (3): 187–200.

Dumont-Leblond, N., Veillette, M., Mubareka, S. et al. (2020). Low incidence of airborne SARS-CoV-2 in acute care hospital rooms with optimized ventilation. *Emerging Microbes & Infections* 9 (1): 2597–2605.

Eccles, R. (2005). Understanding the symptoms of the common cold and influenza. *The Lancet Infectious Diseases* 5 (11): 718–725.

Feng, Y., Zhao, J., Spinolo, M. et al. (2021a). Assessing the filtration effectiveness of a portable ultraviolet air cleaner on airborne SARS-CoV-2 laden droplets in a patient room: a numerical study. *Aerosol and Air Quality Research* 21 (5): 200608.

Feng, Z., Cao, S.J., Wang, J. et al. (2021b). Indoor airborne disinfection with electrostatic disinfector (ESD): numerical simulations of ESD performance and reduction of computing time. *Building and Environment* 200: 107956.

Fernstrom, A. and Goldblatt, M. (2013). Aerobiology and its role in the transmission of infectious diseases. *Journal of Pathogens* 2013: 1–13.

Flerlage, T., Boyd, D.F., Meliopoulos, V. et al. (2021). Influenza virus and SARS-CoV-2: pathogenesis and host responses in the respiratory tract. *Nature Reviews Microbiology* 19 (7): 425–441.

Foster, H.A., Ditta, I.B., Varghese, S., and Steele, A. (2011). Photocatalytic disinfection using titanium dioxide: spectrum and mechanism of antimicrobial activity. *Applied Microbiology and Biotechnology* 90: 1847–1868.

Fujiyoshi, S., Tanaka, D., and Maruyama, F. (2017). Transmission of airborne bacteria across built environments and its measurement standards: a review. *Frontiers in Microbiology* 8: 2336.

Goswami, D.Y. (2003). Decontamination of ventilation systems using photocatalytic air cleaning technology. *The Journal of Solar Energy Engineering* 125 (3): 359–365.

Gozdzielewska, L., Kilpatrick, C., Reilly, J. et al. (2022). The effectiveness of hand hygiene interventions for preventing community transmission or acquisition of novel coronavirus or influenza infections: a systematic review. *BMC Public Health* 22 (1): 1–21.

Gregory, P.H. (1973). *The Microbiology of the Atmosphere*, 2e. Leonard Hall.

Guo, L., Yang, Z., Zhang, L. et al. (2021). Systematic review of the effects of environmental factors on virus inactivation: implications for coronavirus disease 2019. *International Journal of Environmental Science and Technology* 18: 2865–2878.

Hahn, T., Cummings, K.M., Michalek, A.M. et al. (2002). Efficacy of high-efficiency particulate air filtration in preventing aspergillosis in immunocompromised patients with hematologic malignancies. *Infection Control & Hospital Epidemiology* 23 (9): 525–531.

Hall, S., Johnson, P., Bailey, C. et al. (2023). Evaluation of face shields, goggles, and safety glasses as a virus transmission control measure to protect the wearer against cough droplets. *Annals of Work Exposures and Health* 67 (1): 36–49.

Hedworth, H.A., Karam, M., McConnell, J. et al. (2021). Mitigation strategies for airborne disease transmission in orchestras using computational fluid dynamics. *Science Advances* 7 (26): eabg4511.

Humbal, C., Gautam, S., and Trivedi, U. (2018). A review on recent progress in observations, and health effects of bioaerosols. *Environment International* 118: 189–193.

Hwang, G.B., Jung, J.H., Jeong, T.G., and Lee, B.U. (2010). Effect of hybrid UV-thermal energy stimuli on inactivation of *S. epidermidis* and *B. subtilis* bacterial bioaerosols. *Science of the Total Environment* 408 (23): 5903–5909.

Izadyar, N. and Miller, W. (2022). Ventilation strategies and design impacts on indoor airborne transmission: a review. *Building and Environment* 218: 109158.

James, P.T., Kunoor, A., and Rakesh, P.S. (2018). Awareness of health care workers, patients and visitors regarding air borne infection control–a descriptive study from a tertiary care centre in Kerala, southern India. *Indian Journal of Tuberculosis* 65 (2): 168–171.

Kausar, M.A., Khatoon, F., Saleem, M. et al. (2021). Bioaerosols from indoor environment: seasonal variation and impact on human health. *Bioscience Research* 18 (1): 788–794.

Kek, H.Y., Saupi, S.B.M., Tan, H. et al. (2023). Ventilation strategies for mitigating airborne infection in healthcare facilities: a review and bibliometric analysis (1993 to 2022). *Energy and Buildings* 295: 113323.

Kembel, S.W., Jones, E., Kline, J. et al. (2012). Architectural design influences the diversity and structure of the built environment microbiome. *The ISME Journal* 6 (8): 1469–1479.

Kim, K.H., Kabir, E., and Jahan, S.A. (2018). Airborne bioaerosols and their impact on human health. *Journal of Environmental Sciences* 67: 23–35.

Kumar, P., Kausar, M.A., Singh, A.B., and Singh, R. (2021). Biological contaminants in the indoor air environment and their impacts on human health. *Air Quality, Atmosphere & Health* 14 (11): 1723–1736.

Kumari, H., Chakraborti, T., Singh, M. et al. (2020). Prevalence and antibiogram of coagulase negative Staphylococci in bioaerosols from different indoors of a university in India. *BMC Microbiology* 20: 1–14.

Lacey, J. (1991). Aerobiology and health: the role of airborne fungal spores in respiratory disease. *Frontiers in Mycology* 157: 185.

Lacey, M.E. and West, J.S. (2006). *The Air Spora: A Manual for Catching and Identifying Airborne Biological Particles*, 32–33. Dordrecht: The Netherlands.

Larsson, D.J. and Flach, C.F. (2022). Antibiotic resistance in the environment. *Nature Reviews Microbiology* 20 (5): 257–269.

Lee, T., Grinshpun, S.A., Martuzevicius, D. et al. (2006). Relationship between indoor and outdoor bioaerosols collected with a button inhalable aerosol sampler in urban homes. *Indoor Air* 16 (1): 37.

Lee, B.U., Hong, I.G., Lee, D.H. et al. (2012). Bacterial bioaerosol concentrations in public restroom environments. *Aerosol and Air Quality Research* 12 (2): 251–255.

Lee, J.Y., Miao, Y., Chau, R.L. et al. (2023). Artificial intelligence-based prediction of indoor bioaerosol concentrations from indoor air quality sensor data. *Environment International* 174: 107900.

Lemieux, J., Veillette, M., Mbareche, H., and Duchaine, C. (2019). Re-aerosolization in liquid-based air samplers induces bias in bacterial diversity. *Aerosol Science and Technology* 53 (11): 1244–1260.

Li, C.S. and Wen, Y.M. (2003). Control effectiveness of electrostatic precipitation on airborne microorganisms. *Aerosol Science & Technology* 37 (12): 933–938.

Li, J., Cao, J., Zhu, Y.G. et al. (2018). Global survey of antibiotic resistance genes in air. *Environmental Science & Technology* 52 (19): 10975–10984.

Liu, D.T., Phillips, K.M., Speth, M.M. et al. (2022). Portable HEPA purifiers to eliminate airborne SARS-CoV-2: a systematic review. *Otolaryngology–Head and Neck Surgery* 166 (4): 615–622.

Mach, K.J., Salas Reyes, R., Pentz, B. et al. (2021). News media coverage of COVID-19 public health and policy information. Humanities and Social Sciences. *Communications* 8 (1): 1–11.

Mamaghani, A.H., Haghighat, F., and Lee, C.S. (2017). Photocatalytic oxidation technology for indoor environment air purification: the state-of-the-art. *Applied Catalysis B: Environmental* 203: 247–269.

Mandal, J. and Brandl, H. (2011). Bioaerosols in indoor environment-a review with special reference to residential and occupational locations. *The Open Environmental & Biological Monitoring Journal* 4 (1): 83–96.

Mannan, M. and Al-Ghamdi, S.G. (2021). Indoor air quality in buildings: a comprehensive review on the factors influencing air pollution in residential and commercial structure. *International Journal of Environmental Research and Public Health* 18 (6): 3276.

Mathur, P. (2011). Hand hygiene: back to the basics of infection control. *The Indian Journal of Medical Research* 134 (5): 611.

Meadow, J.F., Altrichter, A.E., Kembel, S.W. et al. (2014). Indoor airborne bacterial communities are influenced by ventilation, occupancy, and outdoor air source. *Indoor Air* 24 (1): 41–48.

Menzel, A., Matiu, M., Michaelis, R., and Jochner, S. (2017). Indoor birch pollen concentrations differ with ventilation scheme, room location, and meteorological factors. *Indoor Air* 27 (3): 539–550.

Moelling, K. and Broecker, F. (2020). Air microbiome and pollution: composition and potential effects on human health, including SARS coronavirus infection. *Journal of Environmental and Public Health* 2020: 1646943.

Nath, A., Baruah, N., Nonglait, M.L., and Deka, P. (2023). Biological contaminants in indoor environments of educational institutions. *Aerobiologia* 39 (1): 1–20.

Nazaroff, W.W. (2016). Indoor bioaerosol dynamics. *Indoor Air* 26 (1): 61–78.

Nnachi, R.C., Sui, N., Ke, B. et al. (2022). Biosensors for rapid detection of bacterial pathogens in water, food and environment. *Environment International* 166: 107357.

Park, S., Mistrick, R., and Rim, D. (2022). Performance of upper-room ultraviolet germicidal irradiation (UVGI) system in learning environments: Effects of ventilation rate, UV fluence rate, and UV radiating volume. *Sustainable Cities and Society* 85: 104048.

Peccia, J., Hospodsky, D., and Bibby, K. (2011). New directions: a revolution in DNA sequencing now allows for the meaningful integration of biology with aerosol science. *Atmospheric Environment* 45 (10): 1896–1897.

Pertegal, V., Riquelme, E., Lozano-Serra, J. et al. (2023). Cleaning technologies integrated in duct flows for the inactivation of pathogenic microorganisms in indoor environments: a critical review of recent innovations and future challenges. *Journal of Environmental Management* 345: 118798.

Preston, H.E., Bayliss, R., Temperton, N. et al. (2023). Capture and inactivation of viral particles from bioaerosols by electrostatic precipitation. *Iscience* 26 (9): 107567.

Prussin, A.J. and Marr, L.C. (2015). Sources of airborne microorganisms in the built environment. *Microbiome* 3: 1–10.

Ramírez, A.S., Ramondt, S., Van Bogart, K., and Perez-Zuniga, R. (2019). Public awareness of air pollution and health threats: challenges and opportunities for communication strategies to improve environmental health literacy. *Journal of Health Communication* 24 (1): 75–83.

Recer, G.M. (2004). Long-term use of high-efficiency vacuum cleaners and residential airborne fungal-spore exposure. *Aerobiologia* 20: 179–190.

Redmann, R.K., Beddingfield, B.J., Spencer, S. et al. (2022). A miniaturized electrostatic precipitator respirator effectively removes ambient SARS-CoV-2 bioaerosols. *Viruses* 14 (4): 765.

Roberge, R.J. (2016). Face shields for infection control: A review. *Journal of Occupational and Environmental Hygiene* 13 (4): 235–242.

de Robles, D. and Kramer, S.W. (2017). Improving indoor air quality through the use of ultraviolet technology in commercial buildings. *Procedia Engineering* 196: 888–894.

Singh, A.B. and Babu, C.R. (1982). Survey of atmospheric pollen allergens in Delhi: seasonal periodicity. *Annals of Allergy* 48 (2): 115–122.

Stanley, R.G. and Linskins, H.F. (1974). *Pollen: Biology, Chemistry and Management*. Berlin: Springer-Verlag.

Stockwell, R.E., Ballard, E.L., O'Rourke, P. et al. (2019). Indoor hospital air and the impact of ventilation on bioaerosols: a systematic review. *Journal of Hospital Infection* 103 (2): 175–184.

Swamy, G.S.N.V.K.S.N. (2021). Development of an indoor air purification system to improve ventilation and air quality. *Heliyon* 7 (10): e08153.

Tang, J.W., Li, Y., Eames, I. et al. (2006). Factors involved in the aerosol transmission of infection and control of ventilation in healthcare premises. *Journal of Hospital Infection* 64 (2): 100–114.

Taştan, M. and Gökozan, H. (2019). Real-time monitoring of indoor air quality with internet of things-based E-nose. *Applied Sciences* 9 (16): 3435.

Taylor, E.J. (1988). *Dorland's Medical Dictionary*. Philadelphia: W.B. Saunders Co.

Thompson, W.A.R. (1981). *Black's Medical Dictionary*, 3e. Adam and Charles Black.

Wang, X., Liu, W., Huang, C. et al. (2016). Associations of dwelling characteristics, home dampness, and lifestyle behaviors with indoor airborne culturable fungi: On-site inspection in 454 Shanghai residences. *Building and Environment* 102: 159–166.

Weber, D.J., Kanamori, H., and Rutala, W.A. (2016). 'No touch' technologies for environmental decontamination: focus on ultraviolet devices and hydrogen peroxide systems. *Current Opinion in Infectious Diseases* 29 (4): 424–431.

World Health Organization (2014). *Antimicrobial resistance: global report on surveillance*. World Health Organization.

Wu, J., Alipouri, Y., Luo, H., and Zhong, L. (2022). Ultraviolet photocatalytic oxidation technology for indoor volatile organic compound removal: a critical review with particular focus on byproduct formation and modeling. *Journal of Hazardous Materials* 421: 126766.

Yang, Y., Zhou, R., Chen, B. et al. (2018). Characterization of airborne antibiotic resistance genes from typical bioaerosol emission sources in the urban environment using metagenomic approach. *Chemosphere* 213: 463–471.

Yu, C.W. and Kim, J.T. (2013). Photocatalytic oxidation for maintenance of indoor environmental quality. *Indoor and Built Environment* 22 (1): 39–51.

Zhang, H., Srinivasan, R., and Ganesan, V. (2021). Low cost, multi-pollutant sensing system using raspberry pi for indoor air quality monitoring. *Sustainability* 13 (1): 370.

17

Guidelines for the Biological Contaminants in Indoor Areas

Pradeep Kumar[1,2], Sevaram Singh[3,4], Vinay Mohan Pathak[5,6], Rajiv Kumar Singh[7], Shiwangi Gupta[1], and Rajeev Singh[8]

[1] Department of Environmental Studies, Satyawati College, University of Delhi, Delhi, New Delhi, India
[2] Department of Agricultural and Biosystems Engineering, South Dakota State University, Brookings, SD, USA
[3] Translational Health Science and Technology Institute, NCR Biotech Science Cluster, Faridabad, Haryana, India
[4] Jawaharlal Nehru University, Delhi, New Delhi, India
[5] Department of Microbiology, University of Delhi (South Campus), Delhi, New Delhi, India
[6] Department of Botany and Microbiology, Gurukul Kangri Deemed to be University, Haridwar, Uttarakhand, India
[7] Primary Health Center (PHC), Government of Uttar Pradesh, Pali, Hardoi, Uttar Pradesh, India
[8] Department of Environmental Science, Jamia Millia Islamia, Delhi, New Delhi, India

17.1 Introduction

IAQ is a critical factor that significantly impacts human health and well-being (Kumar et al. 2023a). Among various pollutants, biological contaminants pose a significant threat to occupants of indoor spaces. These contaminants include mold, bacteria, viruses, pollen, dust mites, and other microorganisms that can cause respiratory allergies, infections, and other adverse health effects (Kumar et al. 2021b). In response to these concerns, numerous organizations and regulatory bodies have developed guidelines to assess and manage biological contaminants in indoor air environments (WHO 1988; Rao et al. 1996).

According to the Global Burden of Disease (GBD) effort, exposure to air pollution is predicted to cause 6.67 million more deaths per year due to a variety of conditions, including cancer, diabetes, cardiovascular and respiratory disorders, and other illnesses (HEI 2020). Ambient exposure to particle pollution (4.1 million yearly fatalities, or about 61% of the overall burden) and ambient ozone (O_3) exposure (0.37 million, or about 9% of the total) are responsible for a large portion of this burden. The levels of exposure to ambient air pollution (AAP) that people and populations experience are frequently calculated by integrating data from ground

Airborne Biocontaminants and Their Impact on Human Health, First Edition.
Edited by Rajeev Singh and Anamika Singh.

monitoring, pollution levels measured using satellites, and modelled concentrations using chemical transport models (Pillarisetti et al. 2022, Kumar et al. 2023a).

Numerous biological agents are present in the indoor atmosphere because of humidity and insufficient ventilation (Kumar et al. 2022). Microorganisms like mold, fungus, and bacteria thrive when there is too much moisture on practically all interior surfaces (Heseltine and Rosen 2009). These microorganisms then release spores, cells, fragments, and volatile organic compounds (VOCs) into the air in the home. Furthermore, wetness starts the chemical or biological breakdown of things, which also contaminates the air within buildings. Therefore, it has been proposed that dampness is a potent, reliable predictor of the likelihood of developing asthma and experiencing respiratory symptoms (such as coughing and wheezing). Thus, adopting moisture as the risk indicator would allow us to address the health hazards posed by biological pollutants in indoor air (Heseltine and Rosen 2009). Health risks are the consequence of a complicated series of circumstances that include water infiltration indoors, an abundance of moisture that promotes biological development, physical and chemical breakdown, and the release of potentially harmful biological and chemical agents. This series of events is followed by a discussion of the scientific data that underpins these recommendations (Heseltine and Rosen 2009).

While there are many biological agents for which there are no quantitative guidelines available, recommendations, guidance, or guidelines (all of which are referred to as "guidelines" in this text) have been developed to safeguard health and identify health risks by defining indicators of a safe indoor environment (WHO 2009). As a result, the guidelines outline issues and situations where danger is likely to materialize, but they do not offer suggestions for acceptable amounts of exposure to biological agents (WHO 2009). These recommendations are meant to assist the general public and public health agencies in identifying risks and minimizing hazards. However, they do not offer suggestions for particular measures to regulate IAQ, which must be dealt with at the local or national level, as has previously been done in certain nations (WHO 2009).

These recommendations are meant to be followed by everyone around the globe in order to safeguard public health in a variety of social, economic, and environmental contexts. They also aim to help attain ideal IAQ (WHO 2009). They emphasize construction features that guard against the development of harmful health impacts linked to moisture or mold (WHO 2009). The recommendations include all key demographic groups, are applicable to varying degrees of economic development and climates, and offer workable solutions for lowering health risks associated with moisture and microbiological contamination (WHO 2009).

This chapter provides comprehensive guidelines for managing and controlling biological contaminants in indoor air environments. It discusses the sources, types, and health effects of biological contaminants, as well as strategies for their prevention, detection, and remediation. The chapter emphasizes the importance of maintaining good IAQ and outlines practical measures to minimize exposure to

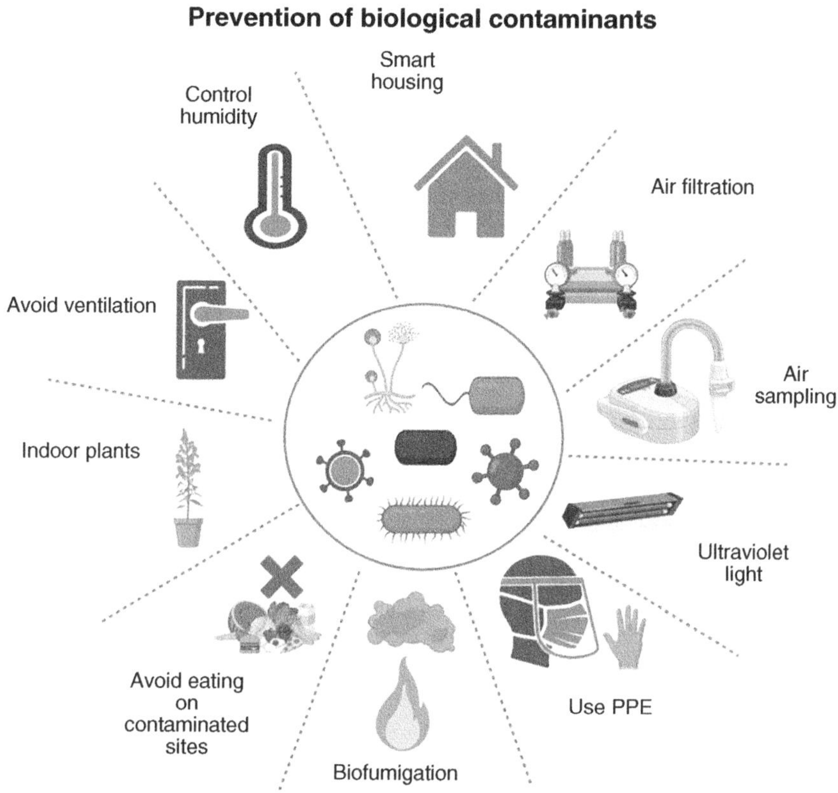

Figure 17.1 General recommendations for reducing the biological contaminants indoors. *Source:* Created with Biorender.

biological agents. It also highlights the significance of regular monitoring and assessment to ensure a healthy and comfortable indoor environment (Figure 17.1). Additionally in this chapter, the necessity for fundamental practices, essential in formulating guidelines and recommendations are discussed.

17.2 Existing Quantitative Regulations and Recommendations from Different Countries and Agencies

Standardized values for the biological pollution of indoor air were helpful to use as references to make an accurate assessment of the data that were obtained. There are no regulations or recommendations regarding the microbiological quality of indoor air at the moment. There are rules and suggestions specific to each nation. Table 17.1 provides the standards and recommendations with quantitative limits for fungal concentrations in indoor air that may be found in the literature.

Table 17.1 Summary of quantitative standards and guidelines in air by various agencies/countries.

S. no.	Recommending agency/ country	Recommendations	Reference
1	Commission of the European Communities (CEC)	Bacteria: 0 undetectable, 1–499 CFU/m^3 low, 500–999 CFU/m^3 medium, >1000 CFU/m^3 high, and 2000 CFU/m^3 very high	ECC (1993)
2	American Industrial Hygiene Association (AIHA)	1000 of viable CFU/m^3 upper limit and fungal spores <500 CFU/m^3 for residential buildings <250 CFU/m^3 for commercial buildings	IRSST (2001) and AIHA (1989)
3	Swedish requirements	500 CFU bacteria and 300 CFU fungal spores in 1 m^3	Abel et al. (2002)
4	Brazil	Airborne microorganisms (especially fungi) should be 750 CFU/m^3	de Aquino Neto and de Góes Siqueira (2000) and Nunes et al. (2005)
5	Hong Kong	Bacteria <1000 CFU/m^3 Excellent <500 CFU/m^3	Lee et al. (2002) and Li et al. (2001)
6	Taiwan	Bacteria $\leq$500 CFU/m^3 and $\leq$1000 CFU/m^3 for fungi	Huang et al. (2013) and Hsu et al. (2012)
7	India	No guideline	Katiyar (2013)
8	Singapore	Bacteria $\leq$500 CFU/m^3	Obbard and Fang (2003)
9	American Conference of Governmental Industrial Hygienists (ACGIH)	Total bacteria $\leq$500 CFU/m^3	ACGIH (1989)
10	WHO (1988)	Fungi counts $\leq$500 CFU/m^3	WHO (1988)
11	Romania	Bacteria $\leq$2500 CFU/m^3 550 CFU/m^3 for fungi In schools, bacteria <1500 CFU/m^3	Moldoveanu (2015)
12	The Korean Indoor Air Quality Control in Public Use Facilities Act	<800 CFU/m^3 for hospitals, nurseries, and public nurseries	MOE (2011) and Lee et al. (2012)

Atypical scenarios, such as hospitals and clean rooms, are not addressed since they involve specialized conditions and cannot be met by conventional building infrastructure due to the needed concentrations.

17.3 Regulatory Frameworks and Standards

Regulatory frameworks and standards play a crucial role in ensuring the reduction of biological contamination in indoor areas. They provide guidelines, requirements, and benchmarks for maintaining a healthy and safe indoor environment. Here are some key regulatory frameworks and standards relevant to this area:

Occupational Safety and Health Administration (OSHA) Standards: OSHA sets and enforces regulations to ensure safe and healthy working conditions in the United States. OSHA's General Duty Clause requires employers to provide a workplace free from recognized hazards, including biological contaminants. Specific standards, such as the Bloodborne Pathogens Standard (29 CFR 1910.1030), address the prevention of exposure to bloodborne pathogens in various occupational settings (Weil 1991).

Environmental Protection Agency (EPA) Guidelines: The EPA provides guidelines and resources for maintaining IAQ and preventing biological contamination in buildings. The IAQ Tools for Schools Program offers guidance on managing and reducing airborne contaminants in educational facilities. The EPA also provides information on mold prevention and remediation in indoor environments (Kubba 2010).

Centers for Disease Control and Prevention (CDC) Recommendations: The CDC offers guidelines and recommendations for preventing and controlling the spread of infectious diseases in various settings, including indoor environments (Chinn and Sehulster 2003). The CDC's Guidelines for Environmental Infection Control in Healthcare Facilities provide comprehensive guidance on infection prevention measures, ventilation, and cleaning protocols.

National Institute for Occupational Safety and Health (NIOSH) Guidelines: NIOSH provides guidelines and resources for protecting workers from occupational exposures to biological agents. The NIOSH publication "Guidelines for Protecting the Safety and Health of Health Care Workers" addresses infection control measures, respiratory protection, and other aspects relevant to healthcare settings (CDC 2018).

International Organization for Standardization (ISO): ISO develops international standards that provide best practices and requirements for various aspects of IAQ and contamination control. ISO 14644 series specifies cleanliness classes for cleanrooms and associated controlled environments. ISO 16890 provides guidelines for testing and classification of air filters used in general ventilation systems (WHO 2014).

Building Codes and Regulations: Local building codes often include requirements related to ventilation, air filtration, and hygiene practices in indoor areas. Codes may specify standards for HVAC systems, exhaust ventilation, and the use of appropriate materials to prevent or mitigate biological contamination.

It is essential to consult the specific regulatory authorities, standard organizations, and local building code agencies in your region or country for the most up-to-date and applicable guidelines. Compliance with these regulatory frameworks and standards helps ensure that appropriate measures are in place to reduce biological contamination and promote the health and well-being of occupants in indoor environments.

17.4 Basic Guidelines for Controlling Biological Contamination

17.4.1 Ventilation

Ensure proper ventilation in indoor spaces to dilute and remove airborne contaminants. This can be achieved through natural ventilation (e.g. opening windows) or mechanical ventilation systems. Maintaining a healthy IAQ is crucial for the well-being and comfort of occupants (WHO 2009). Proper ventilation plays a vital role in achieving this goal. Here are some ventilation practices to help maintain a healthy IAQ:

Natural ventilation: Take advantage of natural airflow by opening windows and doors when weather conditions permit. This allows fresh outdoor air to enter the space and dilute indoor pollutants.

Mechanical ventilation: Utilize mechanical ventilation systems, such as exhaust fans and supply fans, to enhance air circulation and exchange. These systems help remove stale air and introduce fresh outdoor air into the building.

Balanced ventilation: Implement a balanced ventilation system that incorporates both exhaust and supply mechanisms. This ensures a continuous exchange of air while maintaining a balance between indoor and outdoor conditions.

Air filters: Install high-quality air filters in the ventilation system to trap dust, pollen, pet dander, and other airborne particles. Regularly clean or replace the filters according to the manufacturer's instructions (Sikka and Mondal 2022).

Air purifiers: Consider using standalone air purifiers with high-efficiency particulate air (HEPA) filters. These devices help remove pollutants and allergens from the air, improving overall IAQ (Figure 17.2) (Fermo et al. 2021).

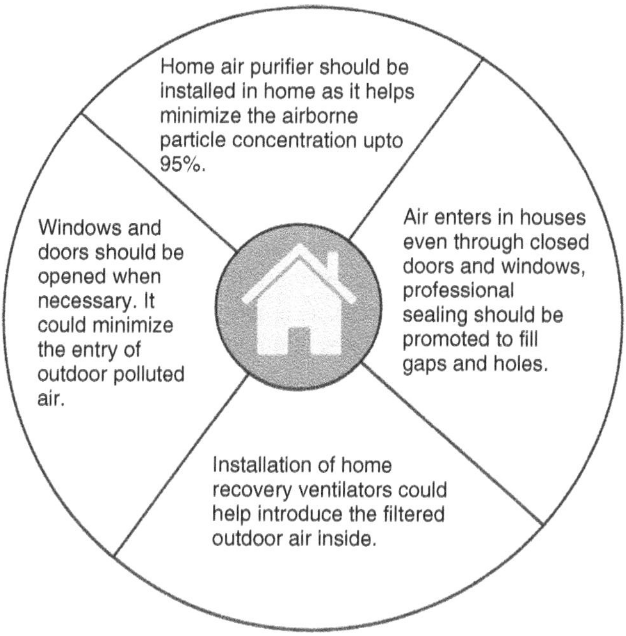

Figure 17.2 Common strategies for improving IAQ.

Regular maintenance: Maintain the ventilation system by scheduling regular inspections and cleanings. This helps prevent the buildup of dirt, dust, and mold, which can negatively impact IAQ (Kubba 2010).

Control humidity: Monitor and control humidity levels within the recommended range (usually between 30% and 50%). High humidity promotes the growth of mold and bacteria, while low humidity can lead to dryness and respiratory issues (Brambilla and Sangiorgio 2020).

Avoid pollutant sources: Identify and minimize or eliminate potential sources of indoor air pollutants, such as smoking, strong chemical cleaners, and VOCs emitted by certain building materials and furnishings (Rumchev et al. 2007).

Proper ventilation design: When designing or renovating a building, consider ventilation needs from the beginning. Ensure an adequate number of vents, properly sized ductwork, and appropriate placement of supply and exhaust openings (Nielsen et al. 2016).

Educate occupants: Raise awareness among occupants about the importance of proper ventilation and its impact on IAQ. Encourage good practices like opening windows, reporting ventilation issues promptly, and maintaining a clean living environment.

Remember that specific ventilation strategies may vary depending on the building type, climate, and local regulations. It is recommended to consult with HVAC professionals or experts in IAQ for personalized advice tailored to your specific circumstances (WHO 2009).

17.4.2 Role of Plants in Reducing the Biological Contaminants

Plants play a significant role in minimizing biological contaminants in indoor areas. Here are some ways in which plants contribute to improving IAQ and reducing biological contaminants (Figure 17.3). It is worth noting that while plants can help minimize biological contaminants in indoor areas, they are not a standalone solution. Proper ventilation, regular cleaning and maintenance, and addressing the root causes of contaminants (e.g. fixing leaks or reducing the use of chemical products) are also crucial for maintaining a healthy indoor environment (Kumar et al. 2023a).

Plants are natural air purifiers. Through a process called phytoremediation, they can absorb and break down harmful gases, such as VOCs, formaldehyde, and benzene, which are commonly found in indoor environments. Plants absorb these pollutants through their leaves and roots, converting them into harmless byproducts (Kumar et al. 2023a).

Plants release moisture through a process called transpiration, which can help regulate indoor humidity levels. Maintaining optimal humidity (between 30% and

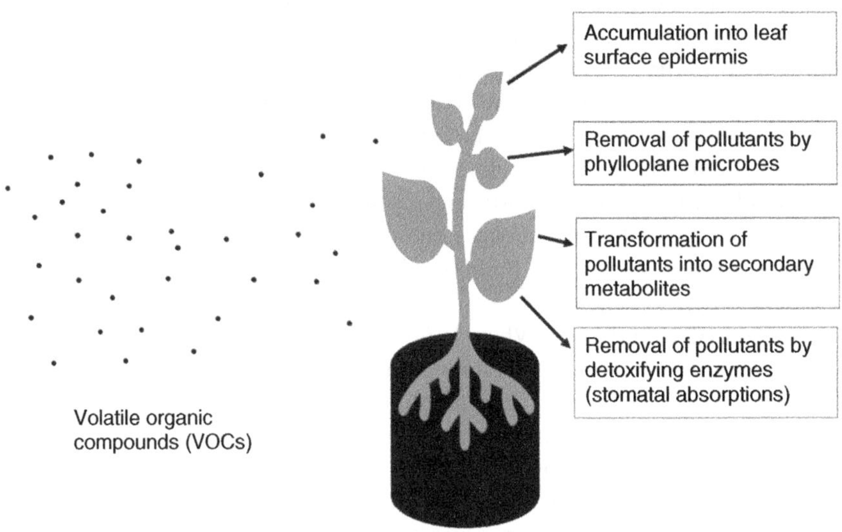

Figure 17.3 Role of plant in removal of VOCs. *Source:* Created with Biorender.com.

60%) can inhibit the growth of biological contaminants like mold, fungi, and bacteria. By increasing humidity in dry environments or decreasing it in damp environments, plants can create an inhospitable environment for these organisms (WHO 2009).

Through photosynthesis, plants absorb carbon dioxide and release oxygen into the air. Increasing oxygen levels can have a positive impact on IAQ and human health (Figure 17.3). Sufficient oxygen promotes alertness, concentration, and overall well-being while reducing the risk of airborne illnesses caused by biological contaminants (Kumar et al. 2023a).

Plants can act as natural filters, trapping airborne particulate matter, including dust, pollen, and spores. The leaves, stems, and other surfaces of plants capture these particles, preventing them from circulating in the air and reducing the potential for respiratory issues and allergies (Kumar et al. 2023a).

Indoor plants have been shown to have a calming effect on individuals and can help reduce stress and improve overall mental well-being. This can indirectly contribute to a healthier indoor environment by promoting a sense of relaxation and creating a more comfortable and pleasant living or working space (Brilli et al. 2018; Kumar et al. 2023b).

17.4.3 Reducing Pollution Sources

Regular cleaning helps minimize the presence of biological contaminants. Vacuum carpets and upholstery frequently remove dust, pet dander, and other allergens (Kumar et al. 2021b). Mop hard floors with damp cloths to capture dust particles effectively. Use microfiber or electrostatic dusters to trap and remove dust from surfaces.

Control pest infestations: Pests like cockroaches, rodents, and insects can contribute to biological contaminants. Take preventive measures to avoid infestations, such as sealing cracks and openings, storing food properly, and keeping a clean environment. If you suspect an infestation, address it promptly by contacting professional pest control services.

Practice proper waste management: Biological contaminants can originate from organic waste materials. Dispose of garbage regularly and ensure it is properly sealed in bins with tight-fitting lids. Clean trash cans frequently to prevent odors and microbial growth.

Avoid smoking indoors: Smoking introduces a range of harmful pollutants, including particulate matter and VOCs, into the indoor environment (Morgan et al. 2022). Establish a strict no-smoking policy indoors to minimize the presence of these contaminants.

Use natural and nontoxic cleaning products: Some cleaning products contain harsh chemicals that can contribute to indoor air pollution. Opt for natural, nontoxic alternatives or make your own cleaning solutions using ingredients like vinegar, baking soda, or lemon juice. Read product labels and choose cleaners with low VOC content (Fermo et al. 2021).

Regularly maintain and clean air filters: If your indoor spaces are equipped with air filtration systems, ensure that the filters are cleaned or replaced regularly according to the manufacturer's instructions. This helps remove particles, including biological contaminants, from the air more effectively (Mata et al. 2022).

Minimize clutter: Excessive clutter in indoor spaces can accumulate dust, provide hiding places for pests, and make cleaning more challenging. Keep your living or working areas well-organized and clutter-free to reduce the potential for biological contaminants to accumulate.

By implementing these strategies, you can minimize pollutant sources and reduce the presence of biological contaminants in your indoor air, creating a healthier and more comfortable environment.

17.4.4 Controlling the Pest Infestations

Controlling pest infestations is crucial for reducing biological contaminants in indoor areas. Pests such as insects, rodents, and other small animals can carry disease-causing microorganisms, contaminate food and water sources, and leave behind droppings and debris that can pose health risks to humans (Baldacchino et al. 2015). Here are some effective methods for pest control to minimize biological contaminants.

Maintain a clean and hygienic indoor environment by regularly cleaning and sanitizing the premises. This includes proper waste management, sweeping, mopping, and disinfecting surfaces. Seal off any cracks, gaps, or openings in walls, floors, doors, and windows to prevent pests from entering the building. Install door sweeps, weather stripping, and screens on windows to create a barrier against pests (Gouge et al. 2015).

Store food items in airtight containers to prevent pests from accessing them. Keep food preparation areas clean and promptly clean up any spills or food waste. Dispose of garbage properly in sealed bins, and regularly empty them to avoid attracting pests (Asnani and Zurbrugg 2007). Maintain cleanliness in trash storage areas and ensure they are well-sealed to prevent pests from accessing the waste. Repair any plumbing leaks, as excess moisture can attract pests like cockroaches and rodents. Ensure proper ventilation and use dehumidifiers in areas prone to dampness to reduce the risk of pest infestations. Conduct regular inspections of the building to identify any signs of pest activity, such as droppings, nests, or gnaw marks. Promptly address any pest-related issues to prevent infestations from spreading (Soh et al. 2023).

Integrated Pest Management (IPM): Implement an IPM program, which combines multiple strategies for effective pest control while minimizing the use of pesticides. IPM focuses on preventive measures, monitoring, and targeted treatments to achieve long-term pest management (Barzman et al. 2015). In severe infestations or cases where DIY methods are insufficient, consider hiring a licensed pest control professional. They have the expertise and specialized equipment to identify and eliminate pests safely and effectively. Remember, it's essential to choose pest control methods that prioritize human and environmental safety while effectively reducing pest infestations and biological contaminants in indoor areas (Barzman et al. 2015).

17.4.5 Humidity Control

Humidity control plays a crucial role in reducing biological contaminants in indoor areas. Regularly monitor humidity levels using a hygrometer or a smart humidity monitor (WHO 2009). Address any issues promptly, such as leaks, condensation, or water damage, to prevent the growth of mold and other biological contaminants. By maintaining optimal humidity levels, you can effectively manage the growth of mold, bacteria, viruses, and dust mites, which can have adverse effects on IAQ and the health of occupants. This range inhibits the growth of most biological contaminants while maintaining a comfortable environment. Mold thrives in environments with high humidity levels (WHO 2009). This is particularly important in areas prone to moisture accumulation, such as bathrooms, basements, and kitchens. Dust mites are microscopic organisms that thrive in humid environments. They are a common allergen and can trigger respiratory issues. Maintaining humidity levels below 50% can help control dust mite populations. To reduce humidity levels, consider using dehumidifiers in areas where moisture is a concern. Dehumidifiers extract excess moisture from the air, helping to maintain optimal humidity and prevent the growth of biological contaminants (Kubba 2010).

Certain viruses and bacteria can survive longer in low-humidity conditions (Lin and Marr 2019). However, excessively high humidity can also support their growth. Maintaining humidity levels within the recommended range helps minimize the survival and spread of airborne pathogens. Remember that different regions and climates may have varying humidity requirements. So, it is essential to consider local recommendations and consult with HVAC professionals for specific guidelines tailored to your location and building type.

17.4.6 Bio-fumigation

Bio-fumigation, also known as biological fumigation or bio-disinfection, refers to the use of biological agents to reduce or eliminate biological contamination in indoor areas (Kumar et al. 2021b). It is an alternative to traditional chemical-based fumigation methods and is gaining recognition as an environmentally friendly and sustainable approach. The role of bio-fumigation in reducing biological contamination in indoor areas involves several mechanisms and benefits:

Pathogen elimination: Bio-fumigation relies on the use of living organisms, such as beneficial bacteria, fungi, or viruses, to target and eliminate harmful pathogens. These biocontrol agents can disrupt the life cycle of pathogens, inhibit their growth, or directly attack them, effectively reducing their populations (Kumar et al. 2021a, b).

Nontoxic approach: Unlike chemical fumigation methods that use harsh and potentially hazardous chemicals, bio-fumigation utilizes naturally occurring or specially selected organisms that are safe for humans, animals, and the environment. This makes it a favorable choice in sensitive indoor environments such as hospitals, schools, food processing facilities, or residential areas (Kusstatscher et al. 2020).

Residual effect: Some bio-fumigation agents can form protective coatings or persist in the environment for extended periods. This residual effect helps prevent the recontamination of treated areas by inhibiting the growth of new pathogens. It provides long-lasting protection and reduces the frequency of fumigation treatments (Kumar and Akhtar 2018).

Targeted action: Bio-fumigation agents can be specifically selected to target particular pathogens or groups of pathogens. This specificity allows for a tailored approach, where the agents are chosen based on the identified contaminants, ensuring effective control of the specific biological hazards present in the indoor area (Morawska et al. 2021).

Reduced resistance development: Chemical fumigation methods can sometimes lead to the development of resistance in pathogens, rendering them ineffective over time. Bio-fumigation, on the other hand, typically involves a combination of multiple biological agents or strains, making it less prone to resistance development. This enhances the long-term effectiveness of the bio-fumigation approach (Sharma 2014).

Eco-friendly solution: Bio-fumigation aligns with sustainable practices and contributes to a healthier environment. It minimizes the release of harmful chemicals into the air, water, and soil, reducing the potential negative impact on ecosystems. Additionally, it promotes the conservation of beneficial organisms and biodiversity, further supporting ecological balance.

It is important to note that while bio-fumigation offers many advantages, its effectiveness may vary depending on the specific pathogens, environmental conditions, and application methods. Proper assessment, selection, and implementation of bio-fumigation strategies should be carried out by trained professionals to ensure optimal results.

17.4.7 Design and Construction of a Building

Designing and constructing a building with the aim of reducing biological contamination in indoor areas involves a combination of architectural, engineering, and operational considerations (Spengler and Chen 2000).

- Implement a ventilation system that provides sufficient fresh air exchange to dilute and remove airborne contaminants. Use high-efficiency air filters to trap particles and allergens. Consider incorporating natural ventilation strategies and ensure proper air distribution throughout the building.
- Prevent moisture buildup and address any water intrusion issues. Moisture can create an ideal environment for mold, bacteria, and other biological contaminants to thrive. Proper waterproofing, insulation, and drainage systems should be in place to prevent dampness and mold growth.
- Choose building materials that are resistant to microbial growth and easy to clean. Nonporous and antimicrobial surfaces, such as stainless steel, glazed tiles, and sealed concrete, can help inhibit the attachment and growth of bacteria and mold (Spengler and Chen 2000).
- Minimize seams, joints, and crevices in surfaces to prevent the accumulation of dirt, moisture, and microbial contamination. Smooth, easy-to-clean finishes should be used in areas where hygiene is critical, such as kitchens, bathrooms, and healthcare facilities.
- Install effective drainage systems in areas prone to water accumulation, such as bathrooms, kitchens, and basements. Proper slope and adequate floor drains will prevent standing water and potential microbial growth.
- Ensure that the building design allows for easy access and maintenance of HVAC systems, filters, and other equipment. Regular cleaning, inspection, and maintenance of these systems are crucial for preventing the buildup and spread of biological contaminants.
- Provide adequate hand hygiene facilities throughout the building, such as handwashing sinks with soap dispensers or hand sanitizer stations. Proper hand hygiene is essential in preventing the spread of pathogens.
- Promote education and awareness among building occupants regarding the importance of hygiene practices, such as handwashing, proper food handling, and respiratory etiquette. Regular training programs can help reinforce good hygiene habits and reduce the risk of contamination.
- Consult with professionals in the field of environmental health, IAQ, and building design to ensure that the building meets the necessary standards and guidelines for reducing biological contamination. Consider seeking certifications such as LEED (Leadership in Energy and Environmental Design) or WELL Building Standard, which incorporate measures to improve IAQ and occupant health (Lee and Kim 2008).

17.4.8 Social and Behavioral Factors

- Educate occupants about the importance of maintaining cleanliness and hygiene in indoor areas.
- Promote awareness about the potential health risks associated with biological contamination.
- Provide information on proper hand hygiene, respiratory etiquette, and personal hygiene practices.
- Establish a routine cleaning schedule for all indoor areas, including floors, surfaces, and frequently touched objects.
- Use appropriate disinfectants recommended for specific pathogens to effectively eliminate microbes.
- Pay special attention to high-touch surfaces such as doorknobs, light switches, faucets, and handrails.
- Ensure proper waste disposal systems are in place, including designated containers for different types of waste (e.g. recyclables, organic waste, hazardous materials).
- Provide clear instructions on waste segregation and disposal to minimize the risk of contamination.
- Encourage individuals to practice good personal hygiene, such as regular handwashing with soap and water for at least 20 seconds.
- Promote the use of hand sanitizers containing at least 60% alcohol when handwashing facilities are not readily available. Encourage individuals to cover their mouth and nose with a tissue or elbow when coughing or sneezing.
- Encourage individuals to stay home when they are sick to prevent the spread of infectious diseases. Promote social distancing measures, such as maintaining a safe distance from others and avoiding crowded spaces (Kumar and Singh 2020).
- Discourage behaviors that can contribute to contamination, such as spitting or improper disposal of tissues (Kumar and Singh 2020).
- Regularly monitor the implementation of hygiene practices and compliance with guidelines.
- Provide feedback and reminders to individuals and staff to ensure ongoing adherence to cleanliness protocols.
- Consider implementing incentives or recognition programs to motivate individuals to maintain good hygiene practices.
- Foster a collaborative environment where individuals feel comfortable reporting contamination concerns or issues.
- Establish effective communication channels to disseminate information and updates related to hygiene practices and contamination control.

17.4.9 Future Trends and Emerging Technologies

Reducing biological contamination in indoor areas is an ongoing challenge, and emerging technologies are constantly being developed to address this issue. Here are some future trends and emerging technologies that hold promise for reducing biological contamination:

Ultraviolet germicidal irradiation (UVGI): UVGI technology utilizes ultraviolet (UV) light to deactivate microorganisms, including bacteria and viruses, in the air and on surfaces. Advances in UVGI systems include the development of more efficient UV-C light sources, integrated sensors for real-time monitoring, and automated disinfection systems (Memarzadeh 2021).

Photocatalytic oxidation (PCO): PCO technology combines UV light and a photocatalyst material to generate reactive oxygen species that can destroy organic contaminants, including pathogens (Chen et al. 2010). Ongoing research focuses on improving PCO reactor design, optimizing catalyst materials, and integrating PCO systems into HVAC and ventilation systems.

Biocide-embedded surfaces: Innovative materials are being developed with biocidal properties that can inhibit the growth and survival of microorganisms on surfaces (Kumar et al. 2021). These surfaces may incorporate antimicrobial agents, such as nanoparticles or coatings that continuously release or activate disinfectants.

Self-cleaning materials: Self-cleaning materials have properties that prevent the adhesion and growth of microorganisms, reducing the need for manual cleaning (Midtdal and Jelle 2013). Examples include surfaces with hydrophobic or antimicrobial coatings, which can repel or kill microbes upon contact.

Internet of Things (IoT) and smart sensors: IoT-enabled devices and smart sensors can monitor indoor environmental conditions, air quality, and occupancy to optimize ventilation, air purification, and cleaning routines (Pantelic et al. 2023). Real-time data collection and analysis enable proactive responses and efficient allocation of resources.

Robotics and automation: Robotic systems equipped with disinfection capabilities can autonomously navigate indoor areas and perform targeted cleaning and disinfection tasks. UV-C disinfection robots, electrostatic sprayers, and autonomous floor cleaners are examples of emerging technologies in this field (Ruan et al. 2021).

Antimicrobial coatings: Antimicrobial coatings are being developed to inhibit the growth and spread of microorganisms on various surfaces, including textiles, walls, and furniture. These coatings can provide long-lasting protection and reduce the risk of cross-contamination (Tan et al. 2019).

Airflow management and ventilation design: Advances in computational fluid dynamics (CFD) modeling and simulation help optimize airflow patterns and ventilation systems to minimize the spread of contaminants. Integrated design approaches consider factors like air distribution, filtration efficiency, and occupancy patterns to improve IAQ (Peng et al. 2020).

Biotechnology and bioengineering: Research in biotechnology and bioengineering is exploring novel methods for combating biological contamination, such as using engineered microorganisms or enzymes to degrade pathogens or biofilms.

It is important to note that while these technologies show promise, their effectiveness, safety, and practicality need to be thoroughly evaluated and validated before widespread implementation. Additionally, a holistic approach that combines multiple strategies, including behavioral practices and regular maintenance, is crucial for effective contamination control in indoor environments.

17.4.10 Risk Assessment and Management

Risk assessment and management for reducing biological contamination in indoor areas involves identifying potential hazards, evaluating the risks associated with them, implementing control measures, and monitoring their effectiveness. Here is a step-by-step guide to the process:

Identify potential hazards: Determine the potential sources of biological contamination, such as bacteria, viruses, fungi, and allergens. Common sources include poor ventilation, water leaks, HVAC systems, human activity, pests, and contaminated surfaces.

Assess risks: Evaluate the likelihood and severity of biological contamination occurring. Consider the susceptibility of occupants, the type of indoor environment (e.g. healthcare facility, office), and existing control measures in place. Identify high-risk areas, such as restrooms, food preparation areas, or areas with high occupant density.

Implement control measures: Develop a comprehensive plan to minimize and control biological contamination. Improve ventilation systems to enhance air circulation and filtration. Establish cleaning and disinfection protocols for surfaces, equipment, and shared spaces. Implement proper waste management procedures. Educate occupants about good hygiene practices, such as handwashing and respiratory etiquette. Consider the use of antimicrobial materials and coatings in high-risk areas (WHO 1988, 2009).

Monitor and review: Regularly inspect and monitor the indoor environment for signs of biological contamination. Conduct air and surface sampling, if necessary, to assess microbial levels. Keep records of cleaning and maintenance activities. Encourage occupants to report any concerns or incidents related to biological contamination. Review control measures periodically to ensure their effectiveness (WHO 1988, 2009).

Response and remediation: Develop a response plan for dealing with incidents of biological contamination. Establish procedures for isolating affected areas, cleaning, disinfection, and restoration. Follow guidelines from regulatory bodies and public health agencies (WHO 1988, 2009).

Continual improvement: Stay updated with the latest research, guidelines, and best practices for reducing biological contamination. Regularly train and educate staff and occupants on infection control measures. Engage with experts or consultants to assess and improve your risk management strategies. Remember that risk assessment and management should be tailored to the specific indoor environment and its associated hazards. Consultation with experts in industrial hygiene, environmental health, or public health can provide valuable guidance in developing an effective strategy (WHO 1988, 2009).

17.5 Conclusion

The conclusion summarizes the key findings from the literature and emphasizes the importance of following guidelines to ensure a healthy indoor environment. It also highlights the need for continued research and collaboration to further improve the guidelines and address emerging challenges. Overall, this chapter provides a comprehensive and up-to-date overview of guidelines for managing biological contaminants in indoor air environments. It may serve as a valuable resource for researchers, professionals, policymakers, and individuals seeking guidance on maintaining healthy IAQ. By following these guidelines, we can create safer and more comfortable indoor spaces for everyone.

References

Abel, E., Andersson, J.V., Dawidowicz, N. et al. (2002). The Swedish key action 'The Healthy Building' – research results achieved during the first three years period 1998–2000, Indoor Air 2002. *Proceedings of the 9th International Conference on Indoor Air Quality and Climate* 2: 996–1001.

AIHA (1989). *Field Guide for the Determination of Biological Contaminates in Environmental American Conference of Governmental Industrial Hygienists (ACGIH). Guidelines for the Assessment of Bioaerosols in the Indoor Environment.* ISBN-13: 978-0936712833.

Asnani, P.U. and Zurbrugg, C. (2007). *Improving Municipal Solid Waste Management in India: A Sourcebook for Policymakers and Practitioners*. World Bank Publications.

Baldacchino, F., Caputo, B., Chandre, F. et al. (2015). Control methods against invasive Aedes mosquitoes in Europe: a review. *Pest Management Science* 71 (11): 1471–1485.

Barzman, M., Bàrberi, P., Birch, A.N.E. et al. (2015). Eight principles of integrated pest management. *Agronomy for Sustainable Development* 35: 1199–1215.

Brambilla, A. and Sangiorgio, A. (2020). Mould growth in energy efficient buildings: Causes, health implications and strategies to mitigate the risk. *Renewable and Sustainable Energy Reviews* 132: 110093.

Brilli, F., Fares, S., Ghirardo, A. et al. (2018). Plants for sustainable improvement of indoor air quality. *Trends in Plant Science* 23 (6): 507–512. https://doi.org/10.1016/j.tplants.2018.03.004.

CDC (2018). *NIOSH Criteria for a Recommended Standard: Occupational Exposure to Heat and Hot Environments*. National Institute on Drug Abuse.

Chen, F., Yang, X., Mak, H.K., and Chan, D.W. (2010). Photocatalytic oxidation for antimicrobial control in built environment: a brief literature overview. *Building and Environment* 45 (8): 1747–1754.

Chinn, R.Y. and Sehulster, L. (2003). Guidelines for environmental infection control in health-care facilities: recommendations of CDC and Healthcare Infection Control Practices Advisory Committee (HICPAC). https://stacks.cdc.gov/view/cdc/11303 (accessed 15 July 2023).

de Aquino Neto, F.R. and de GóesSiqueira, L.F. (2000). Guidelines for indoor air quality in offices in Brazil. *Proceedings of Healthy Buildings. Biological Contamination of Air in Indoor Spaces* 4: 549–554. https://doi.org/10.5772/59727513.

European Collaborative Center (1993). Indoor Air Quality & Its Impact On Man, Report No. 12, Biological Particles in Indoor Environment, Luxembourg.

Fermo, P., Artíñano, B., De Gennaro, G. et al. (2021). Improving indoor air quality through an air purifier able to reduce aerosol particulate matter (PM) and volatile organic compounds (VOCs): experimental results. *Environmental Research* 197: 111131.

Gouge, D.H., Nair, S., Li, S., and Stock, T. (2015). Pest-proofing Your Home. https://repository.arizona.edu/handle/10150/578402 (accessed on: 14 July 2023).

Health Effects Institute (2020). *State of Global Air 2020. Special Report*. Boston, MA: Health Eff. Inst.

Heseltine, E. and Rosen, J. (eds.) (2009). WHO guidelines for indoor air quality: dampness and mould.

Hsu, C.-H., Lu, M.-C., and Huang, D.-J. (2012). Application of chlorine dioxide for disinfection of student health centers. *Environmental Monitoring and Assess* 184: 741–747.

Huang, P.Y., Shi, Z.Y., Chen, C.H. et al. (2013). Microbial contamination in two intensive care units of a medical center in Central Taiwan. *Aerosol and Air Quality Research* 13: 1060–1069.

IRSST (2001). Occupational Health and Safety Research Institute Robert Sauvé: Bioaerosol in Workplace: evaluation, control and prevention guide. Technical guide. www.irsst.qc.ca (accessed 14 July 2023).

Katiyar, V. (2013). Assessment of indoor air micro-flora in selected schools. *Advances in Environmental Research* 2 (1): 61–80.

Kubba, S. (2010). Indoor environmental quality. In: *LEED Practices, Certification, and Accreditation Handbook* (ed. S. Kubba), 211. Elsevier.

Kumar, A. and Akhtar, J. (2018). Integrated disease management for seed production under organic cultivation. In: *Integrated Crop Management Practices* (ed. A.K. Choudhary, R.S. Bana, and V. Pooniya), 121.

Kumar, P. and Singh, R. (2020). Novel coronavirus (severe acute respiratory syndrome coronavirus 2) as threat to general and reproductive health: challenges and research needs. *Journal of Reproductive Healthcare and Medicine 2*: 1–7.

Kumar, P., Kausar, M.A., Singh, A.B., and Singh, R. (2021b). Biological contaminants in the indoor air environment and their impacts on human health. *Air Quality, Atmosphere & Health* 14 (11): 1723–1736.

Kumar, P., Singh, A.B., and Singh, R. (2021a). Seasonal variation and size distribution in the airborne indoor microbial concentration of residential houses in Delhi and its impact on health. *Aerobiologia* 37 (4): 719–732.

Kumar, P., Singh, A.B., and Singh, R. (2022). Comprehensive health risk assessment of microbial indoor air quality in microenvironments. *Plos one* 17 (2): e0264226.

Kumar, P., Singh, A.B., Arora, T. et al. (2023a). Critical review on emerging health effects associated with the indoor air quality and its sustainable management. *Science of the Total Environment* 872: 162163.

Kumar, R., Verma, V., Thakur, M. et al. (2023b). A systematic review on mitigation of common indoor air pollutants using plant-based methods: a phytoremediation approach. *Air Quality, Atmosphere & Health* 16: 1501–1527.

Kumar, S., Roy, D.N., and Dey, V. (2021). A comprehensive review on techniques to create the anti-microbial surface of biomaterials to intervene in biofouling. *Colloid and Interface Science Communications* 43: 100464.

Kusstatscher, P., Cernava, T., and Berg, G. (2020). Using bacteria-derived volatile organic compounds (VOCs) for industrial processes. In: *Bacterial Volatile Compounds as Mediators of Airborne Interactions* (ed. C.-M. Ryu, L. Weisskopf, and B. Piechulla), 305–316.

Lee, J.H., Ahn, K.H., and Yu, I.J. (2012). Outbreak of bioaerosols with continuous use of humidifier in apartment room. *Toxicology Research* 28 (2): 103–106.

Lee, S.C., Li, W.-M., and Ao, C.-H. (2002). Investigation of indoor air quality at residential homes in Hong Kong – case study. *Atmospheric Environment* 36: 225–237.

Lee, Y.S. and Kim, S.K. (2008). Indoor environmental quality in LEED-certified buildings in the US. *Journal of Asian Architecture and Building Engineering* 7 (2): 293–300.

Li, W.M., Lee, S.C., and Chan, L.Y. (2001). Indoor air quality at nine shopping malls in Hong Kong. *Science of the Total Environment* 273 (1–3): 27–40.

Lin, K. and Marr, L.C. (2019). Humidity-dependent decay of viruses, but not bacteria, in aerosols and droplets follows disinfection kinetics. *Environmental Science & Technology* 54 (2): 1024–1032.

Mata, T.M., Felgueiras, F., Martins, A.A. et al. (2022). Indoor air quality in elderly centers: pollutants emission and health effects. *Environments* 9 (7): 86.

Memarzadeh, F. (2021). A review of recent evidence for utilizing ultraviolet irradiation technology to disinfect both indoor air and surfaces. *Applied Biosafety* 26 (1): 52–56.

Morawska, L., Allen, J., Bahnfleth, W. et al. (2021). A paradigm shift to combat indoor respiratory infection. *Science 372* (6543): 689–691.

Midtdal, K. and Jelle, B.P. (2013). Self-cleaning glazing products: a state-of-the-art review and future research pathways. *Solar Energy Materials and Solar Cells* 109: 126–141.

Ministry of Environment (MOE) (2011). Korean Indoor Air Quality Control in Public Use Facilities, etc. *Act. Republic of Korea*. https://elaw.klri.re.kr/eng_mobile/viewer.do?hseq=60420&type=new&key= (accessed 15 July 2023).

Moldoveanu, A.M. (2015). Biological contamination of air in indoor spaces. In: *Current Air Quality Issues* (ed. F. Nejadkoorki), 489–514. IntechOpen.

Morgan, A.L., Torpy, F.R., Irga, P.J. et al. (2022). The botanical biofiltration of volatile organic compounds and particulate matter derived from cigarette smoke. *Chemosphere* 295: 133942.

Nielsen, A.N., Jensen, R.L., Larsen, T.S., and Nissen, S.B. (2016). Early stage decision support for sustainable building renovation – a review. *Building and Environment* 103: 165–181.

Nunes, Z.G., Martins, A.S., Altoe, A.L.F. et al. (2005). Indoor air microbiological evaluation of offices, hospitals, industries, and shopping centers. *Memórias do Instituto Oswaldo Cruz* 100 (4): 351–357.

Obbard, J.P. and Fang, L.S. (2003). Airborne concentrations of bacteria in a hospital environment in Singapore. *Water, Air, & Soil Pollution* 144 (1): 333–341.

Pantelic, J., Son, Y.J., Staven, B., and Liu, Q. (2023). Cooking emission control with IoT sensors and connected air quality interventions for smart and healthy homes: Evaluation of effectiveness and energy consumption. *Energy and Buildings* 286: 112932.

Peng, S., Chen, Q., and Liu, E. (2020). The role of computational fluid dynamics tools on investigation of pathogen transmission: prevention and control. *Science of The Total Environment* 746: 142090.

Pillarisetti, A., Ye, W., and Chowdhury, S. (2022). Indoor air pollution and health: bridging perspectives from developing and developed countries. *Annual Review of Environment and Resources* 47: 197–229.

Rao, C.Y., Burge, H.A., and Chang, J.C. (1996). Review of quantitative standards and guidelines for fungi in indoor air. *Journal of the Air & Waste Management Association* 46 (9): 899–908.

Ruan, K., Wu, Z., and Xu, Q. (2021). Smart cleaner: a new autonomous indoor disinfection robot for combating the covid-19 pandemic. *Robotics* 10 (3): 87.

Rumchev, K., Brown, H., and Spickett, J. (2007). Volatile organic compounds: do they present a risk to our health? *Reviews on Environmental Health* 22 (1): 39–56.

Sharma, N. (2014). Biologicals. In: *Biological Controls for Preventing Food Deterioration* (ed. N. Sharma), 1–25.

Sikka, M.P. and Mondal, M. (2022). A critical review on cleanroom filtration. *Research Journal of Textile and Apparel* 26 (4): 452–467.

Soh, S., Chua, C.H., Neo, Z.W. et al. (2023). Rodent activity in municipal waste collection premises in Singapore: an analysis of risk factors using mixed-effects modelling. *Scientific Reports* 13 (1): 3070.

Spengler, J.D. and Chen, Q. (2000). Indoor air quality factors in designing a healthy building. *Annual Review of Energy and the Environment* 25 (1): 567–600.

Tan, L.Y., Sin, L.T., Bee, S.T. et al. (2019). A review of antimicrobial fabric containing nanostructures metal-based compound. *Journal of Vinyl and Additive Technology* 25 (S1): E3–E27.

Weil, D. (1991). Enforcing OSHA: the role of labor unions. *Industrial Relations: A Journal of Economy and Society* 30 (1): 20–36.

World Health Organization WHO Guidelines for Indoor Air Quality: Dampness and Mould. Geneva: World Health Organization; 2009. 1, Introduction. https://www.ncbi.nlm.nih.gov/books/NBK143944/ (accessed 15 July 2023).

World Health Organization (WHO) (1988). *Indoor Air Quality: Biological Contaminants: European Series, Number No. 31*. Copenhagen: WHO Regional Publication.

World Health Organization (2014). *WHO Guidelines for Indoor Air Quality: Household Fuel Combustion*. World Health Organization.

Index

Airborne Biocontaminants and Their Impact on Human Health, First Edition.
Edited by Rajeev Singh and Anamika Singh.

t

u

v

w

Printed and bound by CPI Group (UK) Ltd, Croydon, CR0 4YY

17/09/2026

14964308-0007